Low Cost
Flip Chip
Technologies

Other Electronics Books from McGraw-Hill

Low Cost Flip Chip Technologies

For DCA, WLCSP, and PBGA Assemblies

John H. Lau
Express Packaging Systems, Inc.

McGraw-Hill

New York   San Francisco   Washington, D.C.   Auckland   Bogotá
Caracas   Lisbon   London   Madrid   Mexico City   Milan
Montreal   New Delhi   San Juan   Singapore
Sydney   Tokyo   Toronto

Library of Congress Cataloging-in-Publication Data

Lau, John H.
    Low cost flip chip technologies : for DCA, WLSCP, and PBGA
    assemblies / John H. Lau.
      p. cm.
    Includes bibliographical references and index.
    ISBN 0-07-135141-8
    1. Multichip modules (Microelectronics)—Design and construction.
   2. Microelectronic packaging. I. Title.

    TK7874.L3168 2000
    621.3815—dc21

                                     99-088307

McGraw-Hill

*A Division of The **McGraw·Hill** Companies*

1 2 3 4 5 6 7 8 9 0 DOC/DOC 9 0 5 4 3 2 1 0

ISBN 0-07-135141-8

*The sponsoring editor for this book was Stephen S. Chapman and the
production supervisor was Sherri Souffrance. It was set in Century
Schoolbook by North Market Street Graphics.*

Printed and bound by R. R. Donnelley & Sons Company.

 This book was printed on recycled, acid-free paper containing a
minimum of 50% recycled, de-inked fiber.

McGraw-Hill books are available at special quantity discounts to use as
premiums and sales promotions, or for use in corporate training programs. For more information, please write to the Director of Special
Sales, Professional Publishing, McGraw-Hill, Inc., Two Penn Plaza, New
York, NY 10121-2298, or contact your local bookstore.

Contents

Chapter 9. Thermal Management of Flip Chip on Board   301

Chapter 10. Wafer-Level Packaging   317

Chapter 13. Wire Bonding Chip (Face-Up) in PBGA Packages   417

Chapter 14. Wire Bonding Chip (Face-Down) in PBGA Packages   465

Foreword

Flip chip technology has been practiced by AT&T (now Lucent Technologies) since the 1950s; Bell Labs called the technology beam-leaded devices. Also, IBM developed the controlled-collapse chip connection (C4) in the early 1960s for its mainframe computer applications. These flip chip devices are mounted on an expansive ceramic substrate where the thermal expansion mismatch between the silicon chip and the ceramic substrate is less critical. Silicone elastomers have been used for many years to underfill the small beam-leaded devices for corrosion protection. However, due to the high cost of the ceramic substrate, it was not popular until the late 1980s, when IBM at Yasu Lab in Japan reported the successful implementation of flip chip on organic substrate that first demonstrated the application of high-performance flip chip on low-cost printed wiring board substrate. Epoxy underfills were used to reduce the thermal expansion mismatch and enhance the thermal fatigue performance of the flip chip on organic structures. Since then, there has been much interest in this low-cost flip chip technology and much research and development work has been sparked in this important area.

Dr. John Lau, a leading electronic packaging expert who has published over 10 books and has made vast contributions to the field of electronic packaging, has undertaken the important task of writing a comprehensive book on low-cost flip chip technology. This book is intended for scientists and engineers who are actively working in this critical area and want to gain a comprehensive view of the technology, as well as for packaging engineers who experience practical problems in the flip chip area. In addition, system engineers who want to maximize the advantage of flip chip technology will find this book extremely helpful and useful.

The book has a total of 16 chapters that cover the whole spectrum of low-cost flip chip technology, from the fundamentals to the latest developments in this field. The reference list is extensive and comprehensive, and could provide a valuable source for reference and in-depth study. Many examples, products, and processes related to flip chip technology are described in detail. If you are interested in flip chip technology, you will find this book to be of great value. I congratulate John in successfully completing this valuable text that benefits the electronic packaging community. I hope you, like myself, find it to be very useful for your R&D in flip chip technology and for further stimulating your work in this area.

C. P. WONG, PH.D.
Fellow, IEEE and Bell Labs
Professor of Materials Science and Engineering
Research Director of the NSF-ERC Packaging
  Research Center
Georgia Institute of Technology

Preface

The past decade has witnessed explosive growth in the research and development efforts devoted to solder-bumped flip chips on low-cost organic substrates as a direct result of the higher requirements for package density and performance and of cost advantages over ceramic substrates. Just like many other new technologies, low-cost solder-bumped flip chips still face many critical issues. In the development of flip chip on board or on substrate in a plastic package, the following (compared with the wire bonding and conventional surface mount technologies) must be noted and understood: the infrastructure of flip chip is not well established; flip chip expertise is not commonly available; wafer bumping is still too costly; bare die/wafer is not commonly available; bare die/wafer handling is not easy; pick and place is more difficult; fluxing is more critical; there are underfill issues; rework is more difficult; solder joint reliability is more critical; inspection is more difficult; flip chip assembly testability is not well established; there are die shrink and expansion issues; there are known-good-die issues; and chip cracking during solder reflow and curing of underfill has been reported.

In the past five years, some of these critical issues have been studied by many experts. Their results have already been disclosed in diverse journals or, more incidentally, in the proceedings of many conferences, symposiums, and workshops whose primary emphases are electrical designs, materials science, manufacturing engineering, or electronic packaging and interconnection. Consequently, there is no single source of information devoted to the state of the art of low-cost flip chip technologies. This book aims to remedy this deficiency and to present, in one volume, a timely summary of progress in all aspects of this fascinating field within the past five years. This book is written for everyone who can quickly learn about the basics and problem-solving methods, understand the trade-offs, and make system-level decisions involving the low-cost flip chip technologies.

This book is organized into sixteen parts. Chapter 1 briefly discusses the IC trends and updates in packaging technology. Chapter 2 presents two popular chip-level interconnects, namely wire bonds and solder bumps. More than 12 different wafer-bumping methods are discussed. Chapter 3 examines the physical and mechanical properties of lead-free solders. Also, more than 100 different lead-free solder alloys in the paste, bar, and wire forms are presented. Chapter 4 discusses the high-density printed circuit board (PCB) and substrates, with a focus on sequential buildup (SBU) technologies with microvias. Some useful charts for designing high-speed circuits are also provided. Chapter 5 presents the flip chip on board (FCOB) with solderless and fluxless materials such as anisotropic conductive adhesive (ACA) and anisotropic conductive film (ACF). Emphasis is placed on the design, materials, processing, and reliability of the ACA and ACF FCOB assemblies.

Next FCOB with solder bumps is discussed. Chapter 6 presents solder-bumped FCOB with the conventional underfills. Both high- and low-melting-point solder alloys are discussed. Chapter 7 presents solder-bumped FCOB with no-flow underfills. Both liquidlike and film-like no-flow underfills are considered. Chapter 8 presents solder-bumped FCOB with imperfect underfills. The effects of various imperfect underfills on the solder joint reliability of the FCOB assemblies are discussed.

Chapter 9 discusses the thermal management of solder-bumped FCOB. The effects of PCB construction, airflow, chip size and power dissipation area, underfill, solder-bump population, and heat sink on the thermal resistance of FCOB assemblies are discussed.

The wafer-level packaging is presented in Chap. 10. From the system makers' point of view, most of the wafer-level packages are just like solder-bumped flip chip, except that the solder bumps of the wafer-level packages are much larger and easier to put on the PCB, and the manufacturers do not have to struggle with the underfill encapsulant. Six different companies' wafer-level packages are briefly discussed in this book. For the other eight different companies' wafer-level packages, please read *Chip Scale Package: Design, Materials, Processes, Reliability, and Applications* (Lau and Lee, McGraw-Hill, 1999).

More than 35 different companies' CSPs have been reported in *Chip Scale Packages*. However, in this book, the applications of solder-bumped flip chip on microvia and via-in-pad (VIP) CSP substrate are presented in Chap. 11 and μBGA on the direct Rambus modules is presented in Chap. 12. The PCB manufacturing, testing, and assembly of the Rambus modules are also discussed.

Next the wire bonding chip on an organic substrate in a plastic ball grid array (PBGA) package is presented. Chapter 13 examines the popcorning effects of face-up PBGA packages by both experimental measurement and the fracture mechanics method. Assembly of a large PBGA with a large plastic quad flat pack (PQFP) directly on the opposite side of the PCB is also presented. Chapter 14 presents a family of face-down PBGA packages. The superb electrical and thermal performance of these packages is demonstrated through simulations and measurements.

The last part of this book presents the solder-bumped flip chip on an organic substrate in a PBGA package. Chapter 15 discusses the design, materials, processing, and reliability of solder-bumped flip chip PBGA packages manufactured by four different companies. The failure analysis of solder-bumped flip chip in a PBGA package is presented in Chap. 16. Emphasis is placed on solder-bump extrusion, die cracking, and delaminations.

For whom is this book intended? Undoubtedly it will be of interest to three groups of specialists: (1) those who are active or intend to become active in research and development of low-cost flip chip technologies; (2) those who have encountered practical flip chip problems and wish to understand and learn more methods for solving such problems; and (3) those who have to choose a reliable, creative, high-performance, robust, and cost-effective packaging technique for their interconnect systems. This book can also be used as a text for college and graduate students who could become our future leaders, scientists, and engineers in the electronics industry.

I hope this book will serve as a valuable source of reference to all those faced with the challenging problems created by the ever increasing IC speed and density and shrinking product size and weight. I also hope that it will aid in stimulating further research and development on electrical and thermal designs, materials, processes, manufacturing, electrical and thermal management, testing, and reliability, and more sound applications of low-cost flip chip technologies in electronics products.

The organizations that learn how to design low-cost flip chip technologies in their interconnect systems have the potential to make major advances in the electronics industry and to gain great benefits in cost, performance, quality, size, and weight. It is my hope that the information presented in this book will assist in removing roadblocks, avoiding unnecessary false starts, and accelerating design, materials, and process development of low-cost flip chip technologies. I am strongly against the notion that electronics packaging and interconnection are the bottleneck of high-speed computing. Rather, I want to consider this as the golden opportunity to make a major contribution to the electronics industry by developing innovative, cost-effective, and reliable low-cost flip chip products. It is an exciting time for low-cost flip chip technologies!

JOHN H. LAU
Palo Alto, California

To my parents, for their encouragement.
To my wife, Teresa Lau, for her support.
To my daughter, Judy Lau, for her understanding.
To those who published peer-reviewed papers, for sharing their knowledge.

Acknowledgments

Development and preparation of *Low Cost Flip Chip Technologies for DCA, WLCSP, and PBGA Assemblies* was facilitated by the efforts of a number of dedicated people at McGraw-Hill and North Market Street Graphics. I would like to thank them all, with special mention to Stephanie Landis of North Market Street Graphics, and Petra Captein and Thomas Kowalczyk of McGraw-Hill for their unswerving support and advocacy. My special thanks to Steve Chapman (executive editor of electronics and optical engineering), who made my dream of this book come true by effectively sponsoring the project and solving many problems that arose during the book's preparation. It has been a great pleasure and a fruitful experience to work with all these people in transforming my messy manuscripts into a very attractive printed book.

The material in this book has clearly been derived from many sources including individuals, companies, and organizations, and I have attempted to acknowledge, in the appropriate parts of the book, the assistance that I have been given. It would be quite impossible for me to express my thanks to everyone concerned (please read the names in the references of each chapter of the book) for their cooperation in producing this book, but I would like to extend due gratitude.

Also, I want to thank several professional societies and publishers for permitting me to reproduce some of their illustrations and information in this book. For example, the American Society of Mechanical Engineers (ASME) Conference Proceedings (*International Intersociety Electronic Packaging Conference*) and Transactions (*Journal of Electronic Packaging*), the Institute of Electrical and Electronic Engineers (IEEE) Conference Proceedings (*Electronic Components and Technology Conference* and *International Electronic Manufacturing and Technology Symposium*) and Transactions (*Components, Packaging, and Manufacturing Technology*), the International Microelectronics and Packaging Society (IMAPS) Conference Proceedings (*International Symposium on Microelectronics*) and Transactions (*International Journal of Microcircuits and Electronic Packaging*), the Surface Mount Technology Association (SMTA) Conference Proceedings (*Journal of Surface Mount Technology*), the National Electronic Packaging Conferences (NEPCON) and Proceedings, the *IBM Journal of Research and Development, Electronic Packaging and Production, Advanced Packaging, Circuits Assembly, Surface Mount Technology, Connection Technology, Solid State Technology, Circuit World, Microelectronics International,* and *Soldering and Surface Mount Technology.*

I want to thank my former employer, Hewlett-Packard, for providing me an excellent working environment that has nurtured me as a human being, fulfilled my job satisfaction, and enhanced my professional reputation. I also want to thank Mr. Terry T. M. Gou (chairman and CEO of Hon Hai Precision Industry Co., Ltd.) for his trust, respect, and support of my work at EPS.

I want to thank my eminent colleagues at Hewlett-Packard and throughout the electronics industry for their useful help, strong support, and stimulating discussions. Also, I want to thank Chris Chang, Chih-Chiang Chen, Jennifer Leong, Ricky Lee (on sabbatical leave from Hong Kong University of Science and Technology), Huabo Chen, Bill Wun, Kuan-Luen Chen, Ray Chen, Yung-Shih Chen, Tony Chen, Tai-Yu Chou, Sally Chung, Livia Hu, Cathy Hung, Kalok Jim (summer intern from Oxford University), Wei Koh, Chien Ouyang, and Frank Wu of EPS for their very helpful support and constructive suggestions. Working and socializing with them has been a privilege and an adventure, and I have learned a lot about life and electronics packaging from them.

I am indebted to Professor C. P. Wong for taking the trouble to read through the contents of the manuscript in the course of preparing his insightful foreword. As one of the key players in low-cost flip chip technologies and in his role as the research director of the NSF-ERC Packaging Research Center at Georgia Institute of Technology, Dr. Wong has particular skills and insight into the material contained in this book, and therefore I greatly appreciate his undertaking to write the foreword.

Lastly, I want to thank my daughter Judy and my wife Teresa for their love, consideration, and patience in allowing me to work over many weekends on this book. Their simple belief that I am making a small contribution to the electronics industry was a strong motivation for me. During this holiday season, thinking that Judy is doing very well in the Department of Engineering Physics at UC Berkeley, and that Teresa and I are in good health, I want to thank God for his generous blessings.

JOHN H. LAU
Palo Alto, California

Integrated Circuit Packaging Trends

1.1 Introduction

The electronics industry is one of the most dynamic, fascinating, and important areas of manufacturing. In a relatively short period of time, it has become the largest and most pervasive manufacturing industry in the developed world.

The invention of the bipolar transistor by John Bardeen, Walter Brattain, and William Shockley at Bell Laboratories (now Lucent Technologies) in 1947 foreshadowed the development of generations of computers yet to come. The invention of the silicon integrated circuit (IC) by Jack Kilby of Texas Instruments in 1958 and by Robert Noyce and Gordon Moore of Fairchild Semiconductor in 1959 stimulated the development of generations of small, medium, large, very large, ultra-large, giga, and yet-to-come scale integrations (SSI, MSI, LSI, VLSI, ULSI, GSI, and . . . SI).

Fifty years ago, the average family probably owned five active electronic devices. Today, the average family owns hundreds of millions of transistors, and it is probable that by the year 2000, the figure will be more than 1 billion. This has produced a global revolution of major proportions that affects the lives of people everywhere.

The IC chip is not an isolated island. It must communicate with other IC chips in a circuit through an input/output (I/O) system of interconnects. Furthermore, the IC chip and its embedded circuitry are delicate, requiring the package to both carry and protect the chip. Consequently, the major functions of the electronics package are: (1) to provide a path for the electric current that powers the circuits on the IC chip; (2) to distribute the signals onto and off of the IC chip; (3) to

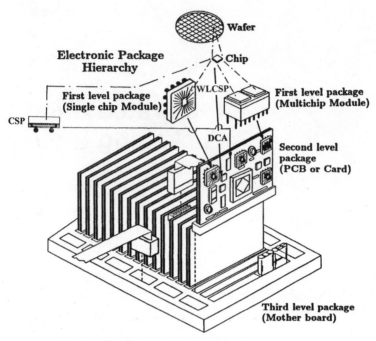

Figure 1.1 The first three levels of electronics packaging.

remove the heat generated by the circuits on the IC chip; and (4) to support and protect the IC chip from hostile environments.

Figure 1.1 shows a schematic representation of electronic package hierarchy. Packaging focuses on how a chip (or many chips) is packaged cost efficiently and reliably.[1–175] Packaging is an art based on the science of establishing interconnections ranging from zero-level packages (i.e., chip-level connections such as gold and solder bumps) to first-level packages (either single- or multichip modules), second-level packages [e.g., printed circuit board (PCB)], and third-level packages (e.g., motherboard). In this chapter, the IC trends will be briefly discussed first. Then the packaging technologies for the ICs will be briefly updated.

1.2 IC Trends

1.2.1 IC density and feature size

The past decade has witnessed an explosive growth in IC density (i.e., number of transistors per chip) and chip size. One of the key reasons for this is the advance of the complementary metal-oxide-semiconductor (CMOS) process (Fig. 1.2), which has very fine feature sizes (Fig. 1.3) and high yields. Today, 0.2-μm sizes are in volume production. According

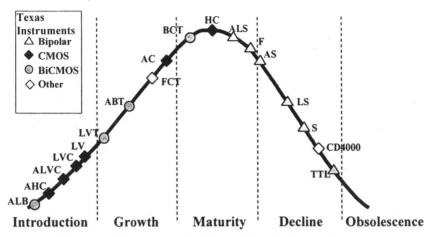

Figure 1.2   IC process technology trends. *(Source: Texas Instruments)*

to the Semiconductor Industry Association (SIA)'s technology road map, 1-cm^2 chips containing 16 million transistors are forecast to be in volume production by the year 2001 (Table 1.1).

1.2.2 IC operating voltage

Because of the explosive growth in portable electronic products, the operating voltage (power supply) on the IC chip has been reduced from

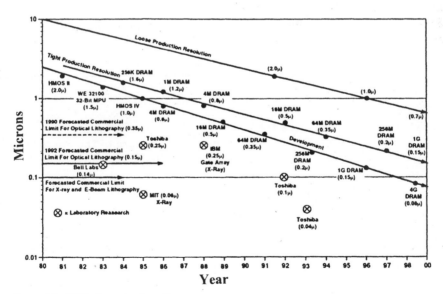

Figure 1.3   IC feature size trends.

TABLE 1.1   SIA's Overall Device Technology Trends

	1997	1999	2001	2006	2012
Memory					
Generation at production					
ramp	64M	256M	1G	4G	64G
Bits/cm^2 at sample/					
introduction	96M	270M	380M	2.2B	17B
Logic (high-volume, cost-					
performance MPU)					
Logic transistors/cm^2	3.7M	6.2M	10M	39M	180M
Logic (low-volume ASIC)					
Transistors/cm^2	8M	14M	16M	40M	100M
Functions/chip					
DRAM bits/chip	267M	1.07G	1.7G	17.2G	275G
Chip size (cm^2)					
DRAM	2.8	4	4.45	7.9	15.8
MPU	3	3.4	3.85	5.2	7.5
ASIC (Maximum litho					
field area)	4.8	8	8.5	10	13
Number of chip I/Os					
Chip-to-package, high-					
performance	1450	2000	2400	4000	7300
Chip-to-package, cost-					
performance	800	975	1195	1970	3585
Number of package pins/balls					
ASIC, high-performance	1100	1500	1800	3000	5500
MPU/controller, cost-					
performance	600	810	900	1500	2700
Chip frequency (MHz)					
On-chip, across-chip clock,					
high-performance	750	1200	1400	2000	3000
Power supply voltage (V)					
Minimum logic voltage	1.8–2.5	1.5–1.8	1.2–1.5	0.9–1.2	0.5–0.6
Power (W)					
High-performance with					
heat sink	70	90	110	160	175

5 V to either 3 or 3.3 V. This figure will eventually drop to 2.5 V, then 1.5 V, as shown in Fig. 1.4 and Table 1.1. One of the key reasons is that the power consumption is proportional to the square of the operating voltage, i.e., the lower the operating voltage, the longer the battery life of the portable electronic products.

1.2.3   Microprocessor, ASIC, DRAM, and SRAM

In most of the electronic products, there are four major IC devices: the microprocessor, the application-specific IC (ASIC), the cache memory, and the main memory. For example, a personal computer usually has one microprocessor; a few cache memories, e.g., fast static random access memory (SRAM); a few ASICs, e.g., video, sound, data path,

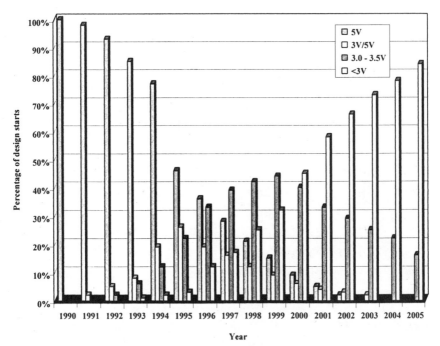

Figure 1.4   Transition of operating voltage from 5 V to less than 3 V.

high-speed memory controller, NuBus controller, and I/O controller; and many system memories, such as read only memory (ROM), which contains permanent code used by software applications, and dynamic random access memory (DRAM) to store the information while the power is turned on.

The microprocessor is the brain of a computing system. Some well-known microprocessors are Intel's complex instruction set computing (CISC)-based microprocessor family (e.g., the Pentiums); IBM, Motorola, and Apple's reduced instruction set computing (RISC)-based PowerPC microprocessors; Hewlett-Packard's RISC-based PA8000; Compaq Computer's RISC-based Alpha chipset; Silicon Graphics' RISC-based MIPS; AMD's CISC-based K5, K6, and K7; and Sun Microsystems' RISC-based UltraSparc. Both RISC-based and CISC-based microprocessors are expected to require package pin counts of over 1000 and to perform at over 500-MHz on-chip clock frequencies (Fig. 1.5).

The SRAMs for cache memories are expected to perform at high speeds similar to those of the microprocessors to prevent system data bottlenecks. Even the ASICs are expected to run at on-chip clock frequencies faster than 300 MHz and to have package pin counts of up to 900. (The ASICs of some telecommunication products need pin counts

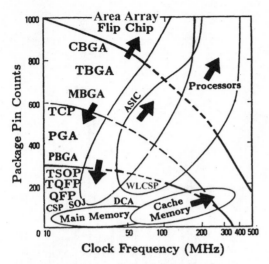

Figure 1.5   IC devices and packaging trends.

of more than 1000.) The DRAM densities are performing in accordance with Moore's law (Figs. 1.6 and 1.7), i.e., the number of transistors per chip doubles for every 18-month period.

1.2.4   Copper interconnects

More than 99 percent of IC chips used in the world today are equipped with Al interconnects. However, IC chips with Cu interconnects are being manufactured and implemented in the products of some well-known companies. For example, IBM is shipping the 400-MHz Cu-based powerPC 750 microprocessors (with 33 percent improvement) and implementing the Cu-pad technology into its S/390, RS/6000, and AS/400 server families. Motorola's Cu-based CMOS process produces the highest-clock-frequency SRAM devices. Apple Computer uses Cu wire interconnects to enhance the performance of its Power Mac G3. Cu-based devices in the 0.18-µm generation are being offered by TSMC; Lucent is implementing these devices in the 0.16-µm generation; and Intel is planning them in the 0.13-µm generation.

The advantages of Cu-based IC devices are: (1) higher conductivity with lower resistance, (2) lower capacitive load on the interconnects, (3) lower power consumption, (4) lower crosstalk, (5) fewer metal layers, (6) fewer manufacturing steps, (7) less fabrication space, (8) shorter production cycle time, (9) fewer defects in IC devices, and (10) better electromigration performance. The disadvantages are: (1) deposition of Cu in layers only several atoms thick is more difficult, (2) contamina-

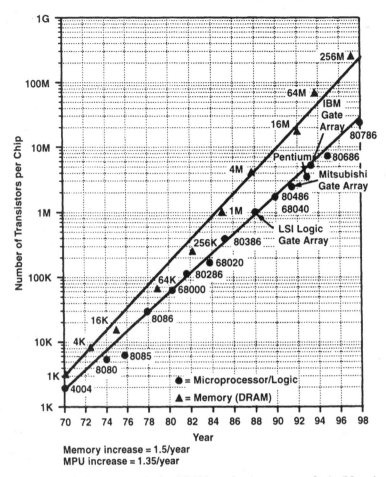

Figure 1.6 IC density trends for DRAMs and microprocessor/logic (Moore's law). *(Source: ICE Corp.)*

tion with the underlying silicon layer will lower transistor performance, (3) migration between Cu lines makes shorts more likely, (4) the equipment is more expensive, and (5) there is a thermal coefficient of expansion mismatch between the Si chip and the Cu pads and lines.

Some well-known equipment companies are also supporting the Cu-based technology. For example, Applied Materials provides the Copper Interconnect Equipment Set Solution (CIESS) for fabricating a complete multilevel copper interconnect structure, and the millenia electra electrochemical plating system (MEEPS) for filling vias and trenches with copper. LAM Research, Novellus, and KLA-Tencor are planning to offer a complete Cu-based production line soon.

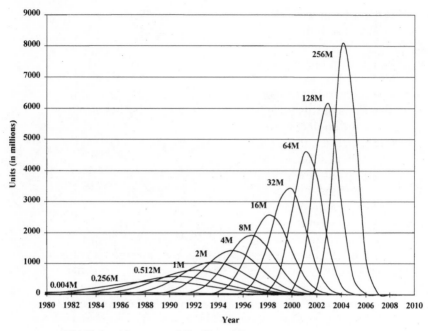

Figure 1.7   DRAM densities are tracking Moore's law.

1.2.5   Moore's Law

Since 1965, Moore's Law has been an unshakable principle for the computer industry: every 18 months, the number of transistors that will fit on a silicon chip doubles. As transistors have been scaled ever smaller, computing performance has risen exponentially while the cost of that power has been driven down. And it has been assumed (or set as a goal) in the semiconductor industry that this rate of progress would hold for at least another 10 to 15 years.

Recently, however, more and more people have reported that insurmountable barriers have forced the repeal of Moore's Law much closer at hand, perhaps early in the coming decade. It is not clear whether the most common type of silicon transistor can be scaled down beyond the generation of chips that will begin to appear in the next couple of years, because semiconductor engineers have not found ways around basic physical limits.

For example, in the current generation of semiconductors, the wires that interconnect transistors are etched as fine as 0.18 µm and the individual insulating layers inside a transistor may be only four or five atoms thick. Semiconductor factories in Japan plan to begin mass production of chips based on 0.13-µm technology early next year, and such chips should be in widespread use within two years. But beyond that generation, the industry's leading researchers acknowledge there remain far more questions than answers. This is because the next step would be 0.1-μm technology, a milestone that in the Moore's Law progression would be expected three to five years from now. But at that scale, transistors will be composed of fewer than 100 atoms, and statistical variations in this lilliputian world are beyond the ability of silicon semiconductor engineers to control.

Without further advances in the miniaturization of silicon-based transistors, hopes for continued progress would have to be based on technologies that are promising but unproved: new materials, new transistor designs, and advances like IBM's silicon-on-insulator and HP's molecular computing. Instead of shrinking today's CMOS technologies, IBM discovered that its silicon-on-insulator technology holds great promise at dimensions of 0.1 μm and smaller. HP and the University of California at Los Angeles invented the Carbon 60 molecule known as the buckyball, in which single molecules act as digital on-off switches.

Thus, in order to extend the reign of Moore's Law, the industry should be prepared and accepting of new technology. Says Randall Isaac, vice president for systems technology and science at IBM's Watson Laboratory in Yorktown Heights, New York, "When a given technology saturates, it is usually replaced by a new one."

1.3 Packaging Technology Update

The major trend in electronic products today is to make them smarter, lighter, smaller, thinner, shorter, and faster, while at the same time more friendly, functional, powerful, reliable, robust, innovative, creative, and inexpensive (Fig. 1.8). As the trend toward miniature and compact products continues, the introduction of products that are more user friendly, that contain a wider variety of functions, and that can be put into a tiny box will provide growth in the market. One of the key technologies that is helping to make these goals for miniaturized and integrated product design possible is electronics packaging and assembly technology, especially low-cost flip chip technology. For example, in 1992 the typical video camera consisted of over 1800 components, with a density of 12 components per cm^2. It occupied 2380 mm^3 and utilized a black-and-white charge-coupled device (CCD) array for viewing. The 1998 video camera consisted only 848 components, with a density of 21 components per cm^2. It occupied only 1598 mm^3 and utilized a color liquid crystal display (LCD) for viewing (Table 1.2). The size and weight of the video camera are expected to be reduced by 50 percent by the year 2002.

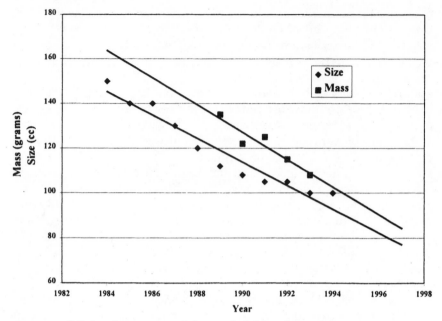

Figure 1.8   Cellular phone mass and size trends (without battery).

Different packaging technologies are required for different semiconductor IC devices and applications.[1–174] Figure 1.5 shows some of the well-known packaging technologies for four of the major IC devices (e.g., ASIC, microprocessor, cache, and system memory).

The dashed lines in Fig. 1.5 demarcate four regions of applications. The first region is for low-pin-count IC devices; the packaging technologies are plastic quad flat pack (PQFP), Swiss outline package (SOP), Swiss outline J-leaded (SOJ), small outline IC (SOIC), plastic leaded chip carrier (PLCC), thin quad flat pack (TQFP), thin small outline package (TSOP), direct chip attach (DCA), chip scale package (CSP), and wafer level CSP (WLCSP).

The packaging technologies such as the tape carrier package (TCP), plastic pin grid array (PPGA), ceramic pin grid array (CPGA), and plastic ball grid array (PBGA) will meet the needs for the second region of applications. For higher-pin-count and higher-performance IC devices (the third region of applications), ceramic ball grid array (CBGA), tape ball grid array (TBGA), and metal ball grid array (MBGA) are cost-effective packaging technologies. For the fourth region of applications (IC devices with very high pin counts and performances), area-array solder-bumped flip chip technology is the solution.

TABLE 1.2   Electronic Product Trend Example: Video Camera

Year Model	1992 NV-S1	1998 DS7
Size		
Volume	2,380 mm^3	1,590 mm^3
Weight	900 g	690 g
Technology	Analog	Digital
Display	Black and white	Color
	CCD array	LCD
Circuit board		
Density	12 components/cm^2	21 components/cm^2
Technology	0.5 mm quad flat pack	CSP
Component technology		
Chips		
1005	115	56
1608	990	29
2124/2520	190	55
Transistor	205	58
Tantalum	130	31
Odd-form	116	60
IC		
SOP/QFP	24	38
CSP	0	10
Total components	1800	848

1.3.1   Area-array flip chip technology

For high-price and high-speed microprocessors and ASICs, the complex IC designs require very high I/O and performance packages. For these types of ICs and subsystems, area-array solder-bumped flip chip technology provides a viable answer to the needs (Fig. 1.5). Usually, these ICs are packaged in a CPGA, a PPGA, a CBGA, a TBGA, or an MBGA in a single-chip format. Recently, PBGAs with solder-bumped flip chips (Fig. 1.9) using the DCA technology have become very popular.

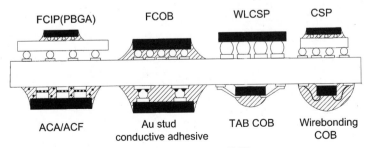

Figure 1.9   Various packaging technologies on PCB.

1.3.2 BGA technology

There are many different kinds of ball grid array (BGA) packages.[3, 5] Depending on their substrates, they are known as CBGA, MBGA, TBGA, and PBGA. Figure 1.5 shows that for high-I/O and high-performance ASICs and microprocessers, CBGA, MBGA, and TBGA could meet the high pin count (>500), power, and clock frequency requirements, but at higher costs.

Due to the fine pitch and pin count limitations of PQFP and the high cost of CBGAs, MBGAs, and TBGAs, the PBGA is cost effective at package pin counts between 250 and 600. The major difference between PQFP and PBGA is that PQFP has a leadframe and PBGA has an organic substrate. The leadframe was standardized for fanning the circuitry more than 15 years ago, while the PBGA's substrate is still custom designed. Today, the leadframe is cheaper than the organic substrate. Thus, in order for the PBGA packages to be popular, the substrate has to be standardized and lower in cost. Again, some system makers employ PBGA packages to house the solder-bumped flip chip by using the DCA technology (Fig. 1.9) for very high pin count (>1000) and performance (see Chap. 15).

1.3.3 TCP

TCP offers smaller pitches, thinner package profiles, and smaller footprints on the PCB. TCP can provide moderate-performance solutions for applications (e.g., ASICs and microprocessors) with package pin counts of up to about 600. It is noted that unless it is a very high volume production, TCP may not be cost effective due to very high development cost of these custom-designed packages. Also, TCPs suffer the same drawbacks (e.g., peripheral chip carrier, long lead length, handling, board level manufacturing yield loss) as the PQFPs.

1.3.4 TSOP and PQFP

Up to 208 pins (0.5-mm pitch and 28-mm body size), 240 pins (0.5-mm pitch, 32-mm body size), and 304 pins (0.5-mm pitch and 40-mm body size), the PQFPs are the most cost-effective packages for surface-mount technology (SMT). They have been used extensively for ASICs and low-performance and low-pin-count microprocessors. Sometimes, they are used to house one or more cache memories. Currently, the price for the PQFP packages is 0.5¢ per pin.

TSOP is a very low-profile plastic package that is specifically designed for SRAM, DRAM, and flash memory (which retains information even when the power is turned off) devices for space-limited applications. Right now, the price for TSOP is less than 1¢ per pin.

1.3.5 CSP and DCA

One of the most cost-effective packaging technologies is DCA. Some examples are shown in Figure 1.9, e.g., wire bonding chip on board,[6] tape automated bonding (TAB) chip on board,[10] flip chip with Au stud and conductive adhesive on board,[4] flip chip with anisotropic conductive adhesive (ACA) on board, flip chip with anisotropic conductive film (ACF) on board, and solder-bumped flip chip on board. However, because of the infrastructure and the cost of supplying the known good die (KGD) and the corresponding fine line and spacing PCB, most in the industry are still working on these issues.

In the meantime, a class of new technology called CSP has surfaced. There are more than 40 different CSPs reported today;[1] most are used for SRAMs, DRAMs, flash memories, and ASICs and microprocessors without high pin counts. The unique feature of most of the CSPs is the use of a substrate (carrier or interposer) to redistribute the very fine-pitch (as small as 0.075 mm) peripheral pads on the chip to much larger-pitch (1, 0.8, 0.75, and 0.5 mm) area-array pads on the PCB.

The advantages of CSP over DCA are that with the carrier (interposer), the CSP is easier to test at speed and burn in for KGD, to handle, to assemble, to rework, and to standardize; it is easier to protect the die and to deal with die shrink; and there are fewer infrastructure constraints. On the other hand, DCA has better electrical performance and less weight, size, and cost.

1.3.6 Wafer-level packaging

Since the publications by Tsukada, et al.[58–60] in 1992, the electronic packaging industry has witnessed an explosive growth in the research and development efforts devoted to solder bumped-flip chips on low-cost PCB or organic substrate,[61–168, 175] even though Citizen was making low-end watches at very high volumes with this technology a few years earlier (Fig. 1.10). Because of the thermal expansion mismatch between the silicon chip and the epoxy substrate, underfill encapsulant is usually needed for solder joint reliability. However, due to the underfill operation, the manufacturing cost is increased and the manufacturing throughput is reduced. In addition, the rework of an underfilled flip chip on PCB is very difficult, if it is not impossible. This further complicates the KGD-related issues.

There is another reason why DCA is not very popular yet. Usually, the pitch and size of the pads on the peripheral-arrayed chips (more than 90 percent of the IC chips used in the world) are very small and pose great demands on the supporting structures (PCB or organic substrate). The high-density and fine-line-width/spacing PCBs with sequential built-up circuits with microvias are not commonly available at reasonable cost yet.

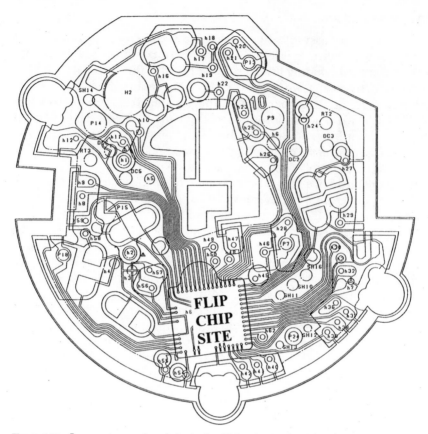

Figure 1.10 Low-cost organic substrate for Citizen's watch with solder-bumped flip chip with underfill.

Meantime, a new class of packaging called wafer-level chip scale package (WLCSP) provides a solution to these problems. (Lau and Lee[1] discuss the WLCSPs of eight different companies.) The unique feature of most WLCSPs is the use of a metal layer to redistribute the very fine-pitch peripheral-arrayed pads on the chip to much larger-pitch area-array pads with much larger solder joints (see Figs. 1.11 and 1.12) on the PCB. With WLCSPs, the demands on the PCB are relaxed, the underfill is not needed, and the KGD issues are much simpler. Thus, WLCSP is welcomed by system makers, because WLCSP is (potentially) lower in cost for high-yield IC chips (a necessary condition) and easier to assemble on PCB with higher yield (just like DCA but without underfill encapsulant).

There are more advantages of WLCSP compared to conventional CSP. Since the package is processed on a whole wafer before the individual

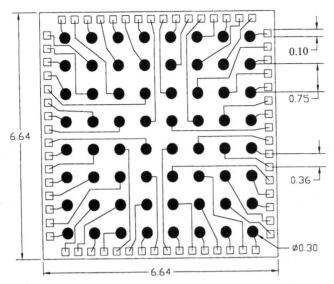

Figure 1.11   WLCSP of a peripheral-arrayed chip.

chips are separated, it may be possible to shorten the logistics chain by bringing packaging and final test back to the wafer FAB. In this case, the wafer would be functionally tested only once—after final packaging. There will be no need for a wafer probe for screening test before packaging, since the whole wafer will be packaged anyway. Also, WLCSP is not only one of the chip scale packages but a real chip size package. Finally, WLCSP enables copper interconnects on the chip. In view of Fig. 1.5,

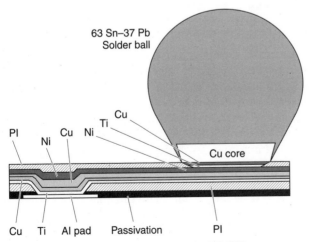

Figure 1.12   Structural cross section of the WLCSP.

WLCSP is most cost effective if the pin count is less than 250 and is perfect for DRAM devices in which the chip is large relative to its pin count. The disadvantages of WLCSP compared to conventional CSP are: (1) high cost for poor-yield IC chips, (2) problems with system makers if the IC chip shrinks, (3) difficulty of test at speed and burn-in at high temperature on a wafer, and (4) higher-density substrates. It should be noted that even though the concept of wafer-level packaging has been used by IBM for its solder-bumped flip chip technology for more than 30 years, nevertheless, now is the exciting time for WLCSP!

It is noted that a new burn-in method that has recently been qualified at Motorola is called *wafer-level burn-in* (WLBI). This is a lower-cost method than die-level burn-in, which allows simultaneous burn-in of all die on a wafer. This process uses a sacrificial metal circuit applied to the wafer for the burn-in and test contacts, and also provides for flip chip requirements. The WLBI circuit design, burn-in process, test strategy, die preparation, and assembly requirements can be found in Beddingfield et al.[175]

1.4   Summary

A brief overview of trends in IC devices (e.g., microprocessor, ASIC, cache memory, and system memory) has been presented. Also, various packaging technologies such as the area-array flip chip, BGA, TCP, CSP, WLCSP, and DCA for the IC devices have been briefly discussed.

The area-array solder-bumped flip chip technology is suitable for very high-I/O and high-performance single-chip modules for the microprocessors and ASICs. CBGAs, MBGAs, and TBGAs are cost-effective packages for microprocessors and ASICs with pin counts over 500 and high power dissipation. Usually, the area-array solder-bumped flip chip is packaged in a CBGA, an MBGA, or a TBGA.

In general, PBGAs are cost effective for packaging the ASICs (and sometimes the microprocessors) with pin counts of 250 to 600 and power dissipations of less than 5 W. In some cases, PBGAs with lower pin counts (less than 200) are also cost effective for housing a cache or a few fast SRAMs. For the solder-bumped flip chip in a PBGA package or the cavity-down thermal-enhanced wire bonding chip in a PBGA package, the back of the chip is attached to a heat spreader, which could be mounted with a heat sink. In that case, its power dissipation could be as high as 50 W and it could be used to house high-power microprocessors and ASICs.

CSPs are usually used to package memory and low-power (less than 1 W), low-pin-count (less than 200) ASICs, even though higher-pin-count (more than 1000), higher-power ICs have been tried in laboratories. Except for a handful of vertically integrated companies, DCAs are still awaiting the time when the KGD and high-density microvia buildup PCBs are commonly available at reasonable costs. The wait should not be too long, and they will come in a few years. While we are waiting, WLCSP is enjoying its honeymoon.

References

1. Lau, J. H., and S. W. R. Lee, *Chip Scale Package, Design, Materials, Process, Reliability, and Applications*, McGraw-Hill, New York, 1999.
2. Lau, J. H., C. P. Wong, J. L. Prince, and W. Nakayama, *Electronic Packaging, Design, Materials, Process, and Reliability*, McGraw-Hill, New York, 1998.
3. Lau, J. H., and Y.-H. Pao, *Solder Joint Reliability of BGA, CSP, Flip Chip, and Fine Pitch SMT Assemblies*, McGraw-Hill, New York, 1997.
4. Lau, J. H., *Flip Chip Technologies*, McGraw-Hill, New York, 1996.
5. Lau, J. H., *Ball Grid Array Technology*, McGraw-Hill, New York, 1995.
6. Lau, J. H., *Chip On Board Technologies for Multichip Modules*, Van Nostrand Reinhold, New York, 1994.
7. Lau, J. H., *Handbook of Fine Pitch Surface Mount Technology*, Van Nostrand Reinhold, New York, 1994.
8. Frear, D., H. Morgan, S. Burchett, and J. Lau, *The Mechanics of Solder Alloy*, Van Nostrand Reinhold, New York, 1994.
9. Lau, J. H., *Thermal Stress and Strain in Microelectronics Packaging*, Van Nostrand Reinhold, New York, 1993.
10. Lau, J. H., *Handbook of Tape Automated Bonding*, Van Nostrand Reinhold, New York, 1992.
11. Lau, J. H., *Solder Joint Reliability, Theory and Applications*, Van Nostrand Reinhold, New York, 1991.
12. Elshabini-Riad, A., and F. Barlow, III, *Thin Film Technology Handbook*, McGraw-Hill, New York, 1998.
13. Garrou, P. E., and I. Turlik, *Multichip Module Technology Handbook*, McGraw-Hill, New York, 1998.
14. Wong, C. P., *Polymers for Electronic and Photonic Applications*, Academic Press, San Diego, CA, 1993.
15. Senthinathan, R., and J. L. Prince, *Simultaneous Switching Noise of CMOS Devices and Systems*, Kluwer Academic Publishers, New York, 1994.
16. Tummala, R. R., E. Rymaszewski, and A. Klopfenstein, *Microelectronics Packaging Handbook*, Chapman & Hall, New York, 1997.
17. Tummala, R. R., and E. Rymaszewski, *Microelectronics Packaging Handbook*, Van Nostrand Reinhold, New York, 1989.
18. Seraphim, D. P., R. Lasky, and C. Y. Li, *Principles of Electronic Packaging*, McGraw-Hill, New York, 1989.
19. Vardaman, J., *Surface Mount Technology, Recent Japanese Developments*, IEEE Press, New York, 1992.
20. Hwang, J. S., *Solder Paste in Electronics Packaging*, Van Nostrand Reinhold, New York, 1989.
21. Hwang, J. S., *Modern Solder Technology for Competitive Electronics Manufacturing*, McGraw-Hill, New York, 1996.
22. Johnson, R. W., R. K. Teng, and J. W. Balde, *Multichip Modules: System Advantages, Major Construction, and Materials Technologies*, IEEE Press, New York, 1991.
23. Sandborn, P. A., and H. Moreno, *Conceptual Design of Multichip Modules and Systems*, Kluwer Academic Publishers, New York, 1994.
24. Nash, F. R., *Estimating Device Reliability: Assessment of Credibility*, Kluwer Academic Publishers, New York, 1993.
25. Gyvez, J. P., *Integrated Circuit Defect-Sensitivity: Theory and Computational Models*, Kluwer Academic Publishers, New York, 1993.

26. Doane, D. A., and P. D. Franzon, *Multichip Module Technologies and Alternatives*, Van Nostrand Reinhold, New York, 1992.
27. Messuer, G., I. Turlik, J. Balde, and P. Garrou, *Thin Film Multichip Modules*, International Society for Hybrid Microelectronics, Silver Spring, MD, 1992.
28. Manzione, L. T., *Plastic Packaging of Microelectronic Devices*, Van Nostrand Reinhold, New York, 1990.
29. Hymes, L., *Cleaning Printed Wiring Assemblies in Today's Environment*, Van Nostrand Reinhold, New York, 1991.
30. Gilleo, K., *Handbook of Flexible Circuits*, Van Nostrand Reinhold, New York, 1991.
31. Engel, P. A., *Structural Analysis of Printed Circuit Board Systems*, Springer-Verlag, New York, 1993.
32. Suhir, E., *Structural Analysis in Microelectronic and Fiber Optics Systems*, Van Nostrand Reinhold, New York, 1991.
33. Matisoff, B. S., *Handbook of Electronic Packaging Design and Engineering*, Van Nostrand Reinhold, New York, 1989.
34. Prasad, R. P., *Surface Mount Technology*, Van Nostrand Reinhold, New York, 1989.
35. Manko, H. H., *Soldering Handbook for Printed Circuits and Surface Mounting*, Van Nostrand Reinhold, New York, 1986.
36. Morris, J. E., *Electronics Packaging Forum*, vol. 1, Van Nostrand Reinhold, New York, 1990.
37. Morris, J. E., *Electronics Packaging Forum*, vol. 2, Van Nostrand Reinhold, New York, 1991.
38. Hollomon, J. K., Jr., *Surface-Mount Technology*, Howard W. Sams & Company, Indianapolis, IN, 1989.
39. Solberg, V., *Design Guidelines for SMT*, TAB Professional and Reference Books, New York, 1990.
40. Hutchins, C., *SMT: How to Get Start*, Hutchins and Associates, Raleigh, NC, 1990.
41. Bar-Cohen, A., and A. D. Kraus, *Advances in Thermal Modeling of Electronic Components and Systems*, vol. 1, Hemisphere Publishing, New York, 1988.
42. Bar-Cohen, A., and A. D. Kraus, *Advances in Thermal Modeling of Electronic Components and Systems*, vol. 2, ASME Press, New York, 1990.
43. Kraus, A. D., and A. Bar-Cohen, *Thermal Analysis and Control of Electronic Equipment*, Hemisphere Publishing, New York, 1983.
44. Harper, C. A., *Handbook of Microelectronics Packaging*, McGraw-Hill, New York, 1991.
45. Pecht, M., *Handbook of Electronic Package Design*, Marcel Dekker, New York, 1991.
46. Hannemann, R., A. Kraus, and M. Pecht, *Physical Architecture of VLSI Systems*, John Wiley & Sons, New York, 1994.
47. Pecht, M., *Integrated Circuit, Hybrid, and Multichip Module Package Design Guidelines*, John Wiley & Sons, New York, 1994.
48. Pecht, M., A. Dasgupta, J. Evans, and J. Evans, *Quality Conformance and Qualification of Microelectronic Packages and Interconnects*, John Wiley & Sons, New York, 1994.
49. Mroczkowski, R., *Electronic Connector Handbook*, McGraw-Hill, New York, 1998.
50. Giacomo, G., *Reliability of Electronic packages and Semiconductor Devices*, McGraw-Hill, New York, 1997.
51. Harman, G., *Wire Bonding in Microelectronics*, International Society for Hybrid Microelectronics, Reston, VA, 1989.
52. Lea, C., *A Scientific Guide to Surface Mount Technology*, Electrochemical Publications, Scotland, 1988.
53. Lea, C., *After CFCs? Options for Cleaning Electronics Assemblies*, Electrochemical Publications, IOM, British isles, 1992.
54. Wassink, R. J. K., *Soldering in Electronics*, Electrochemical Publications, Ayr Scotland, 1989.
55. Pawling, J. F., *Surface Mounted Assemblies*, Electrochemical Publications, Ayr Scotland, 1987.
56. Ellis, B. N., *Cleaning and Contamination of Electronics Components and Assemblies*, Electrochemical Publications, Ayr Scotland, 1986.

57. Sinnadurai, F. N., *Handbook of Microelectronics Packaging and Interconnection Technologies,* Electrochemical Publications, Ayr Scotland, 1985.
58. Tsukada, Y., Y. Mashimoto, T. Nishio, and N. Mii, "Reliability and Stress Analysis of Encapsulated Flip Chip Joint on Epoxy Base Printed Circuit Board," *Proceedings of the 1st ASME/JSME Advances in Electronic Packaging Conference,* pp. 827–835, Milpitas, CA, April 1992.
59. Tsukada, Y., S. Tsuchida, and Y. Mashimoto, "Surface Laminar Circuit Packaging," *Proceedings of IEEE Electronic Components and Technology Conference,* pp. 22–27, San Diego, CA, May 1992.
60. Tsukada, Y., and S. Tsuchida, "Surface Laminar Circuit, A Low Cost High Density Printed Circuit Board," *Proceedings of Surface Mount International Conference,* pp. 537–542, August 1992.
61. Guo, Y., W. T. Chen, and K. C. Lim, "Experimental Determinations of Thermal Strains in Semiconductor Packaging Using Moire Interferometry," *Proceedings of the 1st ASME/JSME Advances in Electronic Packaging Conference,* pp. 779–784, Milpitas, CA, April 1992.
62. Lau, J. H., "Thermal Fatigue Life Prediction of Encapsulated Flip Chip Solder Joints for Surface Laminar Circuit Packaging," ASME Paper no. 92W/EEP-34, ASME Winter Annual Meeting, Anaheim, CA, November 1992.
63. Lau, J. H., T. Krulevitch, W. Schar, M. Heydinger, S. Erasmus, and J. Gleason, "Experimental and Analytical Studies of Encapsulated Flip Chip Solder Bumps on Surface Laminar Circuit Boards," *Circuit World,* **19,** (3): 18–24, March 1993.
64. Tsukada, Y., S. Tsuchida, and Y. Mashimoto, "A Novel Chip Replacement Method for Encapsulated Flip Chip Bonding," *Proceedings of IEEE Electronic Components and Technology Conference,* pp. 199–204, Orlando, FL, June 1993.
65. Powell, D. O., and A. K. Trivedi, "Flip-Chip on FR-4 Integrated Circuit Packaging," *Proceedings of IEEE Electronic Components and Technology Conference,* pp. 182–186, Orlando, FL, June 1993.
66. Wang, D. W., and K. I. Papathomas, "Encapsulant for Fatigue Life Enhancement of Controlled Collapse Chip Connection (C4)," *IEEE Transactions on Components, Hybrids, and Manufacturing Technology,* **16:** 863–867, 1993.
67. Tsukada, Y., "Solder Bumped Flip Chip Attach on SLC Board and Multichip Module," in *Chip On Board Technologies for Multichip Modules,* Lau, J. H., ed., Van Nostrand Reinhold, New York, pp. 410–443, 1994.
68. Wong, C. P., J. M. Segelken, and C. N. Robinson, "Chip on Board Encapsulation," in *Chip On Board Technologies for Multichip Modules,* Lau, J. H., ed., Van Nostrand Reinhold, New York, pp. 470–503, 1994.
69. Suryanarayana, D., and D. S. Farquhar, "Underfill Encapsulation for Flip Chip Applications," in *Chip On Board Technologies for Multichip Modules,* Lau, J. H., ed., Van Nostrand Reinhold, New York, pp. 504–531, 1994.
70. Lau, J. H., M. Heydinger, J. Glazer, and D. Uno, "Design and Procurement of Eutectic Sn/Pb Solder-Bumped Flip Chip Test Die and Organic Substrates," *Proceedings of the IEEE International Manufacturing Technology Symposium,* pp. 132–138, San Diego, CA, September 1994.
71. Wun, K. B., and J. H. Lau, "Characterization and Evaluation of the Underfill Encapsulants for Flip Chip Assembly," *Proceedings of the IEEE International Manufacturing Technology Symposium,* pp. 139–146, San Diego, CA, September 1994.
72. Kelly, M., and J. H. Lau, "Low Cost Solder Bumped Flip Chip MCM-L Demonstration," *Proceedings of the IEEE International Manufacturing Technology Symposium,* pp. 147–153, San Diego, CA, September 1994.
73. Pompeo, F. L., A. J. Call, J. T. Coffin, and S. Buchwalter, "Reworkable Encapsulation for Flip Chip Packaging," *Proceedings of the International Intersociety Electronic Packaging Conference,* pp. 781–787, Maui, HI, March 1995.
74. Suryanarayana, D., J. A. Varcoe, and J. V. Ellerson, "Reparability of Underfill Encapsulated Flip-Chip Packages," *Proceedings of IEEE Electronic Components and Technology Conference,* pp. 524–528, Las Vegas, NV, May 1995.
75. Schwiebert, M. K., and W. H. Leong, "Underfill Flow as Viscous Flow Between Parallel Plates Driven by Capillary Action," *Proceedings of the IEEE International Manufacturing Technology Symposium,* pp. 8–13, Austin, TX, October 1995.

76. Han, S., K. Wang, and S. Cho, "Experimental and Analytical Study on the Flow of Encapsulant During Underfill Encapsulation of Flip-Chips," *Proceedings of IEEE Electronic Components and Technology Conference,* pp. 327–334, Orlando, FL, May 1996.

77. Wun, K. B., and G. Margaritis, "The Evaluation of Fast-Flow, Fast-Cure Underfills for Flip Chip on Organic Substrates," *Proceedings of IEEE Electronic Components and Technology Conference,* pp. 540–545, Orlando, FL, May 1996.

78. Hwang, J. S., *Modern Solder Technology for Competitive Electronics Manufacturing,* McGraw-Hill, New York, 1996.

79. Lau, J. H., "Solder Joint Reliability of Flip Chip and Plastic Ball Grid Array Assemblies Under Thermal, Mechanical, and Vibration Conditions," *IEEE Trans. Component, Packaging, and Manufacturing Technol.,* part B, **19** (4): 728–735, November 1996.

80. Lau, J. H., E. Schneider, and T. Baker, "Shock and Vibration of Solder Bumped Flip Chip on Organic Coated Copper Boards," *ASME Trans., J. Electronic Packaging,* **118:** 101–104, June 1996.

81. Gamota, D., and C. Melton, "Reflowable Material Systems to Integrate the Reflow and Encapsulant Dispensing Process for Flip Chip on Board Assemblies," *IPC-TP-1098,* 1996.

82. Ito, S., M. Kuwamura, S. Sudo, M. Mizutani, T. Fukushima, H. Noro, S. Akizuki, and A. Prabhu, "Study of Encapsulating System for Diversified Area Bump Packages," *Proceedings of IEEE Electronic Components and Technology Conference,* pp. 46–53, San Jose, CA, May 1997.

83. Pascarella, N., and D. Baldwin, "Advanced Encapsulation Processing for Low Cost Electronics Assembly—A Cost Analysis," *The 3rd International Symposium and Exhibition on Advanced Packaging Materials, Processes, Properties, and Interfaces,* pp. 50–53, Braselton, GA, March 1997.

84. Naguyen, L., L. Hoang, P. Fine, Q. Tong, B. Ma, R. Humphreys, A. Savoca, C. P. Wong, S. Shi, M. Vincent, and L. Wang, "High Performance Underfills Development—Materials, Processes, and Reliability," *IEEE 1st International Symposium on Polymeric Electronics Packaging,* pp. 300–306, Norrkoping, Sweden, October 1997.

85. Erickson, M., and K. Kirsten, "Simplifying the Assembly Process with a Reflow Encapsulant," *Electronic Packaging and Production,* pp. 81–86, February 1997,

86. Wong, C. P., M. B. Vincent, and S. Shi, "Fast-Flow Underfill Encapsulant: Flow Rate and Coefficient of Thermal Expansion," *Proc. ASME—Adv. Electronic Packaging,* **19** (1): 301–306, 1997.

87. Wong, C. P., S. H. Shi, and G. Jefferson, "High Performance No Flow Underfills for Low-Cost Flip-Chip Applications," *Proceedings of IEEE Electronic Components and Technology Conference,* pp. 850–858, San Jose, CA, May 1997.

88. Tummala, R., E. Rymaszewski, and A. Klopfenstein, *Microelectronics Packaging Handbook,* Chapman & Hall, New York, 1997.

89. Lau, J. H., C. Chang, and R. Chen, "Effects of Underfill Encapsulant on the Mechanical and Electrical Performance of a Functional Flip Chip Device," *J. Electronics Manufacturing,* **7** (4): 269–277, December 1997.

90. Lau, J. H., and C. Chang, "How to Select Underfill Materials for Solder Bumped Flip Chip on Low Cost Substrates?" *Proceedings of the International Symposium on Microelectronics,* pp. 693–700, San Diego, CA, November 1998.

91. Nguyen, L., C. Quentin, P. Fine, B. Cobb, S. Bayyuk, H. Yang, and S. A. Bidstrup-Allen, "Underfill of Flip Chip on Laminates: Simulation and Validation," *Proceedings of the International Symposium on Adhesives in Electronics,* pp. 27–30, Binghamton, NY, September 1998.

92. Pascarella, N., and D. Baldwin, "Compression Flow Modeling of Underfill Encapsulants for Low Cost Flip Chip Assembly," *Proceedings of IEEE Electronic Components and Technology Conference,* pp. 463–470, Seattle, WA, May 1998.

93. Nguyen, L., P. Fine, B. Cobb, Q. Tong, B. Ma, and A. Savoca, "Reworkable Flip Chip Underfill—Materials and Processes," *Proceedings of the International Symposium on Microelectronics,* pp. 707–713, San Diego, CA, November 1998.

94. Capote, M. A., and S. Zhu, "No-Underfill Flip-Chip Encapsulation," *Proceedings of Surface Mount International Conference,* pp. 291–293, San Jose, CA, August 1998.

95. Capote, M. A., W. Johnson, S. Zhu, L. Zhou, and B. Gao, "Reflow-Curable Polymer Fluxes for Flip Chip Encapsulation," *Proceedings of the International Conference on Multichip Modules and High Density Packaging*, pp. 41–46, Denver, CO, April 1998.

96. Vincent, M. B., and C. P. Wong, "Enhancement of Underfill Encapsulants for Flip-Chip Technology," *Proceedings of Surface Mount International Conference*, pp. 303–312, San Jose, CA, August 1998.

97. Vincent, M. B., L. Meyers, and C. P. Wong, "Enhancement of Underfill Performance for Flip-Chip Applications by Use of Silane Additives," *Proceedings of IEEE Electronic Components and Technology Conference*, pp. 125–131, Seattle, WA, May 1998.

98. Wang, L., and C. P. Wong, "Novel Thermally Reworkable Underfill Encapsulants for Flip-Chip Applications," *Proceedings of IEEE Electronic Components and Technology Conference*, pp. 92–100, Seattle, WA, May 1998.

99. Shi, S. H., and C. P. Wong, "Study of the Fluxing Agent Effects on the Properties of No-Flow Underfill Materials for Flip-Chip Applications," *Proceedings of IEEE Electronic Components and Technology Conference*, pp. 117–124, Seattle, WA, May 1998.

100. Wong, C. P., D. Baldwin, M. B. Vincent, B. Fennell, L. J. Wang, and S. H. Shi, "Characterization of a No-Flow Underfill Encapsulant During the Solder Reflow Process," *Proceedings of IEEE Electronic Components and Technology Conference*, pp. 1253–1259, Seattle, WA, May 1998.

101. Ito, S., M. Mizutani, H. Noro, M. Kuwamura, and A. Prabhu, "A Novel Flip Chip Technology Using non-Conductive Resin Sheet," *Proceedings of IEEE Electronic Components and Technology Conference*, pp. 1047–1051, Seattle, WA, May 1998.

102. Gilleo, K., and D. Blumel, "The Great Underfill Race," *Proceedings of the International Symposium on Microelectronics*, pp. 701–706, San Diego, CA, November 1998.

103. Lau, J. H., C. Chang, T. Chen, D. Cheng, and E. Lao, "A Low-Cost Solder-Bumped Chip Scale Package—NuCSP," *Circuit World*, 24 (3): 11–25, April 1998.

104. Elshabini-Riad, A., and F. Barlow III, *Thin Film Technology Handbook*, McGraw-Hill, New York, 1998.

105. Garrou, P. E., and I. Turlik, *Multichip Module Technology Handbook*, McGraw-Hill, New York, 1998.

106. Lau, J. H., C. Chang, and O. Chien, "SMT Compatible No-Flow Underfill for Solder Bumped Flip Chip on Low-Cost Substrates," *J. Electronics Manufacturing*, 8 (3, 4): 151–164, December 1998.

107. Lau, J. H., and C. Chang, "Characterization of Underfill Materials for Functional Solder Bumped Flip Chips on Board Applications," *IEEE Trans. Components and Packaging Technology*, part A, 22 (1): 111–119, March 1999.

108. Thorpe, R., and D. F. Baldwin, "High Throughput Flip Chip Processing and Reliability Analysis Using No-Flow Underfills," *Proceedings of IEEE Electronic Components and Technology Conference*, pp. 419–425, San Diego, CA, June 1999.

109. Qian, Z., M. Lu, W. Ren, and S. Liu, "Fatigue Life Prediction of Flip-Chips in Terms of Nonlinear Behaviors of Solder and Underfill," *Proceedings of IEEE Electronic Components and Technology Conference*, pp. 141–148, San Diego, CA, June 1999.

110. Wang, L., and C. P. Wong, "Epoxy-Additive Interaction Studies of Thermally Reworkable Underfills for Flip-Chip Applications," *Proceedings of IEEE Electronic Components and Technology Conference*, pp. 34–42, San Diego, CA, June 1999.

111. Lau, J. H., S.-W. Lee, C. Chang, and O. Chien, "Effects of Underfill Material Properties on the Reliability of Solder Bumped Flip Chip on Board with Imperfect Underfill Encapsulants," *Proceedings of IEEE Electronic Components and Technology Conference*, pp. 571–582, San Diego, CA, June 1999.

112. Lau, J. H., C. Chang, and O. Chien, "No-Flow Underfill for Solder Bumped Flip Chip on Low-Cost Substrates," *Proceedings of NEPCON West*, pp. 158–181, February 1999.

113. Tong, Q., A. Savoca, L. Nguyen, P. Fine, and B. Cobb, "Novel Fast Cure and Reworkable Underfill Materials," *Proceedings of IEEE Electronic Components and Technology Conference*, pp. 43–48, San Diego, CA, June 1999.

114. Benjamin, T. A., A. Chang, D. A. Dubois, M. Fan, D. L. Gelles, S. R. Iyer, S. Mohindra, P. N. Tutunjian, P. K. Wang, and W. J. Wright, "CARIVERSE Resin: A Thermally Reversible Network Polymer for Electronic Applications," *Proceedings of IEEE Elec-*

tronic Components and Technology Conference, pp. 49–55, San Diego, CA, June 1999.

115. Wada, M., "Development of Underfill Material with High Valued Performance," *Proceedings of IEEE Electronic Components and Technology Conference,* pp. 56–60, San Diego, CA, June 1999.

116. Houston, P. N., D. F. Baldwin, M. Deladisma, L. N. Crane, and M. Konarski, "Low Cost Flip Chip Processing and Reliability of Fast-Flow, Snap-Cure Underfills," *Proceedings of IEEE Electronic Components and Technology Conference,* pp. 61–70, San Diego, CA, June 1999.

117. Kulojarvi, K., S. Pienimaa, and J. K. Kivilahti, "High Volume Capable Direct Chip Attachment Methods," *Proceedings of IEEE Electronic Components and Technology Conference,* pp. 441–445, San Diego, CA, June 1999.

118. Shi, S. H., and C. P. Wong, "Recent Advances in the Development of No-Flow Underfill Encapsulants—A Practical Approach towards the Actual Manufacturing Application," *Proceedings of IEEE Electronic Components and Technology Conference,* pp. 770–776, San Diego, CA, June 1999.

119. Rao, Y., S. H. Shi, and C. P. Wong, "A Simple Evaluation Methodology of Young's Modulus—Temperature Relationship for the Underfill Encapsulants," *Proceedings of IEEE Electronic Components and Technology Conference,* pp. 784–789, San Diego, CA, June 1999.

120. Fine, P., and L. Nguyen, "Flip Chip Underfill Flow Characteristics and Prediction," *Proceedings of IEEE Electronic Components and Technology Conference,* pp. 790–796, San Diego, CA, June 1999.

121. Johnson, C. H., and D. F. Baldwin, "Wafer Scale Packaging Based on Underfill Applied at the Wafer Level for Low-Cost Flip Chip Processing," *Proceedings of IEEE Electronic Components and Technology Conference,* pp. 950–954, San Diego, CA, June 1999.

122. DeBarros, T., P. Neathway, and Q. Chu, "The No-Flow Fluxing Underfill Adhesive for Low Cost, High Reliability Flip Chip Assembly," *Proceedings of IEEE Electronic Components and Technology Conference,* pp. 955–960, San Diego, CA, June 1999.

123. Shi, S. H., T. Yamashita, and C. P. Wong, "Development of the Wafer Level Compressive-Flow Underfill Process and Its Required Materials," *Proceedings of IEEE Electronic Components and Technology Conference,* pp. 961–966, San Diego, CA, June 1999.

124. Chau, M. M., B. Ho, T. Herrington, and J. Bowen, "Novel Flip Chip Underfills," *Proceedings of IEEE Electronic Components and Technology Conference,* pp. 967–974, San Diego, CA, June 1999.

125. Feustel, F., and A. Eckebracht, "Influence of Flux Selection and Underfill Selection on the Reliability of Flip Chips on FR-4," *Proceedings of IEEE Electronic Components and Technology Conference,* pp. 583–588, San Diego, CA, June 1999.

126. Okura, J. H., K. Drabha, S. Shetty, and A. Dasgupta, "Guidelines to Select Underfills for Flip Chip on Board Assemblies," *Proceedings of IEEE Electronic Components and Technology Conference,* pp. 589–594, San Diego, CA, June 1999.

127. Anderson, B., "Development Methodology for a High-Performance, Snap-Cure Flip-Chip Underfill," *Proceedings of NEPCON WEST,* pp. 135–143, February 1999.

128. Wyllie, G., and B. Miquel, "Technical Advancements in Underfill Dispensing," *Proceedings of NEPCON WEST,* pp. 152–157, February 1999.

129. Crane, L., A. Torres-Filho, E. Yager, M. Heuel, C. Ober, S. Yang, J. Chen, and R. Johnson, "Development of Reworkable Underfills, Materials, Reliability and Proceeding," *Proceedings of NEPCON WEST,* pp. 144–151, February 1999.

130. Gilleo, K., "The Ultimate Flip Chip-Integrated Flux/Underfill," *Proceedings of NEPCON WEST,* pp. 1477–1488, February 1999.

131. Miller, M., I. Mohammed, X. Dai, N. Jiang, and P. Ho, "Analysis of Flip-Chip Packages Using High Resolution Moire Interometry," *Proceedings of IEEE Electronic Components and Technology Conference,* pp. 979–986, San Diego, CA, June 1999.

132. Hanna, C., S. Michaelides, P. Palaniappan, D. Baldwin, and S. Sitaraman, "Numerical and Experimental Study of the Evolution of Stresses in Flip Chip Assemblies During Assembly and Thermal Cycling," *Proceedings of IEEE Electronic Components and Technology Conference,* pp. 1001–1009, San Diego, CA, June 1999.

133. Emerson, J., and L. Adkins, "Techniques for Determining the Flow Properties of Underfill Materials," *Proceedings of IEEE Electronic Components and Technology Conference,* pp. 777–781, San Diego, CA, June 1999.

134. Guo, Y., G. Lehmann, T. Driscoll, and E. Cotts, "A Model of the Underfill Flow Process: Particle Distribution Effects," *Proceedings of IEEE Electronic Components and Technology Conference,* pp. 71–76, San Diego, CA, June 1999.

135. Mercado, L., V. Sarihan, Y. Guo, and A. Mawer, *Proceedings of IEEE Electronic Components and Technology Conference,* pp. 255–259, San Diego, CA, June 1999.

136. Qian, Z., M. Lu, W. Ren, and S. Liu, "Fatigue Life Prediction of Flip-Chips in Terms of Nonlinear Behaviors of Solder and Underfill," *Proceedings of IEEE Electronic Components and Technology Conference,* pp. 141–148, San Diego, CA, June 1999.

137. Gektin, V., A. Bar-Cohen, and S. Witzman, "Thermo-Structural Behavior of Underfilled Flip-Chips," *Proceedings of IEEE Electronic Components and Technology Conference,* pp. 440–447, Orlando, FL, May 1996.

138. Wu, T. Y., Y. Tsukada, and W. T. Chen, "Materials and Mechanics Issues in Flip-Chip Organic Packaging," *Proceedings of IEEE Electronic Components and Technology Conference,* pp. 524–534, Orlando, FL, May 1996.

139. Doot, R. K., "Motorola's First DCA Product: The Gold Line Pen Pager," *Proceedings of IEEE Electronic Components and Technology Conference,* pp. 535–539, Orlando, FL, May 1996.

140. Greer, S. T., "An Extended Eutectic Solder Bump for FCOB," *Proceedings of IEEE Electronic Components and Technology Conference,* pp. 546–551, Orlando, FL, May 1996.

141. Peterson, D. W., J. S. Sweet, S. N. Burchett, and A. Hsia, "Stresses From Flip-Chip Assembly and Underfill: Measurements with the ATC4.1 Assembly Test Chip and Analysis by Finite Element Method," *Proceedings of IEEE Electronic Components and Technology Conference,* pp. 134–143, San Jose, CA, May 1997.

142. Zhou, T., M. Hundt, C. Villa, R. Bond, and T. Lao, "Thermal Study for Flip Chip on FR-4 Boards," *Proceedings of IEEE Electronic Components and Technology Conference,* pp. 879–884, San Jose, CA, May 1997.

143. Ni, G., M. H. Gordon, W. F. Schmidt, and R. P. Selvam, "Flow Properties of Liquid Underfill Encapsulations and Underfill Process Considerations," *Proceedings of IEEE Electronic Components and Technology Conference,* pp. 101–108, Seattle, WA, May 1998.

144. Hoang, L., A. Murphy, and K. Desai, "Methodology for Screening High Performance Underfill Materials," *Proceedings of IEEE Electronic Components and Technology Conference,* pp. 111–116, Seattle, WA, May 1998.

145. Dai, X., M. V. Brillhart, and P. S. Ho, "Polymer Interfacial Adhesion in Microelectronic Assemblies," *Proceedings of IEEE Electronic Components and Technology Conference,* pp. 132–137, Seattle, WA, May 1998.

146. Zhao, J. X. Dai, and P. Ho, "Analysis and Modeling Verification for Thermalmechanical Deformation in Flip-chip Packages," *Proceedings of IEEE Electronic Components and Technology Conference,* pp. 336–344, Seattle, WA, May 1998.

147. Matsushima, H., S. Baba, and Y. Tomita, "Thermally Enhanced Flip-chip BGA with Organic Substrate," *Proceedings of IEEE Electronic Components and Technology Conference,* pp. 685–691, Seattle, WA, May 1998.

148. Gurumurthy, C., L. G. Norris, C. Hui, and E. Kramer, "Characterization of Underfill/Passivation Interfacial Adhesion for Direct Chip Attach Assemblies Using Fracture Toughness and Hydro-Thermal Fatigue Measurements," *Proceedings of IEEE Electronic Components and Technology Conference,* pp. 721–728, Seattle, WA, May 1998.

149. Palaniappan, P., P. Selman, D. Baldwin, J. Wu, and C. P. Wong, "Correlation of Flip Chip Underfill Process Parameters and Material Properties with In-Process Stress Generation," *Proceedings of IEEE Electronic Components and Technology Conference,* pp. 838–847, Seattle, WA, May 1998.

150. Qu, J., and C. P. Wong, "Effective Elastic Modulus of Underfill Material for Flip-Chip Applications," *Proceedings of IEEE Electronic Components and Technology Conference,* pp. 848–850, Seattle, WA, May 1998.

151. Sylvester, M., D. Banks, R. Kern, and R. Pofahl, "Thermomechanical Reliability Assessment of Large Organic Flip-Chip Ball Grid Array Packages," *Proceedings of IEEE Electronic Components and Technology Conference,* pp. 851–860, Seattle, WA, May 1998.

152. Wiegele, S., P. Thompson, R. Lee, and E. Ramsland, "Reliability and Process Characterization of Electroless Nickel-Gold/Solder Flip Chip Interconnect," *Proceedings of IEEE Electronic Components and Technology Conference,* pp. 861–866, Seattle, WA, May 1998.

153. Caers, J., R. Oesterholt, R. Bressers, T. Mouthaan, and J. Verweij, "Reliability of Flip Chip on Board, First Order Model for the Effect on Contact Integrity of Moisture Penetration in the Underfill," *Proceedings of IEEE Electronic Components and Technology Conference,* pp. 867–871, Seattle, WA, May 1998.

154. Roesner, B., X. Baraton, K. Guttmann, and C. Samin, "Thermal Fatigue of Solder Flip-Chip Assemblies," *Proceedings of IEEE Electronic Components and Technology Conference,* pp. 872–877, Seattle, WA, May 1998.

155. Pang, J., T. Tan, and S. Sitaraman, "Thermo-Mechanical Analysis of Solder Joint Fatigue and Creep in a Flip Chip on Board Package Subjected to Temperature Cycling Loading," *Proceedings of IEEE Electronic Components and Technology Conference,* pp. 878–883, Seattle, WA, May 1998.

156. Gopalakrishnan, L., M. Ranjan, Y. Sha, K. Srihari, and C. Woychik, "Encapsulant Materials for Flip-Chip Attach," *Proceedings of IEEE Electronic Components and Technology Conference,* pp. 1291–1297, Seattle, WA, May 1998.

157. Yang, H., S. Bayyuk, A. Krishnan, A. Przekwas, L. Nguyen, and P. Fine, "Computational Simulation of Underfill Encapsulation of Flip-Chip ICs, Part I: Flow Modeling and Surface-Tension Effects," *Proceedings of IEEE Electronic Components and Technology Conference,* pp. 1311–1317, Seattle, WA, May 1998.

158. Liu, S., J. Wang, D. Zou, X. He, and Z. Qian, and Y. Guo, "Resolving Displacement Field of Solder Ball in Flip-Chip Package by Both Phase Shifting Moire Interferometry and FEM Modeling," *Proceedings of IEEE Electronic Components and Technology Conference,* pp. 1345–1353, Seattle, WA, May 1998.

159. Hong, B., and T. Yuan, "Integrated Flow—Thermomechanical and Reliability Analysis of a Low Air Cooled Flip Chip-PBGA Package," *Proceedings of IEEE Electronic Components and Technology Conference,* pp. 1354–1360, Seattle, WA, May 1998.

160. Wang, J., Z. Qian, D. Zou, and S. Liu, "Creep Behavior of a Flip-Chip Package by Both FEM Modeling and Real Time Moire Interferometry," *Proceedings of IEEE Electronic Components and Technology Conference,* pp. 1439–1445, Seattle, WA, May 1998.

161. Lau, J., C. Chang, C. Chen, R. Lee, T. Chen, D. Cheng, T. Tseng, and D. Lin, "Via-In-Pad (VIP) Substrates for Solder Bumped Flip Chip Applications," *Proceedings of Surface Mount International Conference,* pp. 128–136, September 1999.

162. Lau, J. H., "Cost Analysis: Solder Bumped Flip Chip versus Wire Bonding," *IEEE Trans. Electronics Packaging Manufacturing,* January 2000.

163. Lau, J. H., T. Chung, R. Lee, C. Chang, and C. Chen, "A Novel and Reliable Wafer-Level Chip Scale Package (WLCSP)," *Proceedings of the Chip Scale International Conference,* pp. H1–8, September 1999.

164. Lau, J. H., R. Lee, C. Chang, and C. Chen, "Solder Joint Reliability of Wafer Level Chip Scale Packages (WLCSP): A Time-Temperature-Dependent Creep Analysis," ASME Paper No. 99-IMECE/EEP-5, International Mechanical Engineering Congress and Exposition, November 1999.

165. Lau, J. H., and R. Lee, "Effects of Printed Circuit Board Thickness on Solder Joint Reliability of Flip Chip Assemblies with Imperfect Underfill," ASME Paper No. 99-IMECE/EEP-4, International Mechanical Engineering Congress and Exposition, November 1999.

166. Lau, J. H., C. Chang, and R. Lee, "Failure Analysis of Solder Bumped Flip Chip on Low-Cost Substrate," *Proceedings of the International Electronic Manufacturing Technology Symposium,* pp. 457–472, October 1999.

167. Lau, J. H., C. Chang, and C. Chen, "Characteristics and Reliability of No-Flow Underfills for Solder Bumped Flip Chips on Low Cost Substrates," *Proceedings of the International Symposium on Microelectronics,* pp. 439–449, October 1999.

168. Lau, J. H., and R. Lee, "Effects of Underfill Delamination and Chip Size on the Reliability of Solder Bumped Flip Chip on Board," *Proceedings of the International Symposium on Microelectronics,* pp. 592–598, October 1999.

169. Lau, J. H., R. Lee, and H. Chao, "How to Assemble Large PBGAs Reliably with large PQFPs Directly on the Opposite Side?" *Proceedings of the International Electronic Manufacturing Technology Symposium,* pp. 448–456, October 1999.

170. Lau, J. H., C. Chang, C. Chen, H. Chao, and N. Gamini, "Reliable Two-Side Printed Circuit Board Assembly of the Direct Rambus RIMM Modules," ASME Paper No. 99-IMECE/EEP-3, International Mechanical Engineering Congress & Exposition, November 1999.

171. Lau, J., H. Chen, C. Chen, C. Chang, M. Lee, D. Cheng, and T. Tseng, "PCB Manufacturing and Testing of the Direct Rambus RIMM Modules," *Proceedings of Surface Mount International Conference,* pp. 526–533, September 1999.

172. Lau, J. H., T. Chen, and R. Lee, "Effect of Heat-Spreader Sizes on the Thermal Performance of Large Cavity-Down Plastic Ball Grid Array Packages," *ASME Trans., J. Electronic Packaging,* December 1990.

173. Lau, J. H., and R. Lee, "Temperature-Dependent Popcorning Analysis of Plastic Ball Grid Array Package During Solder Reflow with Fracture Mechanics Method," *ASME Trans., J. Electronic Packaging,* March 2000.

174. Lau, J., C. Chen, C. Chang, T. Chen, H. Chen, T. Chen, T. Tseng, and D. Cheng, "Design, Analysis, and Measurement of a Novel Plastic Ball Grid Array Package," *Proceedings of Surface Mount International Conference,* pp. 202–210, September 1999.

175. Beddingfield, C., W. Ballouli, F. Carney, and R. Nair, "Wafer-Level KGD Technology for DCA Applications," *Advanced Packaging,* pp. 26–30, September 1999.

2

Chip-Level Interconnects: Wire Bonds and Solder Bumps

2.1 Introduction

The objective of chip-level connection, also called zero-level packaging, is to provide the required connections between the chip and the next level of interconnect, such as the PCB or the substrate of a package. There are many different chip-level interconnects; the most popular are wire bonds, gold bumps, and solder bumps. Today, more than 90 percent of the chips used have wire bonds. No more than 6 percent of the chips used have Au bumps through tape automated bonding (TAB). (TAB is a very interesting subject; however, it has been presented in Lau[1] and will not be discussed in this book.) Fewer than 4 percent of the chips now used have solder bumps.

Wire bonding of chips on board and on organic substrate in a package has been used for many years.[2-11] As mentioned earlier, more than 90 percent of the IC chips used in the world have wire bonds, because the high-speed automatic wire bonders meet most of the needs of the semiconductor devices for the next-level packaging interconnection. Figure 2.1 shows schematically an example of 0.1-mm-pitch wire bonding on a silicon ship. Today, however, pitches of 0.076 to 0.889 mm are used more and more in prototypes. The diameter of gold wire ranges between 20 and 33 μm, and the corresponding ball diameters are determined to be 2.5 to 4 times the wire diameter. The speed of most of the wire bonders is already four wires or eight bonds per second. The bottleneck in wire bonding is the bonding capillary (Fig. 2.2).

Figure 2.3 shows a photo of wires bonded on an organic substrate in a 1000-pin PBGA package.[12] The wires are bonded in a 50-μm-pitch staggered pattern as shown in Fig. 2.4. In order to prevent wiring short

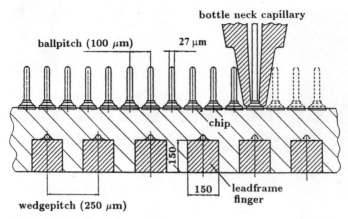

Figure 2.1   Wire bonding on chip.

circuits, the loop heights of the upper and lower wires are not the same, as shown in Fig. 2.5. Figures 2.6 and 2.7 show, respectively, photos of 60-μm-pitch in-line and staggered wire bonds.[13] Wire bonding of chips on board has been examined extensively in Lau[8] and Harman[9] and will only be discussed briefly in this book. Figure 2.8 shows solder bumps on a silicon chip, which will be the focus of this book.

2.2   Wire Bonds Versus Solder Bumps

The past few years have witnessed explosive growth in the research and development efforts devoted to solder-bumped flip chips on low-cost organic substrates as a direct result of the higher requirements of package density, performance, and interconnection and the limitations of wire-bonding technology. In comparison with the popular wire-bonding

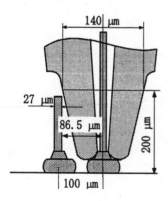

Figure 2.2   Diagram of the bonding capillary during wire bonding.

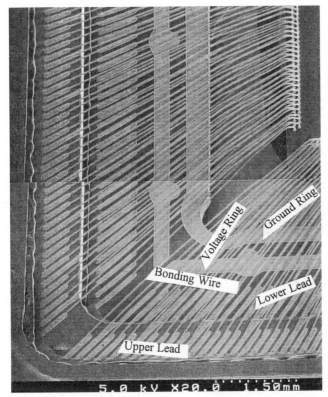

Figure 2.3 Scanning electron microscope (SEM) photograph of wires bonded in fine 50-μm pitch.

face-up technology, flip chip technology provides higher packaging density (greater I/Os) and performance (shorter possible leads, lower inductance, and better noise control), smaller device footprints, and a lower packaging profile. For example, for AC current the inductance for flip chip solder joints is about 0.4 nH, while that for wire bonding is 1.1 nH/mm. For DC current, the resistance of solder bumps is less than 1 mΩ. The advantages and disadvantages of flip chip and wire-bonding technologies are shown in Table 2.1.

Just like many other new technologies, low-cost solder-bumped flip chips still face certain critical issues. In the development of flip chips on board and in a plastic package, the following must be noted and understood in order to obtain the technology's full benefits:

- The infrastructure of flip chips is not well established.

- Flip chip expertise is not commonly available.

- Wafer bumping is still too costly.

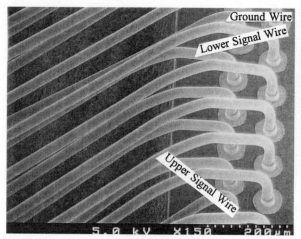

Figure 2.4   Magnified SEM image of the bonding wires at the chip area.

- Bare die/wafer is not commonly available.
- Bare die/wafer handling is not easy.
- Pick and place is more difficult.
- Fluxing is more critical.
- Underfill decreases manufacturing throughput.

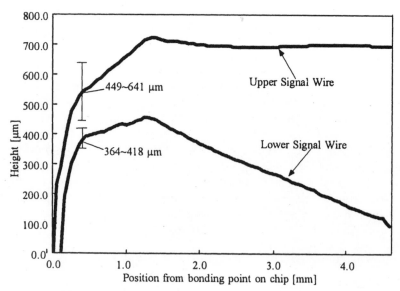

Figure 2.5   Upper and lower signal loop height. The minimum wire distance between the upper and lower signal wires is 31 μm.

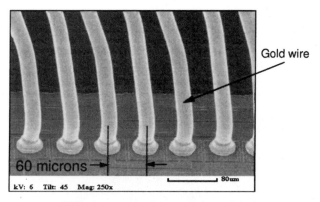

Figure 2.6   SEM photograph of 60-µm in-line pitch device side.

- Rework is more difficult.
- Solder joint reliability is more critical.
- Inspection is more difficult.
- Flip chip assembly testability is not well established.
- Die shrink and expand are troubles for system makers.
- There are known good die issues.
- Chip cracking during solder reflow has been reported.

What is the cost comparison between solder-bumped flip chip and wire bonding? This is a frequently asked question, in addition to the

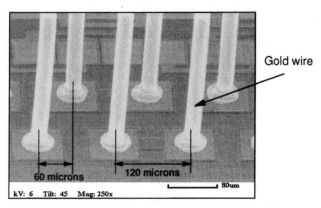

Figure 2.7   SEM photograph of 60-µm staggered pitch device side.

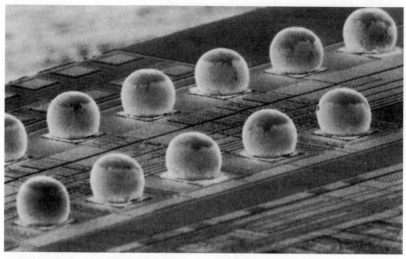

Figure 2.8 SEM photograph of solder bumps on chip.

critical issues just mentioned. The answer is almost impossible to determine. First of all, either flip chip or wire bonding is just an interconnect technology for connecting the IC chip to the next level of interconnection. Second, the impact of these interconnect technologies on the cost of the final product depends on the upstream (e.g., semiconductor design,

TABLE 2.1   Comparison Between Flip Chip and Wire Bonding Technologies

	Flip chip technology	Wire bonding technology
Advantages	■ High density ■ High I/Os ■ High performance ■ Noise control ■ Thin profile ■ SMT compatible ■ Area-array technology ■ Small device footprints ■ Self-alignment	■ Mature technology ■ Infrastructure exists ■ Flexible for new devices ■ Flexible for new bonding patterns
Disadvantages	■ Availability of wafers ■ Availability of dies ■ Availability of KGD ■ Wafer bumping ■ Test and burn-in ■ Underfill encapsulation ■ Additional equipment ■ Additional processes ■ Difficulty of rework after encapsulation ■ Die shrink	■ Availability of wafers ■ Availability of dies ■ Availability of KGD ■ Tests and burn-in ■ I/O limitation ■ Peripheral technology ■ Sequential process ■ Additional equipment ■ Additional processes ■ Difficulty of rework ■ Glob-top encapsulation

materials, and manufacturing)[14-18] and downstream (e.g., substrate and printed circuit board) of this level of interconnect, and the functions of the final product. However, making some assumptions, the cost of solder-bumped flip chip and wire bonding chip on board or on substrate in a plastic package could be analyzed,[19] which is the objective of this section.

2.2.1 Assembly process

Figure 2.9 shows a very simplified assembly process for wire-bonding chip and solder-bumped flip chip on either board or organic substrate. More detailed information on design, materials, processes, reliability, and applications for wire bonding and flip chip technologies can be found in Refs. 2 to 11. It should be noted that there are two ways to do the screening test for solder-bumped flip chip technology. One is to test before wafer bumping. In this case, the probe marks (damages) will be on

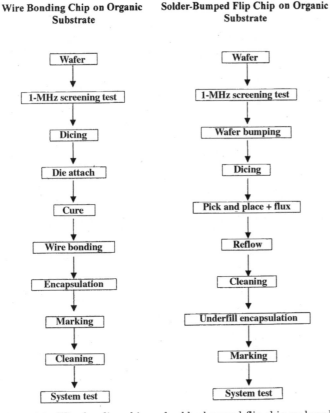

Figure 2.9 Wire bonding chip and solder-bumped flip chip on board process flows.

the pad, which could affect the integrity of the under-bump metallurgy and result in potential effects on long-term reliability. The other screening test method is to test after wafer bumping. This will contaminate the probe needles and result in shorts. Also, solder bumps could be damaged. However, better yield could be obtained because of the better electrical contact between the probe needles and the solder bumps (i.e., the number of false negatives during testing is reduced by better contacts between the probes and the bumps). For a mature wafer-bumping process, the bump yield per wafer (Y_B) is usually very high (over 98 percent).

From a cost point of view, the most important difference between wire bonding and solder-bumped flip chip is that the wire-bonding technology needs gold wire and the solder-bumped flip chip technology needs wafer bumping. The second most important difference between these two technologies is the effect on IC chip yields of the 1-MHz screening test and at-speed/burn-in system tests. Finally, different major equipment is needed for each of these interconnect methods.

2.2.2 Major equipment

The major equipment needed by wire-bonding chip and solder-bumped flip chip on board or on a plastic ball grid array (PBGA) package substrate is shown in Table 2.2. It should be noted that equipment required

TABLE 2.2   Major Equipment for Flip Chip and Wire Bonding Technologies

Flip chip		Wire bonding	
Equipment	Cost	Equipment	Cost
Pick and place + flux	3 @ $500K = $1500K	Die attach	$220K
Reflow oven	$70K	Cure oven	$20K
Clean	$130K	Wire bonders	35 @ $110K = $3850K
Underfill dispenser	5 @ 120K = $600K	Encapsulation	$700K (Transfer mold) or
Vertical cure oven	$100K		5 @ $120K = $600K
X-ray	$300K		(glob-top dispenser)
Total	$2700K		plus vertical cure oven = $100K
		Total	$4790K

- Machine utilization is 7200 hours per year.
- Machine capacity is 12 million chips per year.
- Pick and place + fluxing is assumed to be 6.5 seconds per chip.
- Underfill dispensing is assumed to be 10 seconds per chip (to completely fill the chip).
- Wire bonder speed is assumed to be 4 wires or 8 bonds per second. Also it is assumed that there are 300 pads per chip.
- Glob-top dispensing dam and encapsulant are assumed to be 10 seconds per chip.
- Equipment required by both techniques is not shown.

by both of these technologies is not shown. Also, all equipment is assumed to be utilized 300 days a year, 24 hours a day, and its capacity is assumed to be 12 million chips per year. Each chip has 300 peripheral pads.

It can be seen from Table 2.2 that expensive pick and place + fluxing machinery is necessary for flip chip technology. Even though the wire bonder is cheaper, more wire bonders are needed because their throughput is much smaller (depending on the number of pads on a chip) than that of the pick and place machine (which performs gangbonding on a chip). Consequently, the cost of major equipment for flip chip ($2.7 million) is lower than that for wire bonding ($4.79 million). Also, the manufacturing floor space required for flip chip should be smaller.

It should be noted that if the number of pads per chip is doubled to 600, then 70 wire bonders are needed and the cost of equipment is even higher. On the other hand, if the pads per chip are reduced to only 30, then 4 wire bonders are more than enough to do the job, and in this case the equipment for wire bonding technology is much cheaper than that for flip chip technology. Thus, flip chip technology is cost effective if the pin count and density of the IC devices are high.

2.2.3 Cost of materials

As mentioned earlier, the largest cost differences between wire bonding and flip chip are materials and IC chip yields. The labor/operation costs for these two assembly processes are assumed to be the same.

The wafer. The physically possible number of undamaged chips (N_c) stepped from a wafer (Fig. 2.10) may be given by

$$N_c = \pi \, \frac{[\phi - (1 + \theta)\sqrt{A/\theta}]^2}{4A} \tag{2.1}$$

where

$$A = xy \tag{2.2}$$

and

$$\theta = \frac{x}{y} \geq 1 \tag{2.3}$$

In Eqs. 2.1 to 2.3, x and y are the dimensions of a rectangular chip (mm) with x no less than y; θ is the ratio between x and y; ϕ is the wafer diameter (mm); and A is the area of the chip (mm^2). For example, for a 200-mm wafer with $A = 10 \times 10 = 100$ mm^2, then $N_c = 255$ chips.

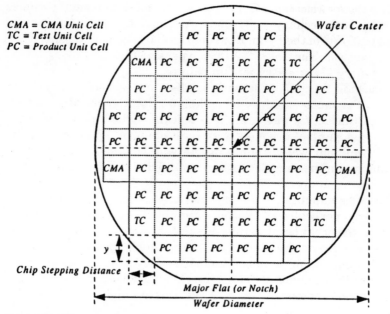

Figure 2.10 Diagram of a wafer.

For a given chip size, the physically possible number of pads (N_p) depends on the pad pitch (p) and pad configuration. The N_p on a chip surface is given as follows.

For area-array pads:

$$N_p = \left(\frac{x}{p} - 1\right)\left(\frac{y}{p} - 1\right) \tag{2.4}$$

For peripheral-array pads:

$$N_p = 2\left(\frac{x}{p} - 1\right) + 2\left(\frac{y}{p} - 1\right) \tag{2.5}$$

For peripheral-staggered-array pads:

$$N_p = 4\left(\frac{x}{p} - 1\right) + 4\left(\frac{y}{p} - 1\right) - 4 \tag{2.6}$$

For example, for a chip with $x = 10$ mm, $y = 8$ mm, and $p = 0.25$ mm, $N_p = 1209$ area-array pads, $N_p = 140$ peripheral-array pads, and $N_p = 276$ peripheral-staggered-array pads. Obviously, area array is the choice for high-density IC and packaging.

Cost per wafer for wafer bumping.   Wafer bumping is the heart of solder-bumped flip chip technology. The cost of wafer bumping is affected by the true yield (Y_T) of the IC chips, the wafer-bumping yield (Y_B), and the good die cost (C_D). The actual wafer-bumping cost per wafer (C_B) is:

$$C_B = C_{WB} + (1 - Y_B)Y_T N_c C_D \qquad (2.7a)$$

where C_B is the actual wafer-bumping cost per wafer, C_{WB} is the wafer-bumping cost per wafer (ranging from \$25 to \$250), Y_B is the wafer-bumping yield per wafer, C_D is the good die cost (not the cost of an individual die on the wafer), N_c is given in Eq. 2.1, and Y_T is the true IC chip yield after at-speed/burn-in system tests. It can be seen that the actual wafer-bumping cost per wafer depends not only on the wafer-bumping cost per wafer but also on the true IC chip yield per wafer, wafer-bumping yield per wafer, and good die cost.

Wafer-bumping yield (Y_B) plays a very important role in wafer bumping. The wafer-bumping yield loss ($1 - Y_B$) could be due to (1) incorrect processes, (2) different materials, (3) bump heights that are too tall or short, (5) not enough shear strength, (6) solder bridging, (7) broken wafer or dies, (8) damaged bumps, (9) missing bumps, and (10) scratching of the wafer. It should be noted that solder bumps are not reworkable. The process has to be done right the first time; otherwise, someone has to pay for it!

For example, if the die size of a 200-mm wafer is 100 mm^2 ($N_c \approx 255$), true IC chip yield per wafer is 80 percent, wafer-bumping yield per wafer is 90 percent, wafer-bumping cost per wafer is \$200, and good die cost is \$100 (e.g., microprocessors), then the actual wafer-bumping cost per wafer is \$2240. For the same size wafer, if the good die cost is \$1 (e.g., memory devices), then the actual wafer-bumping cost per wafer is \$220.40. On the other hand, if the wafer-bumping yield is increased from 90 percent to 99 percent, then the actual cost for bumping the microprocessor wafer is \$404 and the actual cost for bumping the memory wafer is \$202.04. Thus, wafer-bumping yield plays an important role in the cost of wafer bumping and the wafer-bumping operations should strive to make Y_B greater than 99 percent, especially for expensive good dies.

Cost per good die for wafer bumping.   The wafer-bumping cost per good die ($C_{B/D}$) can be determined by

$$C_{B/D} = \frac{C_B}{Y_T N_c} \qquad (2.7b)$$

Figure 2.11 shows the wafer-bumping cost per good die ($C_{B/D}$) for various Y_T values for the 200-mm wafer. It can be seen that $C_{B/D}$ increases as Y_T decreases. Also, it can be shown from Eqs. 2.1 and 2.7 that if the

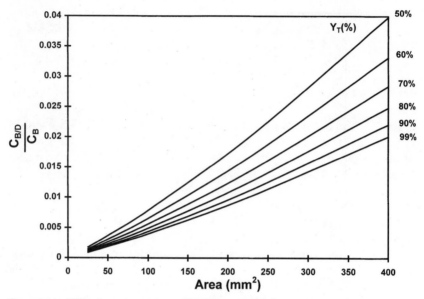

Figure 2.11 Wafer-bumping cost per die (200-mm wafer).

wafer-bumping cost (C_{WB}) of the 200-mm wafer is no more than 2 times of that of the 150-mm wafer, then the wafer-bumping cost per good die is cheaper for the larger wafer (assuming Y_B, Y_T, and good die cost are the same for both wafers). Finally, for example, for a 200-mm wafer with $A = 100$ mm², $N_c \approx 255$. Assuming $C_{WB} = \$100$ and $Y_B = 100$ percent, then $C_{B/D} = \$0.44$ if $Y_T = 0.9$ and $C_{B/D} = \$0.65$ if $Y_T = 0.6$. Thus Y_T plays an important role in wafer bumping. Unless it is mandated for reasons of density, form factor, and/or performance, solder-bumped flip chip technology is not suitable for very low-yield devices.

Cost per pad on a good die for wafer bumping. The wafer-bumping cost per pad ($C_{B/P}$) on a good die can be determined by

$$C_{B/P} = \frac{C_B}{Y_T N_c N_p} \tag{2.8}$$

where N_c is given in Eq. 2.1, N_p is given in Eqs. 2.4, 2.5, or 2.6, C_B is the actual wafer-bumping cost, and Y_T is the true IC chip yield.

Figure 2.12 shows the $C_{B/P}$ for the 200-mm wafer with peripheral-area-array pads. It can be seen that $C_{B/P}$ depends on the chip size, pad pitch, C_B (C_{WB}, Y_B, C_D), and Y_T. The larger the chip size, the higher the $C_{B/P}$, and the larger the pad pitch, the higher the $C_{B/P}$. Also, $C_{B/P}$ is

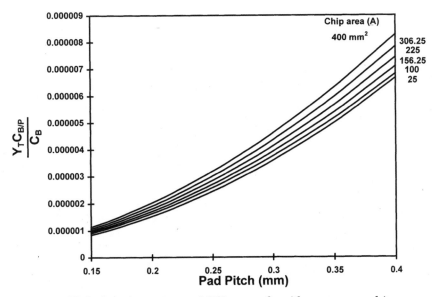

Figure 2.12   Wafer-bumping cost per pad (200-mm wafer with area-array pads).

cheaper for the 200-mm wafer if the C_B of the larger wafer is no more than twice of that of the smaller wafers.

Consider a 200-mm wafer with $A = 10 \times 10 = 100$ mm^2 and $p = 0.25$ mm. Assuming $C_B = C_{WB} = \$200$ ($Y_B = 100\%$) and $Y_T = 0.9$, then $C_{B/P} = \$0.00057$ for area-array pads, $C_{B/P} = \$0.00559$ for peripheral-array pads, and $C_{B/P} = \$0.00283$ for peripheral-staggered-array pads. It can be seen that the cost savings for chips with area-array pads is very compelling.

Cost per die for wire bonding.   The wire-bonding cost per die ($C_{W/D}$) can be determined by

$$C_{W/D} = \frac{Y_I L_A C_W N_p}{Y_T} \qquad (2.9)$$

where C_W is the wire cost (Table 2.3), L_A is the average wire length per pad, Y_I is the initial screening IC chip yield (after 1-MHz screening test), Y_T is the true IC chip yield, and N_p is given by either Eq. 2.5 or Eq. 2.6.

Figure 2.13 shows the $C_{W/D}$ for the chips with peripheral-array pads ($x = y$ has been assumed). It can be seen that $C_{W/D}$ decreases with pad pitch increase and chip size decrease.

TABLE 2.3   Cost of Gold Wire (September 1998)

Diameter (inch)	1,000 (ft)	5,000 (ft)	10,000 (ft)	25,000 (ft)
0.0007	$485.00/LOT	$143.70/Kft	$ 93.70/Kft	$ 60.00/Kft
0.001	$375.00/LOT	$135.30/Kft	$ 96.50/Kft	$ 70.30/Kft
0.0015	$410.00/LOT	$192.10/Kft	$149.70/Kft	$121.10/Kft

99.99 percent gold wire (spool: 2 inch).

Consider a chip with $x = 10$ mm, $y = 8$ mm, and $p = 0.25$ mm. Assuming $L_A = 4.572$ mm, $C_W = \$70.30/\text{Kft} = \$0.00023/\text{mm}$, $Y_I = 0.9$, and $Y_T = 0.6$, then $C_{W/D} = \$0.22$ for peripheral-array pads and $C_{W/D} = \$0.44$ for peripheral-staggered-array pads. If $Y_I = 0.7$ and $Y_T = 0.6$, then $C_{W/D} = \$0.17$ for peripheral-array pads and $C_{W/D} = \$0.34$ for peripheral-staggered-array pads. It can be seen that for wire-bonding technology it is desirable to have $Y_I \rightarrow Y_T$. Otherwise, the expensive gold wire will be bonded on the bad dies.

Cost per pad for wire bonding.   The wire-bonding cost per pad ($C_{W/P}$) can be determined by

$$C_{W/P} = \frac{Y_I L_A C_W}{Y_T} \qquad (2.10)$$

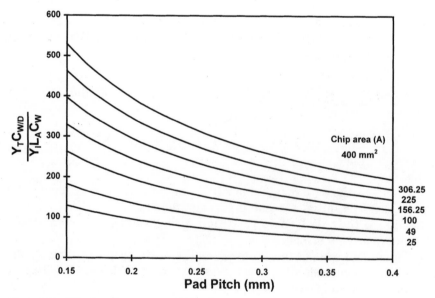

Figure 2.13   Wire bonding cost per die (wafer with peripheral-array pads).

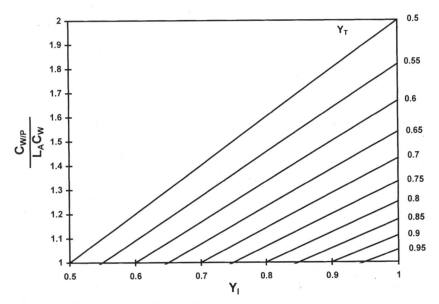

Figure 2.14 Wire bonding cost per pad.

Equation 2.10 is shown in Fig. 2.14 with various values of Y_I and Y_T. It can be seen that the higher the Y_T, the lower the $C_{W/P}$. Also, for a given Y_T, the $C_{W/P}$ increases with a higher Y_I. For example, consider a 200-mm wafer with $A = 100$ mm^2 and $p = 0.25$ mm. Assuming $L_A = 4.572$ mm and $C_W = \$0.000231$/mm, then $C_{W/P} = \$0.001115$ if $Y_T = 0.9$ and $Y_I = 0.95$, and $C_{W/P} = \$0.001056$ if $Y_I = Y_T = 0.9$. By comparing this example with that in the section on wafer-bumping cost per good die, it can be seen that $C_{W/P}$ is smaller than $C_{B/P}$. However, flip chip technology provides the possibility for area-array pads, and in this case $C_{B/P}$ is a few times smaller than $C_{W/P}$.

Cost comparison. The cost ratio (β) between the wafer-bumping cost per die (Eq. 2.7) or pad (Eq. 2.8) and the wire-bonding cost per die (Eq. 2.9) or pad (Eq. 2.10) is given by

$$\beta = \frac{C_B}{Y_I L_A C_W N_p N_c} \tag{2.11}$$

Equation 2.11 is plotted in Fig. 2.15 for the 200-mm wafer with peripheral-array pads. It can be seen that β is larger for larger chip size and larger pad pitch. Also, β is larger for smaller Y_I. When $\beta \geq 1$, this means the costs of solder-bumped flip chip are no less than those of wire bonding.

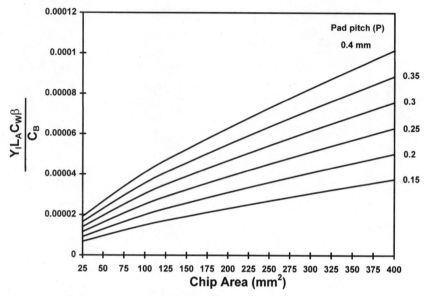

Figure 2.15 Cost ratio between wafer-bumping cost per pad and wire bonding cost per pad (200-mm wafer with peripheral-array pads).

For a 200-mm wafer with $x = y = 10$ mm and $p = 0.25$ mm on a peripheral-staggered-array pad configuration, assuming that $L_A = 4.572$ mm, $C_W = \$0.00023$/mm, $C_B = C_{W/B} = \$100$ ($Y_B = I$), and $Y_I = 0.9$, then $\beta = 1.35$. This means that from materials, labor, and operation points of view, flip chip technology is more expensive than wire-bonding technology. It should be remembered that the major equipment cost for wire bonding is higher than that for flip chip (1.77 times). Thus, flip chip is still cheaper. If we reduce the pitch to $p = 0.16$ mm and keep everything else the same, then $\beta = 0.85$. That means flip chip is cheaper, even without counting on the major equipment cost advantage. However, for the peripheral-array pad configuration, wire bonding is cheaper.

If we reduce the chip size to $x = y = 5$ mm, then $\beta = 0.63$ for $p = 0.25$ mm on peripheral-staggered-array pads and $\beta = 0.39$ for $p = 0.16$ mm on peripheral-staggered-array pads. Thus, in this case, flip chip is cheaper even with peripheral-array pads. It should be pointed out, as shown in the section on wafer-bumping cost per die, that flip chip could be even cheaper if the area-array pad configuration is used. Unfortunately, it cannot be compared with wire bonding, since the latter cannot use area-array pads.

2.2.4 Summary

The cost analysis for wire-bonding chip and solder-bumped flip chip on board or on a PBGA substrate has been presented. Useful equations for determining the wafer-bumping and wire-bonding costs per die and per pad have also been provided. Furthermore, useful charts relating the important variables such as the wafer size, IC chip yields, wafer-bumping yield, chip dimensions, and costs are provided for engineering practice convenience. Some important results are summarized in the following list.

- The major equipment cost of wire bonding is higher than that of flip chip.
- Wire bonding occupies more manufacturing floor space than flip chip does.
- Wafer-bumping cost per die increases as Y_T decreases.
- Wafer-bumping cost per die is cheaper for larger wafers.
- Wafer-bumping cost per pad is higher for larger chip sizes.
- Wafer-bumping cost per pad is higher for larger pad pitches.
- Wafer-bumping cost per pad is cheaper for larger wafers.
- Wafer-bumping cost per pad is very low for area-array chips.
- Wire-bonding cost per die decreases as pad pitch increases.
- Wire-bonding cost per die decreases as chip size decreases.
- Wire-bonding cost per pad increases with higher ratios of Y_I and Y_T.
- Area-array solder-bumped flip chip is cheaper than wire bonding.
- For peripheral-array chip, wire bonding is cheaper than flip chip if the chip size and pad pitch are very large. Otherwise, flip chip is cheaper.
- In order for solder-bumped flip chip to be cost competitive with wire bonding, the wafer-bumping yield should be higher than 99 percent, especially for high-cost dies.
- For low-yield dies, actual wafer-bumping cost may be too high for flip chip applications, unless it is compensated for by performance, density, and form factor.

2.3 Wafer Bumping with Solders

Wafer bumping is the heart of solder-bumped flip chip technology. Today, there are many different ways to put the tiny solder bump (usually no larger than 100 μm) on the wafer. In this chapter, only the 12 best-known methods are discussed.

The most important task in wafer bumping is to put down the ball-limiting metallurgy (BLM) or under-bump metallurgy (UBM) on the aluminum pads of the wafer. Some of the UBMs for different bump materials are shown in Table 2.4. The UBM defines the region of terminal metallurgy on the top surface of the chip that is wetted by the solder.

2.3.1 Evaporation method

The most mature wafer-bumping method is evaporation, which has been perfected by IBM for more than 30 years. A detailed description of this method is beyond the scope of this book; interested readers are encouraged to read Refs. 20 to 46. However, some of the most important features of this method are briefly described in the following paragraphs.

Figure 2.16 shows the evaporation process. It can be seen that it requires two metal masks, as shown in Fig. 2.17. These masks are made of molybdenum, whose thermal coefficient of expansion is very close to that of silicon. Also, molybdenum has excellent long-term dimensional stability at high temperatures. The first metal mask with smaller openings is for the deposition through evaporation of the UBM (chrome/copper/gold), as shown in Fig. 2.18. The second metal mask with larger openings is for the deposition through evaporation of the

TABLE 2.4   Some Well-Known UBMs for Gold, Copper, Aluminum, Solder, and Nickel Bumps

Bump	UBM	Process
Gold	Cr-Cu	Electroplating
	Ti-Pd	Electroplating
	Ti-W	Electroplating
	Ti-Pt	Electroplating
Copper	Cr-Cu	Electroplating
	Al-Ni	Electroplating
Aluminum	Ti	Evaporating
	Cr	Evaporating
Solder	Cr-Cu-Au	Evaporating/printing
	Ni-Cu	Electroplating/printing
	Ti-Cu	Electroplating/printing
	TiW-Cu	Electroplating/printing
	Ni-Au	Electroless + printing
	Al-NiV-Cu	Sputtering + printing
Nickel	Ni	Electroless Ni/Au

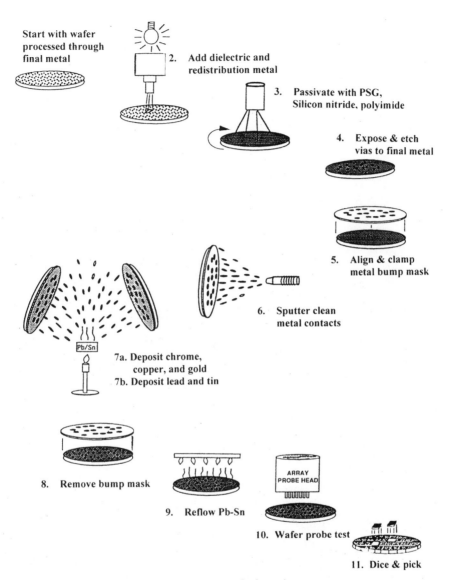

Start with wafer processed through final metal

2. Add dielectric and redistribution metal

3. Passivate with PSG, Silicon nitride, polyimide

4. Expose & etch vias to final metal

5. Align & clamp metal bump mask

6. Sputter clean metal contacts

Pb/Sn

7a. Deposit chrome, copper, and gold
7b. Deposit lead and tin

8. Remove bump mask

9. Reflow Pb-Sn

ARRAY PROBE HEAD

10. Wafer probe test

11. Dice & pick

Figure 2.16   IBM's well-known evaporation wafer-bumping process.

lead-tin solder. After reflow, nice-looking truncated spherical solder bumps can be obtained, as shown in Fig. 2.17. Evaporation bumping is a dry solder buildup process.

It should be noted that because the evaporation rate of the Pb is higher than that of the Sn, this method is usually applied to high-Pb solder bumps such as the 95wt%Pb-5wt%Sn and 97wt%Pb-3wt%Sn.

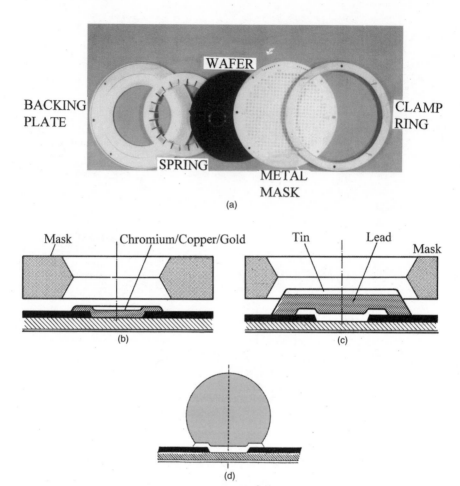

Figure 2.17 Metal mask technology. (*a*) Tooling for alignment of mask to wafer. (*b*) Masking and evaporation of chromium/copper/gold. (*c*) Masking and evaporation of lead and tin. (*d*) Reflowed solder bump.

IBM calls these controlled collapse chip connection (C4) bumps. Since two metal masks and the evaporation process are needed, this method could be expensive.

Recently, Motorola developed an E-3 (evaporated extended eutectic) wafer-bumping process[47] that also uses the evaporation method. It creates a bump with a mostly "pure" Pb column and a small amount of pure Sn at the tip. It is not reflowed prior to die attachment. Since unreflowed Pb is very soft, shipping of these bumps poses a great challenge. Interested readers are encouraged to read Chap. 6 of Lau and Pao.[5]

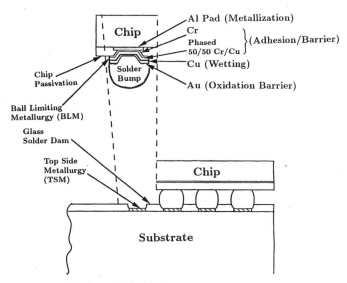

Figure 2.18   IBM's well-known Cr-Cu-Au UBM.

2.3.2 Electroplating method[5,8]

The second most mature method in wafer bumping is electroplating, which is considered a wet solder buildup process. Figure 2.19 shows the bumping process. The most common UBM with electroplating method is either Ti-Cu or TiW-Cu. They are sputtered over the entire surface of the wafer—0.1 to 0.2 µm of Ti or TiW first, followed by 0.3 to 0.8 µm of Cu. A 40-µm layer of resist is then overlaid on the Ti-Cu or TiW-Cu and a solder bump mask is used to define the bump pattern as shown in Fig. 2.19. The openings in the resist are 7 to 10 µm wider than the pad openings in the passivation layer. A 6- to 25-µm layer of Cu is then plated over the UBM, followed by electroplating Sn-Pb solder in any combination. This is done by applying a static or pulsed current through the plating bath with the wafer as the cathode. In order to plate enough solder to achieve a final solder ball height of 100 µm, the solder is plated over the resist coating by about 15 µm to form a mushroom shape. The resist is then removed and the Ti/Cu or TiW/Cu is stripped off with a hydrogen peroxide etch. The wafer is then reflowed, which creates smooth truncated spherical solder bumps due to surface tension.

This method is very popular today because it can plate solders in any combinations, such as the 63wt%Sn-37wt%Pb eutectic solder. One of the drawbacks is bump height nonuniformity. Usually, due to the

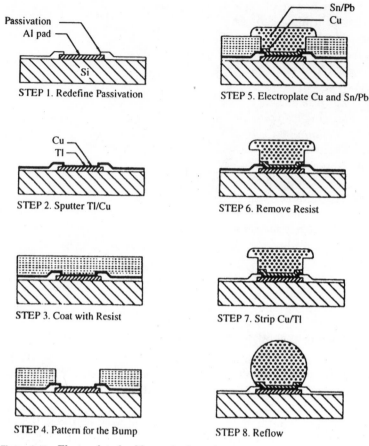

Figure 2.19 Electroplated solder wafer-bumping process.

applied electrical current at the boundary of the wafer, i.e., current density variations across the wafer during the electroplating process, the solder bumps near the edge of the wafer are taller than those near the center of the wafer. (For experimental data, please see pages 233 to 248 of Lau and Pao.[5]) The other drawback is plating time (1 to 2 min/μm). This method is cheaper than the evaporation method, since only one bump mask is needed and the process is simpler.

2.3.3 Under-bump metallurgy (UBM)

One of the key reasons why solder-bumped flip chip technology is not commonly used today is because of the high cost of wafer bumping.

Thus, in the past few years many new and potentially low-cost wafer-bumping methods have been proposed. Almost all of these methods require a UBM: the most common are Ni-Au and Al-NiV-Cu. Since the latter has been discussed in Chap. 31 of Lau and Lee,[3] in this section only the Ni-Au UBM is discussed.

The key process in making the Ni-Au UBM is shown in Fig. 2.20. It can be seen that it is a zincate process and very simple. After the wafer is cleaned, it is put into the zincate solution for less than 3 min and then rinsed in deionized (DI) water. (Very often, double-zincating is performed.) Then the wafer is put into the Ni solution for electroless plating (8 to 10 percent NiP, 85°C for 12 to 15 min) to about 5 μm thick, rinsed in DI water, immersed into the gold solution (80°C for 10 min), and then rinsed again in DI water. The function of this very thin Au layer is to prevent the Ni from oxidizing and prolong the wafer solderable shelf life. Figure 2.21 shows the Ni-Au UBM on a silicon chip.[48]

The quality and uniformity of electroless Ni-Au UBM depend on the careful monitoring of each process step and the tight contamination control (every 2 h) of the zincate, Ni, and Au solution tanks. The Al-Ni adhesion and electrical resistance at the interface are strongly affected by the cleaning and activation processes.

The advantages of the Ni-Au UBM are: (1) its costs are low, (2) it is compatible with eutectic solders, (3) it has high solder wetability, (4) Al-

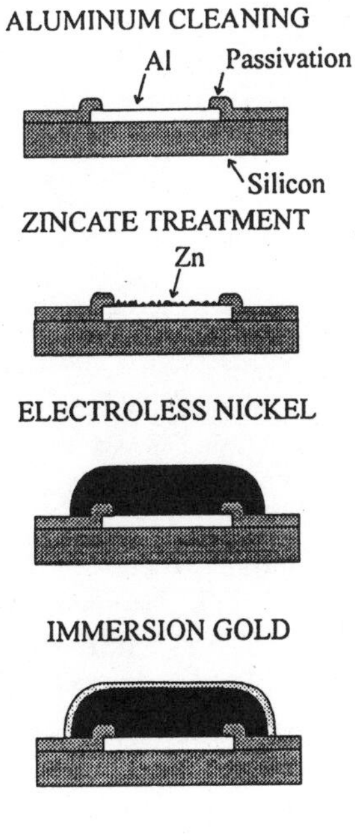

ALUMINUM CLEANING

ZINCATE TREATMENT

ELECTROLESS NICKEL

IMMERSION GOLD

Figure 2.20 Schematic of the electroless Ni-Au bumping process (not to scale).

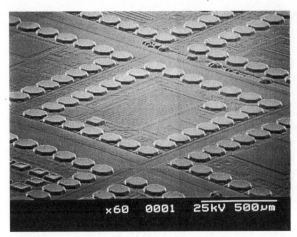

Figure 2.21   SEM photo of electroless Ni-Au UBM about 5 μm thick.

Ni adhesion is more than adequate for SMT, and (5) Ni-Sn intermetallic growth is adequate for solder thermal fatigue life.

It should be emphasized that since all the potential low-cost wafer-bumping methods discussed in Secs. 2.3.4 through 2.3.12 require UBM, logistics and infrastructure become big factors in the cost picture. Usually, truly low cost can be achieved if and only if the wafer bumpers also do UBM and at the same site. Otherwise, there would be many hidden costs.

2.3.4   Stencil printing method[49-53]

Wafer bumping by the stencil printing method is not only potentially low in cost but is also welcomed by system manufacturers, because stencil printing of solder paste is one of their most important steps in SMT. The only limitation is the very fine pitch of the pads on the chip. Today, stencil printing solder paste on the wafer in volume production is limited to 150 μm.[49]

Figure 2.22 shows the key process steps for stencil printing solder paste on the wafer.[50, 51] The first step is to put down the electroless Ni-Au UBM on the Al pads as shown in Figs. 2.23a and b.[52] The second step is to stencil print the 63Sn-37Pb solder paste onto the UBM as shown in Fig. 2.23c[52] and Figs. 2.24a and b. Figure 2.24b is the Ishikawa cause-and-effect (fishbone) diagram for solder paste printing. It can be seen that there are eight major categories (operator, environment, printing

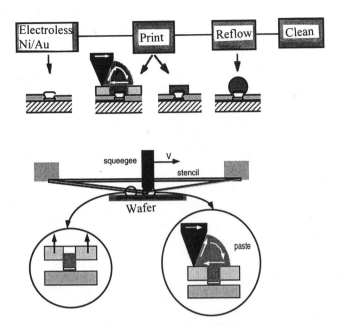

Figure 2.22   Wafer bumping by stencil printing paste process steps.

parameter, printer, stencil, wafer, squeegee, and solder paste) that influence the quality of solder paste printing on wafer. This diagram helps us to sort out the potential causes for solder paste printing defects and to organize the causes into defined categories to yield high-quality solder paste printing.

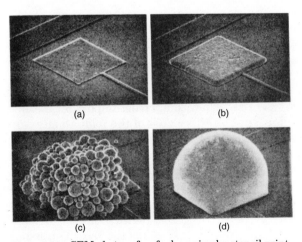

Figure 2.23   SEM photos of wafer bumping by stencil printing method. (*a*) Al pad; (*b*) 5-μm Ni-Au UBM; (*c*) solder paste; (*d*) reflowed bump.

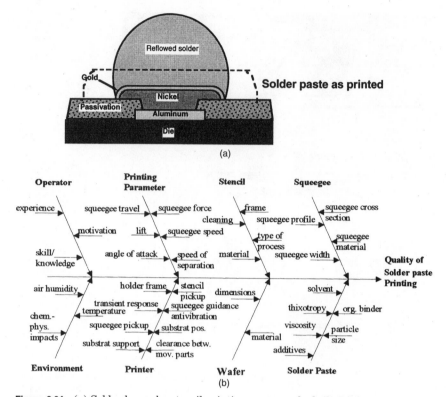

Figure 2.24   (a) Solder bump by stencil printing paste method. (b) Ishikawa (fishbone) diagram for quality solder paste printing.

The next step is to reflow the solder paste on the wafer into nice-looking solder bumps, as shown in Figs. 2.23d and 2.24a. Finally, the wafer is cleaned with DI water (Fig. 2.25). Some typical cross sections are shown in Figs. 2.26a and b.[50] It can be seen that they are very good-looking solder bumps. Also, the Ni-Au UBM looks very good. Energy dispersive spectrum (EDS) results obtained by Li et al.[50] show that a thin Ni-Sn intermetallic layer is formed at the solder-UBM interface. Also, these solder bumps are almost free of voids and have good wetting on the Ni-Au UBM.

The shear strength of the stencil printed solder bumps under certain reflow and aging conditions is also measured by Motorola and is shown in Fig. 2.27. It can be seen that there is no shear strength degradation even after 1000 h of storage at 150°C and 85°C/85% relative humidity (RH) of samples reflowed 11 times.

Figure 2.28a shows a solder bump on a Ni-Au UBM chip. The solder volume (V) of the truncated sphere solder bump with diameter = $2R$,

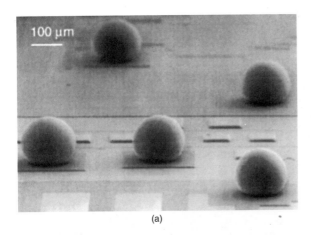

(a)

(b)

Figure 2.25   Stencil-printed solder bumps after reflow and cleaning. (*a*) Reflowed bumps on a 300-μm pitch device. (*b*) Solder bumps on a 200-μm pitch device.

based on the UBM diameter (D) and bump height (H), can be estimated with good accuracy by the equation shown in Fig. 2.28a.[50] If the wetting angle (θ) is known, then the solder volume (V) of a solder bump with height (H), radius (R), and UBM diameter (φ) can be obtained from the equation in Fig. 2.28b.[53]

Figures 2.29a to c show the cross sections of untested DCA on Cu-Ni-Au PCB. These assemblies are subjected to the qualification test conditions shown in Table 2.5. All the requirements are met by Motorola. Figures 2.30a and b show the cross sections of the failure sample (liquid-to-liquid test for –55 to +125) at 1500 cycles. It can be seen that the delamination between the underfill and solder mask causes the solder joint cracks. (Also see Figs. 2.31a to c.)

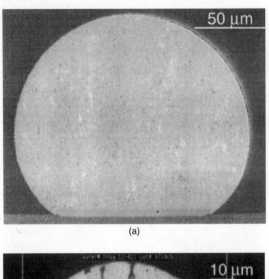

(a)

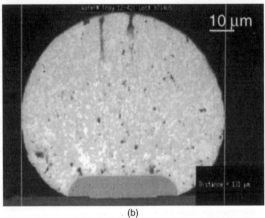

(b)

Figure 2.26 Solder bump cross sections for different pitch devices. (a) 300-μm pitch device, (b) 200-μm pitch device.

In order to achieve a successful stencil printing method for wafer bumping, two things must be noted—the paste material and the stencil. The rheology of solder paste is critical for sufficient paste deposit without bridges. Active rosin flux containing halides should be avoided. Paste containing chemical components that create difficult-to-remove residues should not be used. The type 4 paste is adequate for stencil printing on wafer bumping even though the finer-mesh type 5 and type 6 pastes deposit more consistent solder volume. However, the latter two pastes are not commonly available at reasonable cost.

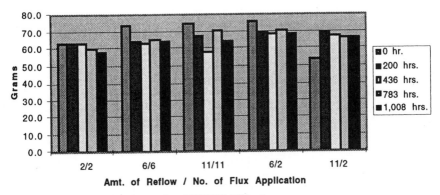

Figure 2.27   Bump shear force after 85°C/85% RH aging.

The electroformed stencil is better than the chemical-etched and laser-cut stencils. Although the chemical-etched stencils are cheaper, because of accuracy and dimension limitations they cannot meet the fine-pitch wafer-bumping requirement. Although the laser-cut stencils are accurate, the sequential processing could make costs prohibitive for wafer bumping.

2.3.5   Solder jet printing method[54–56]

The schematic of a demand-mode ink-jet printing system, also called solder jet technology, developed by MicroFab Technologies, Inc. is shown in Fig. 2.32.[54–56] Solder jet technology is based on piezoelectric demand-mode ink-jet printing technology[54–59] and is capable of producing and placing molten solder droplets 25 to 125 μm in diameter at rates of up to 2000 per second.

In demand-mode ink-jet printing systems, a volumetric change in the fluid is induced either by the displacement of a piezoelectric material that is coupled to the fluid or by the formation of a vapor bubble in the ink caused by heating a resistive element. This volumetric change causes pressure transients to occur in the fluid, and these are directed so as to produce a drop that issues from an orifice. A droplet is created only when it is desired in demand-mode systems. Demand-mode ink-jet printing systems produce droplets that are approximately equal to the orifice diameter of the droplet generator.

The droplet generator is incorporated into a printhead design suitable for integration into a prototype platform. Key features of the printhead include: (1) a heated inert environment localized to the tip of

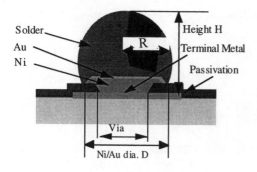

$$V = \pi / 6 \text{ x } H \text{ x } (H^2 + 3 \ (D/2)^2)$$
Bump diameter $2R = (D^2 + H^2) / H$

(a)

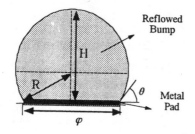

$$V = (\pi /24) \times \varphi^3 \times \{[(\cos \theta + 1) \times (2 - \cos \theta)]$$
$$/ [\sin \theta \times (1 - \cos \theta)]\}$$

$$H = R + R \cos \theta = [\varphi \times (1 + \cos \theta)] / 2 \sin \theta$$

(b)

Figure 2.28 (a) Solder bump volume approximation. (b) Solder bump volume calculation with wetting angle.

the droplet generator and impact area of the substrate, (2) separate heaters for the solder reservoir and droplet generator, (3) vertical dispensing capability, and (4) the ability to deposit solder droplets while the printhead is in motion.

The solder jet printhead has been integrated with Universal Instruments to build the solder jet research platforms (Fig. 2.33a) that have been delivered to many companies to demonstrate fundamental capabilities in the positioning and deposition of solder droplets and to perform initial process development experiments. Figure 2.33b shows the Drop-on-Demand Metal Jet™ developed by MPM and MicroFab Technologies, Inc. Figures 2.34a and b show the solder bumps as jetted and after reflow. It can be seen that the 100-μm solder bumps look very good. The UBM on the Al pads of these chips is electroless Ni-Au, as

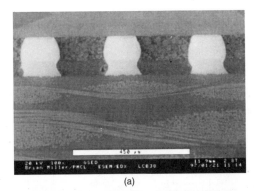

(a)

(b)

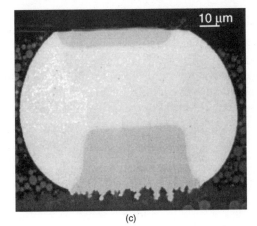

(c)

Figure 2.29 Cross-sectional views of solder bump assembly. (*a*) Overview of the assembly cross section. (*b*) Bump assembly cross section for the 300-μm pitch device. (*c*) Bump assembly cross section for the 200-μm pitch device.

TABLE 2.5   Reliability Evaluation Test Requirements

Test	Condition
Operating life	408 hr, 150°C
Power temperature cycle	1000 c, –40/+125°C
Temperature cycle	1000 c, –40/+150°C
High-temperature bake	1000 hr, 150°C
Temperature cycle	1000 c, –55/+125°C
Temperature shock	1000 c, –55/+125°C
Temperature/ humidity bias	1008 hr, 85°C/85%RH 15-V bias output pins

shown in Fig. 2.34c. The shear test results (based on the test conditions recommended on pages 239 to 241 in Lau and Pao[5]) obtained by MPM is about 35 g. The failure mode is shear fracture in the solder bump and not near the UBM, as shown in Fig. 2.34d.

2.3.6   Fly-through solder jet printing method[60]

As discussed in the previous section, the drop-on-demand (DOD) solder jet technology utilizes the DOD solder ball generators along with relatively high-frequency mechanical positioning. In this section, another approach called fly-through solder joint printing (also called Continuous Stream Metal Jet™) as shown in Figs. 2.35a and b and Fig. 2.36 is presented by Rapid Analysis and Development Co. and MPM Corp. The essential components are as follows.[60]

- *Solder ball generator and integrated stream angular control system:* produces uniform droplets from a constant-speed capillary stream with well-controlled breakup times relative to the disturbance waveform applied to the stream to initiate breakup. The stream from which the droplets are formed originates from a nozzle placed precisely at the center of rotation of a spherical angular control system that is used to remotely adjust the angular orientation of the system with no translational motion of the stream's origin.

- *Pressure control system:* maintains a constant effective driving pressure for the stream.

- *Charging system:* accomplishes accurate and repeatable droplet charging by means of a charging electrode and inductive charging prior to and during droplet separation from the capillary stream.

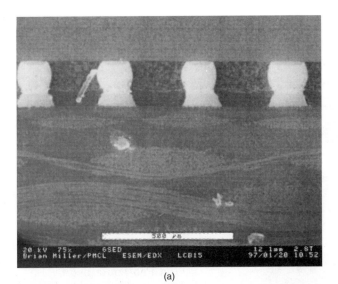

(a)

(b)

Figure 2.30   Liquid-to-liquid temperature cycling (–55 to +125°C), failed at 1500 cycles. (*a*) Overview of the interconnect failure modes. (*b*) Close-up view of the solder joint failure mode.

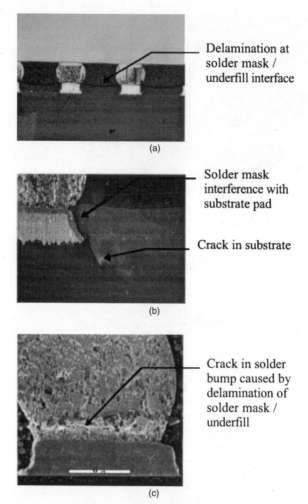

Delamination at
solder mask /
underfill interface

(a)

Solder mask
interference with
substrate pad

Crack in substrate

(b)

Crack in solder
bump caused by
delamination of
solder mask /
underfill

(c)

Figure 2.31 (*a*) Delamination between solder mask and
underfill. (*b*) Cracks in substrate and between solder mask
and underfill. (*c*) Cracks in solder joint.

- *Electrostatic deflection system:* permits control of the relative angular position of the plane defined by the deflected solder droplets relative to the traverse direction.

- *Environmental control system:* rejects oxygen from the vicinity of the capillary stream breakup and shields the solder droplets traveling to the target from oxygen and environmental flow disturbances; also shields the deposition sites from oxygen.

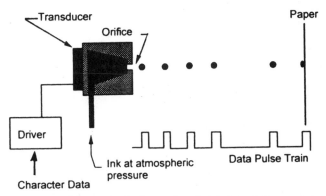

Figure 2.32   Schematic of a demand-mode ink-jet printing system.

- *Droplet stream positioning and deflection calibration system:* stream centering system permits accurate deflection calibration and null positioning of the droplet stream to a fixed home (undeflected) location.

- *Collection system:* collects solder droplets not used for printing.

Some successful printing using the system shown in Figs. 2.35 and 2.36 is demonstrated by the results in Fig. 2.37. It should be noted that the production machines for this technology have already been made. Right now, the Continuous Stream Metal Jet is up to 44,000 solder drops per second.

There are other solder jetting methods, such as the one demonstrated by Sandia National Laboratories[61] and the one designed by IBM, known as micro dynamic solder pump (MDSP).[62] However, these have not been made into commercial machines yet.

2.3.7   Micropunching method[63]

Figure 2.38 schematically shows an overview of the micropunching mechanism developed by NEC.[63] The solder tape is supplied from a spool and rolled up by a motor-driven spool. A micropunch is driven by an electrical actuator and a displacement enlarging mechanism. A micropunch and die set blanks a thin solder tape and forms a small cylindrical piece. Figure 2.39 shows the punched solder pieces for four kinds of solder materials. Figure 2.40 shows the 95wt%Pb-5wt%Sn solder pieces that are punched by different diameters and thicknesses.

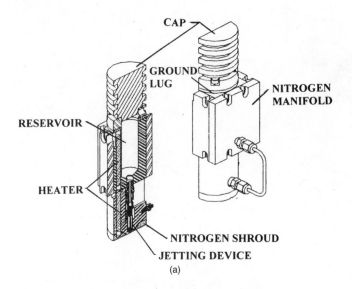

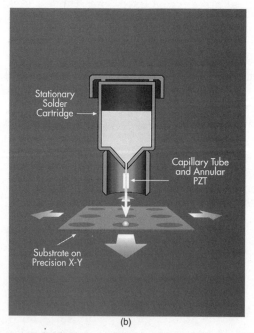

Figure 2.33 (*a*) Solder jet printhead employed on Universal Instruments' Solder Jet platforms. (*b*) Drop-on-Demand Metal Jet.

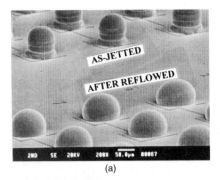

(a)

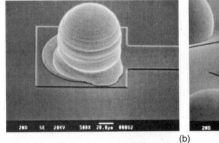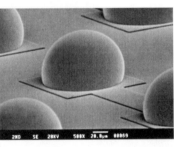

(b)

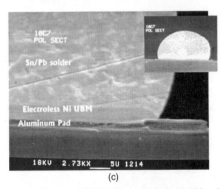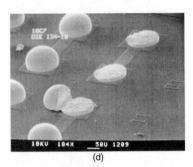

(c)   (d)

Figure 2.34 (*a*) SEM photo showing solder bumps as jetted and reflowed. (*b*) High-magnification views of as-jetted and reflowed bumps. (*c*) Cross section of a reflowed bump showing the UBM. (*d*) Sheared bump fractures (no failure at the UBM).

The characteristic of micropunching technology are as follows.[63]

- The technology can be applied to various kinds of solders.
- Punching size diameter can be selected from 50 to 200 μm.
- The developed punching machine can create 10 bumps per second.
- The structure of the machine is simple and the price is low.
- Solder tape thickness should be less than punching diameter.

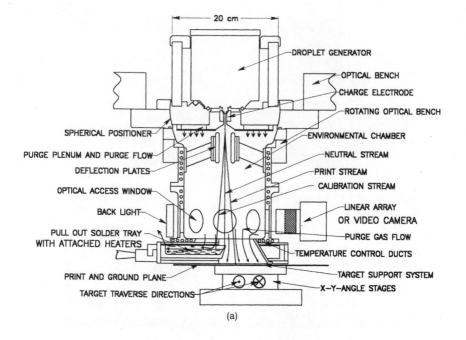

(a)

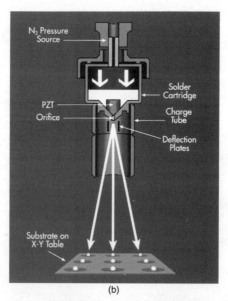

(b)

Figure 2.35 (a) Cross section of RAD fly-through printing system. (b) Continuous Stream Metal Jet.

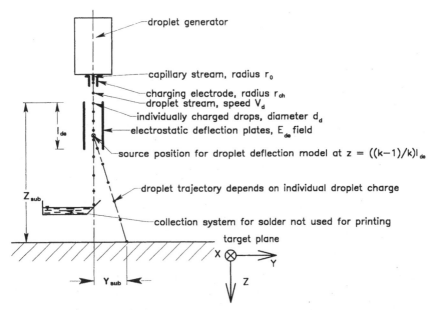

droplet generator

capillary stream, radius r_0

charging electrode, radius r_{ch}
droplet stream, speed V_d

individually charged drops, diameter d_d

electrostatic deflection plates, E_{de} field

source position for droplet deflection model at $z = ((k-1)/k)l_{de}$

droplet trajectory depends on individual droplet charge

collection system for solder not used for printing

target plane

Figure 2.36   Illustration of fly-through printing with important nomenclature defined.

There are two different wafer-bumping processes for micropunch technology, namely the direct bump-forming method and the transfer-ring bump-forming method. Figure 2.41 shows the direct bump-forming process. First, the flux is dispensed on the entire wafer. The solder pieces are punched and placed on the wafer directly. Then the whole wafer is reflowed and cleaned. Figure 2.42 shows the solder bumps on

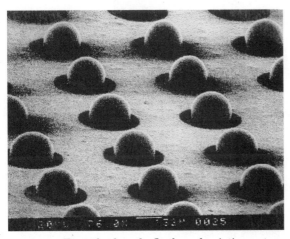

Figure 2.37   Examples from the fly-through printing system.

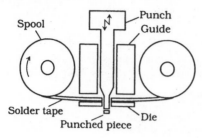

Figure 2.38   Overview of the micro-punching system.

a wafer with different kinds of solders, sizes, and pad pitches. They look very good.

The micropunching technology by transferring bump-forming process is shown in Fig. 2.43. It can be seen that, instead of the solder pieces being punched on the wafer directly, they are placed on a temporary plate laminated with a special adhesive sheet that loses adhesion on heating (Fig. 2.44a). Then, the temporary plate is aligned with the fluxed wafer and a pressure (10 to 20 mN per bump) and a temperature (<120°C) are applied to transfer the solder pieces to the wafer (Fig. 2.44b). Then the entire wafer is reflowed and cleaned (Fig. 2.44c).

The concept of the micropunching technology is very good. However, the infrastructure requirements for supplying the solder tape at a reasonable cost and the micropunching machine with the required accu-

Sn63/Pb37
ϕ 150× t 100 μm

Pb95/Sn5
ϕ 150× t 100 μm

Au80/Sn20
ϕ 150× t 100 μm

Sn96.5/Ag3.5
ϕ 80× t 50 μm

Figure 2.39   SEM photos of the punched solder pieces for four kinds of solders.

ϕ80$\times$ t 50 μm ϕ50$\times$ t 50 μm

ϕ80$\times$ t 40 μm ϕ50$\times$ t 40 μm

Figure 2.40 SEM photos of the 95Pb-5Sn solder pieces punched with different diameters and thicknesses.

ϕ80$\times$ t 30 μm ϕ50$\times$ t 30 μm

racy are great challenges. Also, speed (10 bumps per second) could be an issue.

2.3.8 Molten solder injection method

At Yasu, Japan, IBM developed a wafer-bumping method that uses molten solder injection. Figure 2.45 shows the concept and operation pro-

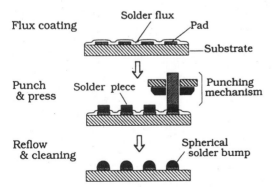

Figure 2.41 Direct solder bump-forming process.

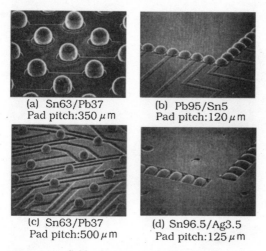

(a) Sn63/Pb37
Pad pitch:350 μm

(b) Pb95/Sn5
Pad pitch:120 μm

(c) Sn63/Pb37
Pad pitch:500 μm

(d) Sn96.5/Ag3.5
Pad pitch:125 μm

Figure 2.42   Solder bumps created by the direct solder bump-forming process.

cess. The head of this machine has a solder reservoir and a metal mask attached to its bottom. The mask has small holes whose pattern is the same as the pad pattern on the wafer. The head comes down to touch the wafer after the alignment is adjusted. When the reservoir is pressurized, the solder in the reservoir is squeezed out from the holes in the mask and

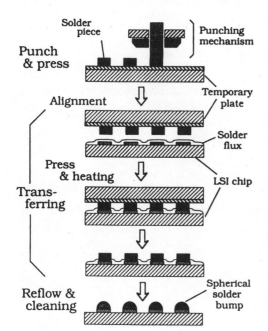

Punch & press

Solder piece

Punching mechanism

Alignment

Temporary plate

Solder flux

Press & heating

LSI chip

Transferring

Reflow & cleaning

Spherical solder bump

Figure 2.43   Transfer solder-bump forming process.

(a) Solder pieces
    on
temporary plate

(b) Solder pieces
    after
    transferring

(c) Solder bumps
    after reflow
    & cleaning

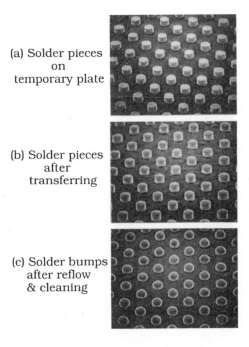

Figure 2.44 Solder bumps created by the transfer solder bump-forming process.

the solder touches the pad on the wafer. After the solder wets the pads, the increased pressure in the reservoir balances with the sum of the surface tensions of the depressed solder columns between the mask and wafer. When the pressure in the reservoir is released, the solder in the columns is pushed back into the holes by the surface tension of the molten solder itself. The solder columns are split when the condition reaches the limit to keep its form. The upper part is drawn back into the reservoir. The lower part remains on the pad and forms the solder bump. The solder bump volume is given by the equation shown in Fig. 2.45.

The concept is great. However, the pad pitch could be a limitation. It works well for 0.25-mm-pitch wafers.

2.3.9 SuperSolder method[64, 65]

The invention disclosure (No. 10941045) of low-cost solder-bumped flip chip application with SuperSolder was filed by the author with Hewlett-Packard in 1994. (The SuperSolder material is made by Furukawa Electric Co., Ltd.) Since then, Furukawa has matured it into a commercial product. SuperSolder (SS) technology works by chemically generating solder through a substitution reaction between tin and organolead salt. A paste composed of tin powder, organolead salt, and flux (SS paste) is heated, undergoing the reactions shown in Fig. 2.46.[64, 65]

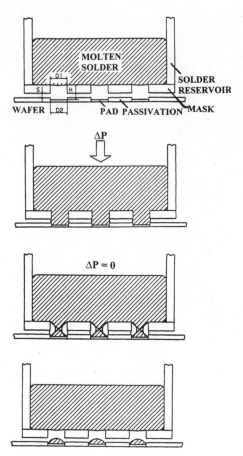

V = solder bump volume = $\pi H D_2^3/12(D_1 + D_2)$
H = height from Al-pad surface to upper surface of mask
D_1 = diameter of mask opening
D_2 = Al-pad diameter

Figure 2.45 Wafer bumping by IBM's molten solder injection process.

The application of SS for wafer bumping is shown in Fig. 2.47. It can be seen that the process steps are very simple. First, the SS paste is applied to the entire wafer or just the entire pad region. Stencils or masks are not needed since SS is not applied to individual pads. Then the whole wafer is heated and cleaned. The solder bump height is about 20 to 25 μm. Multiple SS paste applications are possible to reach the desirable bump height. However, this will increase manufacturing time and cost.

Figure 2.48 shows the secondary ion mass spectrometry (SIMS) measurement results of the oxidation near the solder bump surface using

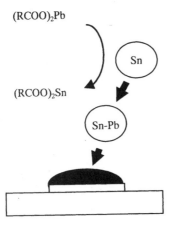

$(RCOO)_2Pb$

$(RCOO)_2Sn$

Step 1   Disassociation of organolead salt:
$(RCOO)_2Pb \rightarrow 2RCOO^- + Pb^{2+}$
(heat)

Step 2   Substitution of tin by lead:
$Pb^{2+} + Sn \rightarrow Pb + Sn^{2+}$
(heat)

Step 3   Diffusion of deposited lead into metallic tin and
alloying:
$Sn + Pb \rightarrow Sn\text{-}Pb$ alloy
(heat)

Step 4   Formation of organotin salt:
$Sn^{2+} + 2RCOO^- \rightarrow (RCOO)_2Sn$
(cool)

Figure 2.46 SuperSolder process. A paste composed of
tin powder, organolead salt, and flux (SS paste) is heated
and undergoes the reaction steps.

the stencil printing, electroplating, and SS methods. It can be seen that
the oxidation in SS solder bumps is lower than that in stencil printing
and electroplating solder bumps. This is because in applying the SS
technology, solder is melted and supplied within a reducing atmo-
sphere. As a result, SS technology is ideal for forming solder bumps
with optimal mountability. The solder supply volume (V) and the solder
supply variation (ΔV) are given in Fig. 2.49. It can be seen that for the
larger value of t in SS solder supply, the larger solder supply volume V
can be attained; however, at the same time the solder supply variation
ΔV also increases.

In order to avoid multiple applications of SS and still reach sufficient
bump height, Furukawa and Motorola recommended increasing the

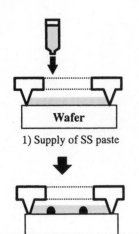

1) Supply of SS paste

2) Heating for supplying solder

Figure 2.47 SuperSolder application steps.

3) Washing off residue

electroless Ni-Au (5 μm) UBM to a 20-μm Ni-Au stud. (The diameter of the Ni-Au UBM is 113.5 μm.) In this case, with just one SS application, solder bump heights of 96 μm could be achieved (Figs. 2.50 and 2.51). It should be pointed out that for such a large Ni-Au stud, the passivation may crack (see Sec. 5.2.2 for more details). Right now, SS paste costs more than the ordinary pastes.

In order to make high-lead solder bumps, such as the C4 bumps, SMT compatible, a 63Sn-37Pb eutectic cap can be made using SS technology (Fig. 2.52). Figure 2.53 shows a cross section of a C4 bump with eutectic cap made by Hewlett-Packard in the 1994 invention disclosure with the SuperSolder provided by Furukawa.

2.3.10   Microball mounting method[66, 67]

Solder ball mounting on PBGA package[7] and CSP[3.5] substrates has been used for 10 and 6 yr, respectively. The ball sizes range from 0.3 to 0.76 mm. In recent years, micro gold (Fig. 2.54) and solder balls are becoming available. In this section, a production method for the micro solder balls

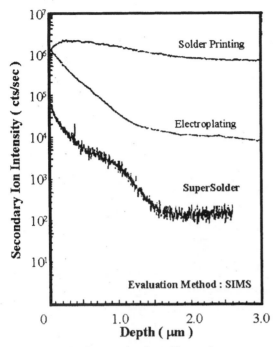

Figure 2.48   Oxidation near the solder surface.

and a gang-bonding method for forming the solder bumps on individual chips or wafers developed by Nippon Steel Corporation[66,67] are presented.

The micro solder balls are made by melting metal pieces previously prepared to a constant weight. The extremely uniform diameters and high sphericity are achieved by controlling the weight of the pieces pre-

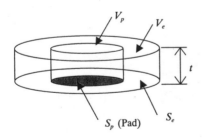

$$V = V_p + x \cdot V_e = t \cdot S_p + x \cdot (t \cdot S_e)$$

$$\Delta V = V_{max} - V_{min}$$
$$= \left\{ t \cdot S_p + (x + \Delta x) \cdot (t \cdot S_e) \right\} - \left\{ t \cdot S_p + (x - \Delta x) \cdot (t \cdot S_e) \right\}$$
$$= 2 \cdot \Delta x \cdot t \cdot S_e$$

where $0 \le x \pm \Delta x \le 1$

Figure 2.49   SuperSolder supply volume model.

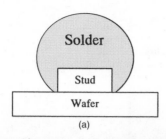

(a)

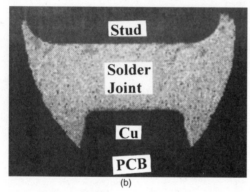

(b)

Figure 2.50   (a) Wafer with Ni-Au stud and SuperSolder. (b) Wafer assembled on PCB.

cisely and by using the surface tension of the molten metals. Micro solder balls 35 to 100 μm in diameter can be made for any solder composition, such as ternary, lead-free, and eutectic solder alloys. Figure 2.55 shows the ball size distribution of micro solder balls 100 μm in diameter.

The process of assembling the micro solder balls on a chip or wafer is shown in Fig. 2.56. It can be seen that the micro solder balls are sucked

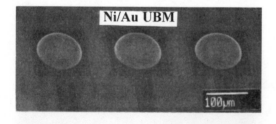

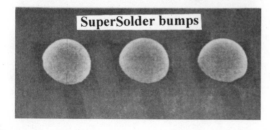

Figure 2.51   Examples of Ni-Au UBM and SuperSolder bumps.

63Sn-37Pb
Eutectic Cap

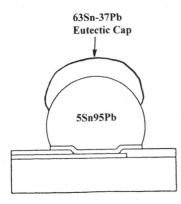

5Sn95Pb

Figure 2.52 Schematic of high melt bump with eutectic cap.

onto through-holes in the arrangement plate by reducing the pressure on the other side of the plate. The through-holes are located at positions corresponding to those of the pads of the individual chip or wafer. It is possible to form bumps on a variety of chips or wafers by selecting this arrangement plate fixed on the bonding head. These arranged balls are transferred to the pads on the chip or wafer in order to form the solder bumps.

The detailed micro solder ball arrangement and transferring process is shown in Fig. 2.56: (*a*) bouncing the micro solder balls uniformly in the ball container by vibration; (*b*) lowering the bonding head with the arrangement plate toward the container and arranging the micro solder balls at the positions of the through-holes by means of vacuum suc-

High Pb
Solder

63Sn/Pb

Figure 2.53 Cross section of a eutectic capped C4 solder bump with SuperSolder.

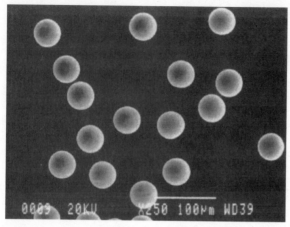

Figure 2.54   SEM image of micro gold balls (45 μm in diameter).

tion; (c) removing the excess balls adhered to the arranged balls or to the plate by vibrating the bonding head at a small amplitude; (d) inspecting the arrangement of the balls; (e) aligning the position of the balls with that of the pads on the chip or wafer; (f) transferring the balls onto the pads.

It is noted that the micro solder balls tend to adhere to each other due to (1) van der Waals forces; (2) surface adsorption of water; (3) electrostatic force; and (4) surface contamination. In order to resolve these issues, Nippon Steel Corporation has developed a micro solder ball mounter (for individual chips only), which is schematically shown in Fig. 2.57. Unique features of this mounter are the system to eliminate excess balls, the ball bouncing system, and the inspection CCD camera system. The complete process flow of micro solder bumping is shown in Fig. 2.58.

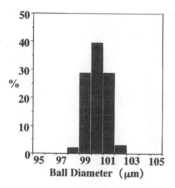

Figure 2.55   Ball size distribution of micro solder balls 100 μm in diameter.

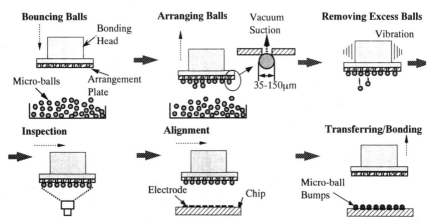

Figure 2.56   Flow chart of the microball arranging and transferring process.

Figure 2.59 shows the solder bumps, bump height distribution, and shear strength distribution of the solder bumps on a chip with 2025 pads on a 220-μm pitch. (The initial micro solder ball diameter is 150 μm.) It can be seen that the solder bumps, bump heights, and shear strengths are very uniform. Figure 2.60 shows the solder bumps on a chip with 200 pads on a 140-μm pitch. (The initial micro solder ball diameter is 80 μm.)

A high-speed automatic microball mounter for wafer bumping on a chip in a wafer is under development. Nippon Steel Corporation estimates that the cycle time should be less than 10 s for a chip in a wafer. The cost picture for micro solder ball bumping is not clear at this moment.

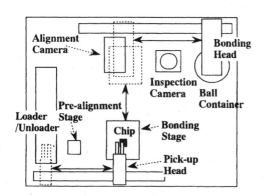

Figure 2.57   Schematic diagram of microball mounter.

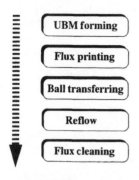

Figure 2.58   Process flow of micro solder bumping.

2.3.11   Tacky Dots™ method

The Tacky Dot™ film is a DuPont® invention[68] produced by coating a proprietary photopolymer adhesive between a polyimide (Kapton®) carrier film and a Mylar® cover sheet. Texas Instruments (TI) joined with DuPont to codevelop Tacky Dots[69, 70] because of its perceived

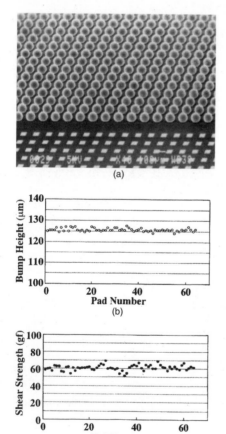

Figure 2.59   (a) SEM image, (b) bump height distribution, and (c) shear strength distribution of area-arrayed micro solder bumps (initial ball diameter: 150 µm; number of bumps: 45 × 45 = 2025; bump pitch: 220 µm).

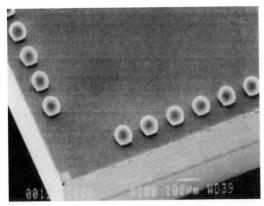

Figure 2.60 Micro solder bumps with 140-mm pitch (initial ball diameter: 80 μm; number of bumps: 200).

advantages as a potential wafer bumping process: (1) compatibility with any metal alloy or thermoplastic composition; (2) independence from the applied UBM; (3) high throughput; (4) fine pitch; (5) cleanliness; and (6) potential low cost.

Tacky Dots is a proprietary technology system that incorporates a film of photoimageable adhesive coated on a support. The Tacky Dots assembly flow is shown in Fig. 2.61. The Tacky Dot film is imaged using

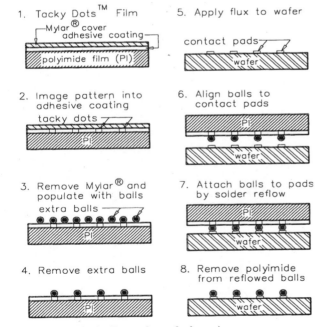

Figure 2.61 Tacky Dots™ for wafer bumping.

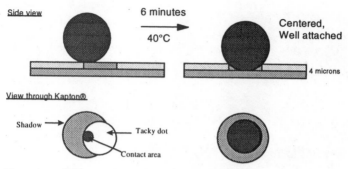

Figure 2.62   Self-centering of spheres in Tacky Dots.

a photo tool and ultraviolet (UV) light to form a pattern of Tacky Dots. The dot size is customized to hold only one particle per dot. For solder wafer bumping, the Tacky Dots are coated on Kapton and populated with conductive solder spheres of certified quality, using equipment to execute the process. The pattern, populated with the desired conductive composition, is aligned with a receiving substrate such as wafer, and the solder conductive particles are reflowed and transferred as solder bumps.

One of the important features of the Tacky Dots technology is that it is self-centering, i.e., the solder spheres are drawn to the center of the dot (Fig. 2.62). The dot size and coating thickness can be designed to perfectly match a solder sphere of a specific size.

Automated population can be accomplished by populating either single, discrete sheets or by populating discrete patterns sequentially on a continuous web. Figures 2.63 and 2.64 show, respectively, the basic elements of the fully automated population module for a single pattern of Tacky Dots mounted on a flex frame and for a roll of continuous film that has been preimaged with a sequential series of Tacky Dot patterns.

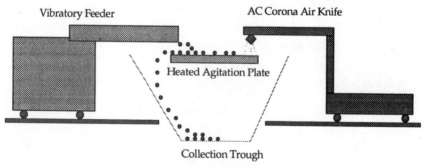

Figure 2.63   Basic elements of EPM-I population module.

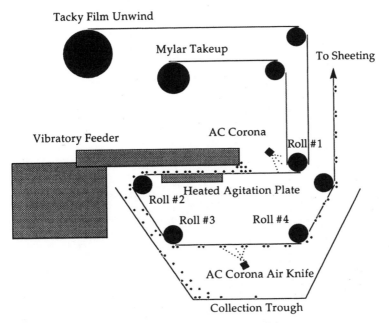

Tacky Film Unwind

Mylar Takeup

To Sheeting

Vibratory Feeder

AC Corona

Roll #1

Heated Agitation Plate

Roll #2

Roll #3   Roll #4

AC Corona Air Knife

Collection Trough

Figure 2.64   Basic elements of EPM-II population module.

The thermal coefficient of expansion for Tacky Dot Kapton film is on the order of 12 ppm/°C; that for silicon is about 2.5 ppm/°C. For 63Sn-37Pb solder and a 100-mm wafer, the Tacky Dot pattern is shrunk by 0.15 percent, allowing for the growth that will occur in the Kapton film going from 25 to 183°C. Thus, the room temperature alignment of the scaled Tacky Dot pattern is done so that all balls at the same distances from the center are equally off their corresponding targets, as shown in Fig. 2.65. As the reflow profile proceeds, the film and wafer heat up and

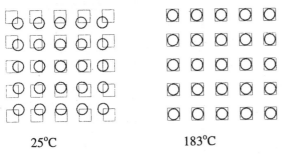

25°C   183°C

Figure 2.65   Scaled Tacky Dots pattern laid over corresponding target site pattern at 25°C and at 183°C (scale exaggerated).

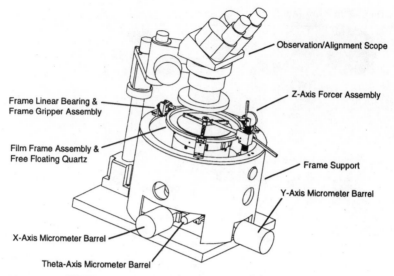

Figure 2.66 WRTP-4 prototype equipment for wafer bumping with Tacky Dots process.

grow linearly with temperature, until at 183°C the patterns match and the transfer takes place.

The Tacky Dots wafer-bumping process prototype equipment is schematically shown in Fig. 2.66.[69] It performs three major functions: film control assembly, reflector assembly, and temperature control. A

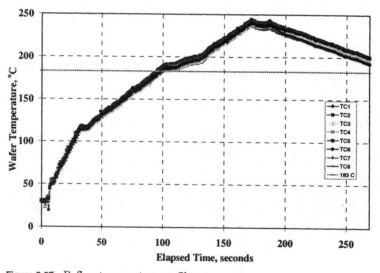

Figure 2.67 Reflow temperature profile.

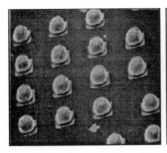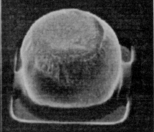

Figure 2.68 Once-reflowed Tacky Dots solder bumps.

typical solder reflow temperature profile obtained by using the proto-type equipment and Tacky Dots is shown in Fig. 2.67.

Figures 2.68 and 2.69 show Tacky Dots solder bumps that have been reflowed one time. It is noted that the solder bump has a flat top sur-face. This is because the solder balls are solidified, with the base of the balls wetting the contact pad on the wafer and the tops of the balls con-strained by the Tacky Dot carrier film. As a result, the array of trans-ferred solder balls will all have flat tops that are coplanar and parallel to the wafer, regardless of the uniformity of the original solder ball. This flat top will disappear if it is reflowed one more time (Fig. 2.70). The UBMs of these chips are TiW-Cu, which works well with the Tacky Dots solder bumps.

The Tacky Dots technology provides a cost-effective, precise, flexible, and environmentally friendly wafer-bumping technique. It is simple, yet flexible in bump size, pitch, solder composition, and compatibility with various UBMs. One of the challenges is peeling off the Mylar cover sheet from the adhesive layer. This tends to generate a static charge and affects the handling of the tiny spheres. In addition, an agi-tated solder sphere bath tends to oxidize the solder sphere surface, which further aggravates the sensitivity toward static.

Figure 2.69 Cross section of once-reflowed Tacky Dots solder bump.

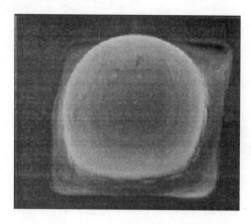

Figure 2.70   Twice-reflowed Tacky Dots solder bump.

2.3.12   Solder bumps on PCB method

Do we have to put the solder bumps on the chip/wafer in order to use the solder-bumped flip chip technology? No! However, this is more convenient for the system makers. They would consider the solder-bumped flip chip as just another SMT component and pick and place it on the PCB.

Can we put the solder bumps on the PCB and not on the chip/wafer in order to use the flip chip technology? Yes! However, in most cases the system makers hate it, because they may have only one or a few flip chip components on the PCB and they have to pay someone to put the solder bumps on the PCB. This not only adds cost but inconvenience.

Figure 2.71 shows a way of stencil printing to make the solder bumps on PCB. Usually, printing must be repeated several times to reach the required solder bump height. The other way is by electroplating the Sn-Pb on the copper pads of the PCB. However, it will take a very long plating time to achieve certain solder bump heights on just a few flip chip sites in a large PCB. Thus, in general, it is more cost effective and convenient to put solder bumps on the wafer.

In order to make the high-lead solder bumps (e.g., C4) SMT compatible, however, a small amount of 63Sn-37Pb solder is put on the copper pads of the PCB (Fig. 2.72). For copper pads with pitches higher than 175 µm, stencil printing works fine; otherwise, electroplating[71] or the molten solder injection method[72] is needed.

2.3.13   Solder bumps on Al pads without UBM[73–76]

It is reemphasized that the low-cost wafer bumping methods discussed in Secs. 2.3.4 through 2.3.12 require UBM. Thus, logistics and infrastructure become big factors in judging the total cost of wafer bumping.

1. Manufacturing an FR-4 board with a solder mask

Cu   Solder mask

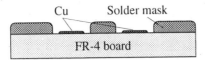

2. Printing solder paste into the solder mask

3. Reflowing the solder

4. Flattening the solder deposits

Figure 2.71   Stencil printing paste on board to either make solder bumps or to coat a thin layer of 63Sn-37Pb solder such that C4 is SMT compatible.

Usually, true low cost can be achieved if the bumpers also do UBM at the same site. Otherwise, there would be many hidden costs.

In this section, a very unique method to solder-bump the die without UBM is presented.[73–76] Figure 2.73 illustrates some of the fundamental motions of the solder bump formation. The movement for bonding is provided by the conventional stud bump bonding machine, e.g., Shinkawa SBB-1. After the solder wire is inserted into a ceramic capillary, a solder ball 110 μm in diameter is formed at the tip of the wire by arc discharge in an $Ar^+10\%H_2$ gas flow (1.5 l/min) introduced by means of two tubes

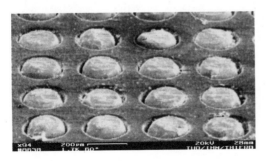

Figure 2.72   PCB with solid solder deposits.

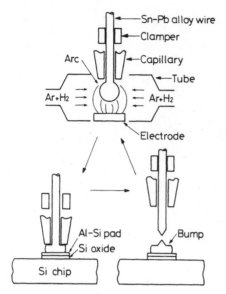

Figure 2.73 Some fundamental motions for the solder bump formation without UBM.

(inner diameter = 6 mm) positioned opposite each other. The discharge conditions are: spark voltage = 2.5 kV; discharge voltage = 60 V; current = 100 mA; time = 1.2 ms. The solder wire functions as an anode. Figure 2.74 shows the solder (Sn-Pb-Sb-Zn-Ag-Cu-Ni) alloy ball formed by arc discharge, and Fig. 2.75 shows the Auger spectra from the solder alloy ball surface before (a) and after (b) Ar ion etching for 120 s. It can be seen that Zn and O are detected near the ball surface.

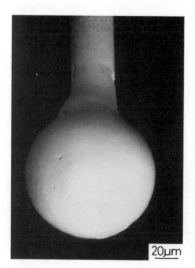

Figure 2.74 SEM photo of the solder alloy ball formed by arc discharge.

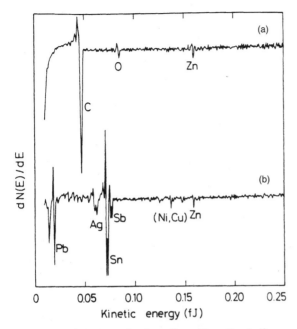

Figure 2.75   Auger spectra from the solder alloy ball surface before (a) and after (b) Ar ion etching.

The ball bonding conditions on the Al-1%Si film pad (0.8 μm), which is prepared by vapor deposition on a Si chip with a Si oxide layer of 0.5 μm, are as follows: applied power = 0.55 to 1.49 W; time = 35 ms for ultrasonic agitation; force = 0.39 to 0.51 N; chip temperature = 125°C. After the solder ball is bonded directly on the Al pad, the capillary is raised, then clamped and further raised, causing the solder wire to break at the ball-wire interface. Figure 2.76 shows the change in the shear force at various applied powers of ultrasonic agitation for the Sn-Pb-Sb-Zn-Ag-Cu-Ni alloy bump before and after reflow at 230°C.

The solder ball deformation $(2 \ln D/D_0)$ is fixed at 0.5 by adjusting the bonding force of the solder ball over the range of ultrasonic power from 0.55 to 1.49 W. In this examination, the D_0 and D showing the diameter of the initial ball and the bonded bump are 110 and 140 μm, respectively. The mean shear force of bonded and reflowed bumps attains a maximum value of 0.95 and 0.59 N, respectively, within a small range at 0.71 W. The bond areas at 0.71 W estimated from the areas of residual solder are 7.9×10^{-3} mm^2 in the as-bonded state and 6.4×10^{-3} mm^2 in the reflowed state. This results in the difference in the shear forces between bonded and reflowed states. Figure 2.77 shows the bonded

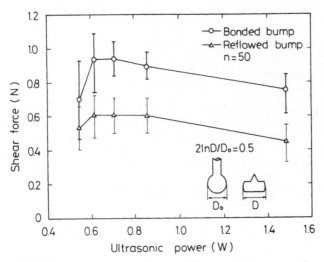

Figure 2.76 Change in the shear force with ultrasonic power for the solder alloy bump before and after reflow at 230°C.

bump shape (a) on Al-1%Si pads and the soldered interconnection and (b) before underfill encapsulation in the flip chip assembly.

The idea is great and it could be very low in cost. However, the throughput could be a big issue, since the solder bumps are formed one by one.

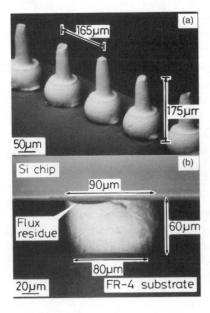

Figure 2.77 SEM photo of the bonded bump shape (a) on Al-1Si pads and the soldered interconnection and (b) before encapsulation in the flip chip assembly.

2.4   Alpha Particle

It is well known that alpha particles from thorium and uranium impurities in packaging raw materials such as lead-bearing solder alloys do not cause permanent damage to the chip but are sufficiently energetic to produce soft errors (data loss)—random, nonrecurring single-bit errors. Alpha particles produce soft errors by penetrating through the lead-bearing solder joint, UBM, Al pad, passivation, silicon, and *pn* junction and generating carriers as a result of giving up kinetic energy as they slow down. The carriers generated by alpha penetration through the junction distort the electric field and also generate charges.

The alpha emission flux is different for different materials. For example, the alpha emission from ceramic ranges from 0.1 to 1 $\alpha/cm^2 \cdot$ h; the values for silicon are less than 0.1 $\alpha/cm^2 \cdot$ h; and the figures for lead-bearing solder alloys vary greatly and are much higher than those from ceramic. Low-alpha lead-bearing solder alloys are available at a premium cost for applications that require purities as high as the order of parts per billion (ppb). One part per billion is equivalent to a flux of about 0.001 $\alpha/cm^2 \cdot$ h.

There is a critical or escape depth, measured from the surface, below which alpha particles are internally absorbed by the emitting material and have no chance of retaining sufficient energy to emerge at the surface. For example, the escape depth for ceramic is on the order of 50 μm; for silicon the figure is 25 μm; and for lead-bearing solder alloys it is less than 10 μm.

Alpha particles have a limited range of penetration that varies with the energy of the particle and with the densty and average mass number of the material. For example, in air, an alpha particle with ~5 MeV of energy may be stopped after traveling 5 to 6 cm. However, in polymers, the particle's range may be on the order of 0.005 to 0.01 cm. Thus, chemically inert materials such as silicone room-temperature vulcanized (RTV) rubber can be used as coatings to create barriers to alpha particles that prevent them from reaching the devices. Since the rubber itself has an average alpha activity of 0.001 $\alpha/cm^2 \cdot$ h and an alpha energy range of 2 to 8 MeV, a minimum of 2 mil of silicone rubber over the active surface of the chip is an effective barrier to alpha particles emitted from the packaging materials.

Alpha particles from lead-bearing solder alloys have been known to cause soft errors in memory devices. The alpha counts of 1 $\alpha/cm^2 \cdot$ h from solders containing lead are marginally acceptable today. However, this figure needs to be reduced to less than 0.01 $\alpha/cm^2 \cdot$ h for low-voltage and high-speed DRAM devices, since the lead-containing solder joints are the principal source of soft error. For more information about alpha particles, readers are encouraged to consult the proceedings of the Low-Alpha Lead Symposium.[77]

2.5 Wafer Bumping with Solderless Materials

The focus of this chapter is wafer bumping with solders. Discussions of wafer bumping with stud bumps can be found in Chap. 12 of Lau[6] and Chap. 23 of Lau and Lee.[3] Material on wafer bumping with Au, Cu, and Ni-Au materials can be found in Chap. 5 of this book.

Acknowledgments

The author would like to thank M. Cho, S. Kang, Y. Kwon, D. Jang, and N. Kim of Samsung; Y. Kaga of Furukawa; G. Hotchkiss, G. Amador, L. Jacobs, R. Stierman, S. Dunford, and P. Hundt of TI; A. Beikmohamadi, A. Cairncross, J. Gantzhorn, B. Quinn, and M. Saltzberg of DuPont; C. Kallmayer, H. Oppermann, S. Ankock, R. Azadeh, R. Aschenbrenner, J. Kloeser, K. Heinricht, E. Jung, L. Lauter, A. Ostmann, and H. Reichl of Technical University of Berlin and Fraunhofer Institute IZM; M. Suwa, T. Miwa, Y. Tsutsumi, and Y. Shirai of Hitachi; Y. Tsukada, S. Ouimet, and M. Paquet of IBM; T. Ogashiwa, T. Arikawa, and A. Inoue of Tanaka and Tohoku University; E. Muntz, M. Orme, G. Pham-Van-Diep, and R. Godin of RAD and MPM; D. Hayes, D. Wallace, M. Boldman, and R. Marusak of MicroFab; S. Wiegele, L. Li, P. Thompson, J. Zhang, R. Lee, and E. Ramsland of Motorola; P. Elenius, J. Leal, J. Ney, D. Stepniak, and S. Yeh of Flip Chip Technologies and Delco; and Y. Kato, Y. Ueoka, E. Kono, and E. Hagimoto of NEC for sharing their important and useful technologies with the industry.

References

1. Lau, J. H., *Handbook of Tape Automated Bonding,* Van Nostrand Reinhold, New York, 1992.
2. Tummala, R., E. Rymaszewski, and A. Klopfenstein, *Microelectronics Packaging Handbook,* 2d ed., Chapman-Hall, New York, 1997.
3. Lau, J. H., and R. Lee, *Chip Scale Package, Design, Materials, Process, Reliability, and Applications,* McGraw-Hill, New York, 1999.
4. Lau, J. H., C. Wong, J. Prince, and W. Nakayama, *Electronic Packaging: Design, Materials, Process, and Reliability,* McGraw-Hill, New York, 1998.
5. Lau, J. H., and Y. H. Pao, *Solder Joint Reliability of BGA, CSP, Flip Chip, and Fine Pitch SMT Assemblies,* McGraw-Hill, New York, 1997.
6. Lau, J. H., *Flip Chip Technologies,* McGraw-Hill, New York, 1996.
7. Lau, J. H., *Ball Grid Array Technology,* McGraw-Hill, New York, 1995.
8. Lau, J. H., *Chip on Board Technologies for Multichip Modules,* Van Nostrand Reinhold, New York, 1994.
9. Harman, G., *Wire Bonding in Microelectronics, Materials, Processes, Reliability, and Yield,* McGraw-Hill, New York, 1997.

10. Garrou, P. E., and I. Turlik, *Multichip Module Technology Handbook,* McGraw-Hill, New York, 1998.
11. Elshabini-Riad, A., and F. D. Barlow, III, *Thin Film Technology Handbook,* McGraw-Hill, New York, 1998.
12. Suwa, M., T. Miwa, Y. Tsutsumi, and Y. Shirai, "Development of a 1000-pin Fine-pitch BGA for High Performance LSI," *Proceedings of IEEE Electronic Components and Technology Conference,* pp. 430–434, June 1999.
13. Ouimet, S., and M. Paquet, "Overmold Technology Applied to Cavity Down Ultrafine Pitch PBGA Package," *Proceedings of IEEE Electronic Components and Technology Conference,* pp. 458–462, May 1998.
14. Dehkordi, P., K. Ramamurthi, D. Bouldin, H. Davidson, and P. Sandborn, "Impact of Packaging Technology on System Partitioning: A Case Study," *Proceedings of the IEEE Multichip Module Conference,* pp. 144–149, Santa Cruz, CA, February 1993.
15. Dehkordi, P. and D. Bouldin, "Design for Packagability: The Impact of Bonding Technology on the Size and Layout of VLSI Dies," *Proceedings of the IEEE Multichip Module Conference,* pp. 153–159, Santa Cruz, CA, February 1993.
16. Singh, P., and D. Landis, "Optimal Chip Sizing for Multi-Chip Modules," *IEEE Trans. Components, Packaging, and Manufacturing Technology,* part B, **17** (3): 369–375, August 1994.
17. Sandborn, P., M. Abadir, and C. Murphy, "The Tradeoff Between Peripheral and Area Array Bonding of Components in Multichip Modules," *IEEE Trans. Components, Packaging, and Manufacturing Technology,* part A, **17** (2): 249–255, June 1994.
18. Dehkordi, P., and D. Bouldin, "Design for Packageability, Early Consideration of Packaging from a VLSI Designer's Viewpoint," *IEEE Computer,* 76–81, April 1993.
19. Lau, J. H., "Cost Analysis: Solder Bumped Flip Chip versus Wire Bonding," *IEEE Trans. Electronics Packaging Manufacturing,* January 2000.
20. Davis, E., W. Harding, R. Schwartz, and J. Corning, "Solid Logic Technology: Versatile, High Performance Microelectronics," *IBM J. Res. Dev.,* 102–114, April 1964.
21. Totta, P. A., and R. P. Sopher, "SLT Device Metallurgy and Its Monolithic Extension," *IBM J. Res. Dev.,* 226–238, May 1969.
22. Goldmann, L. S., and P. A. Totta, "Chip Level Interconnect: Solder Bumped Flip Chip," in *Chip On Board Technologies for Multichip Modules,* Lau, J. H., ed., Van Nostrand Reinhold, New York, pp. 228–250, 1994.
23. Goldmann, L. S., R. J. Herdzik, N. G. Koopman, and V. C. Marcotte, "Lead Indium for Controlled Collapse Chip Joining," *Proceedings of the IEEE Electronic Components Conference,* pp. 25–29, 1977.
24. Totta, P., "Flip Chip Solder Terminals," *Proceedings of the IEEE Electronic Components Conference,* pp. 275–284, 1971.
25. Goldmann, L. S., "Geometric Optimization of Controlled Collapse Interconnections," *IBM J. Res. Dev.,* 251–265, May 1969.
26. Goldmann, L. S., "Optimizing Cycle Fatigue Life of Controlled Collapse Chip Joints," *Proceedings of the 19th IEEE Electronic Components and Technology Conference,* pp. 404–423, 1969.
27. Goldmann, L. S., "Self Alignment Capability of Controlled Collapse Chip Joining," *Proceedings of the 22nd IEEE Electronic Components and Technology Conference,* pp. 332–339, 1972.
28. Shad, H. J., and J. H. Kelly, "Effect of Dwell Time on Thermal Cycling of the Flip Chip Joint," *Proceedings of the ISHM,* pp. 3.4.1–3.4.6, 1970.
29. Hymes, I., R. Sopher, and P. Totta, "Terminals for Microminiaturized Devices and Methods of Connecting Same to Circuit Panels," U.S. Patent 3,303,393, 1967.
30. Karan, C., J. Langdon, R. Pecararo, and P. Totta, "Vapor Depositing Solder," U.S. Patent 3,401,055, 1968.
31. Seraphim, D. P., and J. Feinberg, "Electronic Packaging Evolution," *IBM J. Res. Dev.,* 617–629, May 1981.
32. Miller, L. F., "Controlled Collapse Reflow Chip Joining," *IBM J. Res. Dev.,* 239–250, May 1969.

33. Miller, L. F., "A Survey of Chip Joining Techniques," *Proceedings of the 19th IEEE Electronic Components and Technology Conference*, pp. 60–76, 1969.
34. Miller, L. F., "Joining Semiconductor Devices with Ductile Pads," *Proceedings of ISHM*, pp. 333–342, 1968.
35. Norris, K. C., and A. H. Landzberg, "Reliability of Controlled Collapse Interconnections," *IBM J. Res. Dev.*, 266–271, May 1969.
36. Oktay, S., "Parametric Study of Temperature Profiles in Chips Joined by Controlled Collapse Technique," *IBM J. Res. Dev.*, 272–285, May 1969.
37. Bendz, D. J., R. W. Gedney, and J. Rasile, "Cost/Performance Single Chip Module," *IBM J. Res. Dev.*, 278–285, 1982.
38. Blodgett, A. J., Jr., "A Multilayer Ceramic Multichip Module," *IEEE Trans. Components, Hybrids, and Manufacturing Technology*, 634–637, 1980.
39. Fried, L. J., J. Havas, J. Lechaton, J. Logan, G. Paal, and P. Totta, "A VLSI Bipolar Metallization Design with Three-Level Wiring and Area Array Solder Connections," *IBM J. Res. Dev.*, 362–371, 1982.
40. Clark, B. T., and Y. M. Hill, "IBM Multichip Multilayer Ceramic Modules for LSI Chips—Designed for Performance Density," *IEEE Trans. Components, Hybrids, and Manufacturing Technology*, 89–93, 1980.
41. Blodgett, A. J., and D. R. Barbour, "Thermal Conduction Module: A High-Performance Multilayer Ceramic Package," *IBM J. Res. Dev.*, 30–36, 1982.
42. Dansky, A. H., "Bipolar Circuit Design for a 5000-Circuit VLSI Gate Array," *IBM J. Res. Dev.*, 116–125, 1981.
43. Howard, R. T., "Optimization of Indium-Lead Alloys for Controlled Collapse Chip Connection Application," *IBM J. Res. Dev.*, 372–389, 1982.
44. Lau, J. H., *Solder Joint Reliability: Theory and Applications*, Van Nostrand Reinhold, New York, 1991.
45. Lau, J. H., *Thermal Stress and Strain in Microelectronics Packaging*, Van Nostrand Reinhold, New York, 1993.
46. Frear, D., H. Morgan, S. Burchett, and J. Lau, *The Mechanics of Solder Alloy Interconnects*, Van Nostrand Reinhold, New York, 1994.
47. Greer, S., "An Extended Eutectic Solder Bump for FCOB," *Proceedings of IEEE Electronic Components and Technology Conference*, pp. 546–551, May 1996.
48. Kallmayer, C., H. Oppermann, S. Ankock, R. Azadeh, R. Aschenbrenner, and H. Reichl, "Reliability Investigations for Flip-Chip on Flex Using Different Solder Materials," *Proceedings of IEEE Electronic Components and Technology Conference*, pp. 303–310, May 1998.
49. Elenius, P., J. Leal, J. Ney, D. Stepniak, and S. Yeh, "Recent Advances in Flip Chip Wafer Bumping Using Solder Paste Technology," *Proceedings of IEEE Electronic Components and Technology Conference*, pp. 260–265, June 1999.
50. Li, L., S. Wiegele, P. Thompson, and R. Lee, "Stencil Printing Process Development for Low Cost Flip Chip Interconnect," *Proceedings of IEEE Electronic Components and Technology Conference*, pp. 421–426, May 1998.
51. Wiegele, S., P. Thompson, R. Lee, and E. Ramsland, "Reliability and Process Characterization of Electroless Nickel-Gold/Solder Flip Chip Interconnect Technology," *Proceedings of IEEE Electronic Components and Technology Conference*, pp. 861–866, May 1998.
52. Kloeser, J., K. Heinricht, E. Jung, L. Lauter, A. Ostmann, R. Aschenbrenner, and H. Reichl, "Low Cost Bumping by Stencil Printing: Process Qualification for 200 μm Pitch," *Proceedings of International Symposium on Microelectronics*, pp. 288–297, October 1998.
53. Cho, M., S. Kang, Y. Kwon, D. Jang, and N. Kim, "Flip-Chip Bonding on PCB with Electroless Ni-Au and Stencil Printing Solder Bump," *Proceedings of SMTA International Conference*, pp. 159–164, September 1999.
54. Hayes, D., and D. Wallace, "Solder Jet Printing for Wafer Bumping and CSP," *Proceedings of SMA International Conference*, pp. 508–513, September 1997.
55. Hayes, D., and D. Wallace, "Solder Jet Printing: Wafer Bumping and CSP Applications," *Chip Scale Rev.*, 75–80, September 1998.

56. Hayes, D., D. Wallace, M. Boldman, and R. Marusak, "Picoliter Solder Droplet Dispensing," *Int. J. Microcircuits and Electronic Packaging,* **16** (3): 173–180, 1993.
57. Waldvogel, J., and D. Poulikakos, "Solidification Phenomena in Picoliter Size Solder Droplet Deposition on a Composite Substrate," *Int. J. Heat and Mass Transfer,* **40**: 295–309, 1997.
58. Waldvogel, J., D. Poulikakos, D. Wallace, and R. Marusak, "Transport Phenomena in Picoliter Size Solder Droplet Dispensing on a Composite Substrate," *ASME Trans. J. Heat Transfer,* **118**: 148–156, 1996.
59. Pham-Vam-Diep, G., R. Smith, and R. Godin, "An Investigation of Precision, Continuous Solder Jet Printing for CSP Solder Ball Deposition," *Proceedings of NEPCON West,* 842–858, February 1997.
60. Muntz, E., M. Orme, G. Pham-Van-Diep, and R. Godin, "An Analysis of Precision, Fly-Through Solder Jet Printing," *Proceedings of International Symposium on Microelectronics,* pp. 671–680, October 1997.
61. Frear, D., F. Yost, D. Schmale, and M. Essien, "Area Array Jetting Device for Ball Grid Arrays," *Proceedings of Surface Mount International Conference,* pp. 41–46, September 1997.
62. Schiesser, T., E. Menard, T. Smith, and J. Akin, "Micro Dynamic Solder Pump: An Innovative Liquid Solder Dispense Solution to FCA and BGA Challenges," *Proceedings of NEPCON West,* pp. 1680–1687, February 1995.
63. Kato, Y., Y. Ueoka, E. Kono, and E. Hagimoto, "Solder Bump Forming using Micro Punching Technology," *Proceedings of IEEE Japan IEMTS,* pp. 117–120, December 1995.
64. Kaga, Y., and J. Zhang, "Low Cost Wafer Bumping with Madkless Process," *Proceedings of Surface Mount International Conference,* pp. 265–269, August 1998.
65. Kaga, Y., T. Amano, M. Kohno, and Y. Zhang, "Solder Bumping Through Super Solder," *Proceedings of IEEE IEMTS,* pp. 1–4, October 1995.
66. Shimokawa, K., E. Hashino, Y. Ohzcki, and K. Tatsumi, "Micro-Ball Bump for Flip Chip Interconnections," *Proceedings of IEEE Electronic Components and Technology Conference,* pp. 1472–1476, May 1998.
67. Shimokawa, K., K. Tatsumi, E. Hashino, Y. Ohzeki, M. Konda, and Y. Kawakami, "Micro-Ball Bump Technology for Fine-Pitch Interconnections," *Proceedings of 1st IEMT/IMC Conference,* pp. 105–109, 1997.
68. Cairncross, A., and K. Klabunde, "Method and Product for Particle Mounting," U.S. Patent 5,356,751, October 1994.
69. Hotchkiss, G., G. Amador, L. Jacobs, R. Stierman, S. Dunford, P. Hundt, A. Beikmohamadi, A. Cairncross, J. Gantzhorn, B. Quinn, and M. Saltzberg, "Tacky Dots™ Transfer of Solder Spheres for Flip Chip and Electronic Package Applications," *Proceedings of IEEE Electronic Components and Technology Conference,* pp. 434–441, May 1998.
70. Beikmohamadi, A., A. Cairncross, J. Gantzhorn, B. Quinn, M. Saltzberg, G. Hotchkiss, G. Amador, L. Jacobs, R. Stierman, S. Dunford, and P. Hundt, "Tacky Dots™ Technology for Flip Chip and BGA Solder Bumping," *Proceedings of IEEE Electronic Components and Technology Conference,* pp. 448–453, May 1998.
71. Coombs, C., *Printed Circuits Handbook,* McGraw-Hill, New York, 1996.
72. Tsukada, Y., "Solder Bumped Flip Chip Attach on SLC Board and Multichip Module," in *Chip on Board Technologies for Multichip Modules,* Lau, J. H., ed., Van Nostrand Reinhold, New York, pp. 410–443, 1994.
73. Ogashiwa, T., T. Arikawa, and A. Inoue, "Development of Reflowable Sn-Pb Alloy Bump for Al Pad," *Proceedings of IEEE Electronic Components and Technology Conference,* pp. 664–669, May 1997.
74. Ogashiwa, T., T. Arikawa, H. Murai, A. Inoue, and T. Masumoto, "Reflowable Sn-Pb Bump Formation on Al Pad by a Solder Bumping Method," *Proceedings of IEEE Electronic Components and Technology Conference,* pp. 1203–1208, May 1995.
75. Ogashiwa, T., T. Masumoto, H. Shigyo, Y. Murakami, A. Inoue, and T. Masumoto, "Direct Solder Bump Formation Technique on Al Pad and Its High Reliability," *Jpn. J. Appl. Physics,* **31**: 761–767, March 1992.

76. Ogashiwa, T., C. Kamada, A. Inoue, and T. Masumoto, "Solder Bump Formation for Flip-Chip Interconnection by Ball Bonding Method," *Proceedings of the 6th International Microelectronics Conference,* pp. 228–234, May 1990.
77. Lawrence Livermore National Laboratory, *Proceedings of Low-Alpha Lead Symposium,* February 1997.

Chapter 3

Lead-Free Solders

3.1 Introduction

Low-cost tin-lead solders have been used as joining materials in the electronics industry for many years.[1-11] In the past decade, on average approximately 20,000 tons of lead each year have been used globally among electronics manufacturers. This amount is about 7 percent of total lead consumption worldwide. The unique physical and mechanical properties of tin-lead solders have facilitated printed circuit board (PCB) assembly choices that have fueled creative advances in packaging developments, such as solder-bumped flip chip,[1-4] ball grid array (BGA) packages,[2, 3, 5] and chip scale packages (CSPs).[1-3] For these packaging technologies, Sn-Pb solder is the electrical and mechanical "glue" of the PCB assemblies.

In the past few years, different bills have been introduced in the U.S. Congress to ban lead from a wide variety of uses, including solders. The reasons are, among others: (1) lead and its compounds are ranked among the top 10 hazardous materials and (2) lead is the number one environmental threat to children. Recently, many major electronics companies, national laboratories, universities, research organizations, and solder vendors worldwide have responded by initializing research programs to eliminate lead from solders.[12-42]

3.2 Worldwide Efforts on Lead-Free Solders

In Japan, some electronics manufacturers have announced voluntary plans to reduce their use of lead in solders. For example, Hitachi aims to reduce its use of lead by half during 1999 compared to 1997 levels and to stop using lead solders by 2001 (Sn-Bi-Ag alloys are Hitachi's favorite). Matsushita aims to stop using lead solders by 2001 (Matsushita uses Sn-Ag-Bi-x alloys). Toshiba plans to eliminate lead in its mobile phones by 2000 (Sn-Ag-Cu alloys are Toshiba's choice). NEC intends to reduce lead use in solders by 50 percent by 2002 compared to its usage in 1997 (Sn-Ag and Sn-Ag-Cu alloys are NEC's choice). NTT intends to purchase only equipment safe for the environment (i.e., no lead or cadmium) by 2001. Sony wants to reduce its usage of lead solders but has not yet set a target date. (Sony developed the Sn-2Ag-4Bi-0.5Cu-0.1Ge solder for its own products.) It is interesting to point out that "green" products sell! For example, Matsushita's market share of its lead-free MiniDisc player jumped from 4.6 to 15 percent in 6 months in Japan.

In North America, the world's first lead-free circuit telephone was produced by NORTEL Networks in 1997.[36] NORTEL used eutectic 99.3Sn-0.7Cu to replace Sn-Pb solder for both surface mount and through-hole components. The European Union proposes to ban all lead in electronic products by the year 2004 and wants to be more "green."

Table 3.1 shows some of the solder alloys available in paste, bar, or wire form for different melting temperatures. (The numbers indicate percentages of the elements, with tin making up the balance.) For example, the lead-free solder alloys with melting points below 180°C are Sn-58Bi (138°C), Sn-52In (118°C), Sn-2.8Ag-20In (178°C), and Sn-50In (118 to 125°C). The lead-free solder alloys with melting points between 180 and 200°C are Sn-9Zn (198.5°C), Sn-8Zn-3Bi (189 to 199°C), and Sn-20Bi-10In (143 to 193°C). The lead-free solder alloys with melting points between 200 and 230°C are Sn-3.5Ag (221°C), Sn-2Ag (221 to 226°C), Sn-0.7Cu (227°C), Sn-3.5Ag-3Bi (206 to 213°C), Sn-7.5Bi-2Ag (207 to 212°C), Sn-3.8Ag-0.7Cu (217°C), Sn-3.4Ag-4.8Bi (211°C), and Sn-2Ag-0.8Cu-0.5Sb (216 to 222°C). Finally, the lead-free solder alloys with melting points above 230°C are Sn-5Sb (232 to 240°C), Au-20Sn (280°C), and Sn-25Ag-10Sb (365°C).

The National Center for Manufacturing Sciences (NCMS)[37] has done an extensive study, selecting 7 alloys from an initial list of 79 as drop-in replacements for tin-lead solders. These alloys are Sn-3.5Ag, Sn-58Bi, Sn-3Ag-2Bi, Sn-2.6Ag-0.8Cu-0.5Sb, Sn-3.4Ag-4.8Bi, Sn-2.8Ag-20In, and Sn-3.5Ag-0.5Cu-1Zn. The screening criteria are (1) eliminating toxic elements; (2) eliminating most material families with very minor composition differences; (3) eliminating materials with economic, manufacturability, reliability, and availability limitations; and (4) application of judgement criteria based on those alloys that pass the preceding three screening criteria. It should be pointed out that Sn-58Bi and Sn-2.8Ag-20In are included for manufacturability and reliability assessments even though they both fail economic and availability criteria. These seven alloys are subjected to manufacturability assessment and reliability testing. Both surface-mount and plated-through-hole tech-

TABLE 3.1 Lead-Free Solder Alloys Available as Paste, Bar, or Wire

	Paste	Bar	Wire
Melting point below 180°C	Sn-58Bi Sn-42In Sn-50In Sn-52In	Sn-30Bi Sn-40Bi Sn-58Bi Sn-52Bi Sn-50In Sn-52In Sn-57Bi-2In	Sn-40Bi Sn-58Bi Sn-52Bi Sn-50In Sn-52In Sn-57Bi-2In
Melting point between 180 and 200°C	Sn-8Zn-3Bi Sn-Zn-Bi Sn-Zn-Bi-In Sn-8.8In-27.6Zn Sn-20In-2.8Ag Sn-10.5In-2Ag-0.5Sb Sn-9.5Bi-0.5Cu	Sn-10Zn Sn-9Zn	Sn-9Zn
Melting point between 200 and 230°C	Sn Sn-0.7Cu Sn-1Cu Sn-2.5Ag Sn-3.5Ag Sn-4Ag Sn-5Ag Sn-3.4Ag-1.2Cu Sn-3.4Ag-1.2Cu-3.3Bi Sn-3.5Ag-0.7Cu Sn-4Ag-0.5Cu Sn-4.7Ag-1.7Cu Sn-Ag-Cu-Ge Sn-2.6Ag-0.8Cu-0.5Sb Sn-Ag-Bi Sn-1.5Ag-2Bi-0.5Cu Sn-3.4Ag-4.8Bi Sn-3Ag-2Bi	Sn Low Alpha Sn Sn-2Ag Sn-2.5Ag Sn-3Ag Sn-3.5Ag Sn-3.6Ag Sn-3.65Ag Sn-4Ag Sn-4.6Ag Sn-5Ag Sn-3.5Ag-0.5Cu Sn-12Zn Sn-0.7Cu Sn-1Cu Sn-1.7Cu-4.7Ag Sn-3.4Ag-1.2Cu Sn-2.5Ag-0.8Cu-0.5Sb Sn-Ag-Cu-Ge Sn-0.7Cu-1.5Sb Sn-3Bi-5Sb	Sn Sn-2.5Ag Sn-3Ag Sn-3.5Ag Sn-3.65Ag Sn-4Ag Sn-5Ag Sn-0.7Cu-1.5Sb Sn-3.5Ag-0.5Cu Sn-4.7Ag-1.7Cu Sn-2.5Ag-0.8Cu-0.5Sb Sn-1Sb Sn-0.7Cu Sn-12Zn
Melting point above 230°C	Sn-1Sb Sn-3Sb Sn-5Sb 80Au-20Sn Sn-3Cu Sn-25Ag-10Sb	Sn-2Sb Sn-2.5Sb Sn-3Sb Sn-5Sb Sn-3Sb-1Cu Sn-2Sb-1Cu-2Ag-2Zn Sn-4Sb-2Cu Sn-7.5Sb-0.5Cu Sn-1Sb-3Cu-0.5Ag Sn-5Sb-3.5Cu-1Ni-0.25Ag Sn-20Zn Sn-25Zn Sn-30Zn Sn-40Zn Sn-9.5Bi-0.5Cu	Sn-5.5Ag Sn-6Ag Sn-10Ag Sn-8Ag-0.1In Sn-10Ag-3Cu Sn-25Ag-10Sb Sn-0.5Ag-4Cu Sn-0.1Ag-0.2Ni-4Cu Sn-0.01Ag-0.2Ni-3Cu-5Sb Sn-56Ag-22Cu-17Zn Sn-95Ag Sn-2Sb Sn-2.5Sb Sn-3Sb Sn-5Sb

TABLE 3.1   Lead-Free Solder Alloys Available as Paste, Bar, or Wire *(Continued)*

Paste	Bar	Wire
	Sn-2Cu	Sn-6Sb
	Sn-3Cu	Sn-7.5Sb-3.5Cu
	Sn-4.5Cu	Sn-2Cu
	Sn-3Cu-0.5In	Sn-3Cu
	Sn-3.9Cu-0.1Ag	Sn-1Cu-3Sb
	Sn-4Cu-0.5Ag	Sn-2Cu-0.8Sb-0.2Ag
	Sn-4Cu-0.8Ag	Sn-3Cu-0.5Ag-1Sb
	Sn-2Cu-0.8Sb-0.2Ag	Sn-3.9Cu-0.1Ag
	Sn-4.7Cu-0.3Se	Sn-6.8Cu-0.1Si-0.1Ag
	Sn-6.8Cu-0.1Si-0.1Ag	Sn-0.5Cu-0.5Ag-3Sb-1.5Zn
	80Au-20Sn	Sn-4.7Cu-0.3Se
	Sn-5.6Ag	Sn-15Zn
	Sn-10Ag	Sn-20Zn
	Sn-10Ag-3Cu	Sn-25Zn
	Sn-95Ag	Sn-30Zn
	Sn-56Ag-22Cu-17Zn	Sn-40Zn
	Sn-25Ag-10Sb	Sn-20Zn-5Al
		80Au-20Sn

nologies are considered. The Sn-37Pb solder is similarly evaluated to establish a baseline.

During the manufacturability assessment, it is found that none of the candidate alloys wets as well as the benchmark on four of the five PCB surface finishings (imidazole, immersion tin, nickel-palladium, and palladium). For the fifth finishing (nickel-gold), all candidate alloys wet comparably to the benchmark. Thus, it appears that some modifications will be required in PCB design and process development in order to accommodate use of the lead-free solders.

Based on the results of the NCMS reliability (temperature cycling) tests,[37] the Sn-58Bi, Sn-3.4Ag-4.8Bi, and Sn-3.5Ag lead-free solders appear very promising based on low failure rates. Also, the first two alloys perform better than the Sn-37Pb solder in certain surface-mount applications. However, the Sn-3.5Ag solder performs equivalently to the Sn-37Pb solder at 0 to 125°C and worse than the Sn-37Pb solder at −55 to +125°C. This is rather surprising, since Sn-3.5Ag solder has been used in the past to improve product reliability at higher temperatures. Table 3.2 summarizes the findings of the NCMS report, even though there is no drop-in replacement disclosed.

It should be pointed out that[38] if lead-free solders containing silver are improperly disposed of and contact groundwater, the solders could render that groundwater unsafe to drink according to United States Environmental Protection Agency (USEPA) standards. In that case, lead-free solders containing silver could create hazardous environments or increase the cost of waste treatments.

TABLE 3.2 Brief Summary of NCMS Lead-Free Solder Report

Solder alloys	Liquidus and solidus temperatures	Applications	Evidence for recommendation
Sn-58Bi	139°C (eutectic)	Consumer electronics, telecommunications	Surface-mount technology: better fatigue life than eutectic tin-lead for both thermal cycling ranges; less fatigue damage than eutectic tin-lead seen in surface-mount cross sections. Through-hole technology: mixed results, with fatigue life for CPGA-84 better than eutectic tin-lead; for CDIP-20, worse than eutectic tin-lead.
Sn-3.4Ag-4.8Bi	211°C (eutectic)	Consumer electronics, telecommunications, aerospace, automotive	Surface-mount technology: longer fatigue life than eutectic tin-lead at 0 to 100°C; no failures in 1206 resistors up to 6673 cycles; fatigue life equivalent to eutectic tin-lead at –55 to +125°C; less fatigue damage than eutectic tin-lead seen in surface mount cross sections. Through-hole technology: most joints show failure by fillet lifting.
Sn-3.5Ag	221°C (eutectic)	Consumer electronics, telecommunications, aerospace, automotive	Surface-mount technology: fatigue life equivalent to eutectic tin-lead at 0 to 100°C; worse than eutectic tin-lead at –55 to +125°C. Through-hole technology: less susceptible than other high-tin solders to fillet lifting, but results with Sn-2.6Ag-0.8Cu-0.5Sb indicate that through-hole reliability still may be compromised.

3.3   Physical and Mechanical Properties of Lead-Free Solders

Today, 96.5wt%Sn-3.5wt%Ag (or simply 96.5Sn-3.5Ag) and 42wt%Sn-58wt%Bi (42Sn-58Bi) are two of the most promising lead-free solders. Due to its higher melting temperature, 96.5Sn-3.5Ag solder is suitable for high-temperature environments such as automotive applications. On the other hand, for low-temperature applications such as flexible circuits and smart card assemblies, 42Sn-58Bi solder is a good candidate due to its low melting temperature. It should be noted from a cost perspective that Pb is many times cheaper than the other elements used in lead-free solder alloys, as shown in Table 3.3.

In this section, the thermal coefficient of expansion (TCE), storage modulus, moisture uptake, and melting point of 96.5Sn-3.5Ag and 42Sn-58Bi solders provided by different solder vendors are determined, respectively, by thermal mechanical analysis (TMA), dynamic mechanical analysis (DMA), thermogravimetric analysis (TGA), and a differential scanning calorimeter (DSC).[42] These physical and mechanical properties are very important for the understanding (through finite element and failure analysis) of the solders' behaviors and their responses when they are used in a given packaging system.

3.3.1   Determining melting temperature with a DSC[42]

The objective of the differential scanning calorimeter (DSC) is to measure the amount of energy (heat) absorbed or released by a sample as it is heated, cooled, or held at a constant (isothermal) temperature. The instrument consists of two independent furnaces, one for the sample and one for the reference (Fig. 3.1). When an exothermic or endothermic change occurs in the sample material, energy is applied to or removed from one or both furnaces to compensate for the energy change occurring in the sample. Since the system is always directly measuring energy flow to or from the samples, the DSC can directly

TABLE 3.3   Costs of Elements Used in Lead-Free Solder Alloys

Element	U.S.$ per ton
Ag (silver)	159,646
Bi (bismuth)	7,055
Cu (copper)	1,598
In (indium)	275,000
Pb (lead)	488
Sb (antimony)	1,325
Sn (tin)	5,590
Zn (zinc)	952

Cost in November 1998.

Figure 3.1   Specimen setup for DSC.

measure melting temperature, glass transition temperature, temperature at onset of crystallization, and temperature at onset of curing. The kinetic software enables analysis of a DSC peak to obtain specific kinetic parameters that characterize a reaction process.

Any material reaction can be represented by the following equation:

$$A \xrightarrow{k} B + \Delta H$$

where A is the material before reaction, B is the material after reaction, ΔH is the heat absorbed or released, and k is the Arrhenius rate constant.

The Arrhenius equation is given by:

$$k = Z \exp(-E_a/RT)$$

where Z is the preexponential constant, E_a is the activation energy of the reaction, R is the universal gas constant (8.314 j/°C/mole), and T is the absolute temperature in degrees Kelvin.

The rate of reaction (dx/dt) can be directly measured by the DSC and can be expressed as:

$$dx/dt = k(1 - x)^n$$

where dx/dt is the rate of reaction, x is the fraction reacted, t is the time, k is the Arrhenius rate constant, and n is the order of reaction.

Combining all the preceding equations and assuming nth-order reaction kinetics and a constant program rate, activation energy, and preexponential constant yields:

$$dx/dt = Z \exp(-E_a/RT)(1-x)^n$$

The fraction reacted (x) is directly related to the fractional area of the DSC reaction peak. The kinetic parameters Z, E_a, and n are determined by using an advanced multilinear regression method (MLR).

In this study, the solders are put into an aluminum pan (weight ~ 10 mg) and then put into DSC equipment. A thermal scan is carried out at a heating rate of 2°C/min, ranging from 40 to 260°C. Figure 3.2 shows the typical heat flow–versus–temperature (endothermic) curve of the solders from Vendor A, and Figure 3.3 shows these curves for solders from Vendor B. As expected, there is no reaction observed during the melting of solders.

Table 3.4 summarizes the average melting temperatures of these solders from Vendor A and Vendor B. (Since all the solders under consideration are eutectic solders, they only have a single melting point, i.e., the solidus temperature is the same as the liquidus temperature.) It can be seen that the melting temperature (corresponding to the peak of the heat flow) of 63Sn-37Pb solder is about 183°C, that of 96.5Sn-3.5Ag is about 221°C, and that of 42Sn-58Bi is about 138.89°C (slightly higher than 138°C). These values are very close to those reported in the

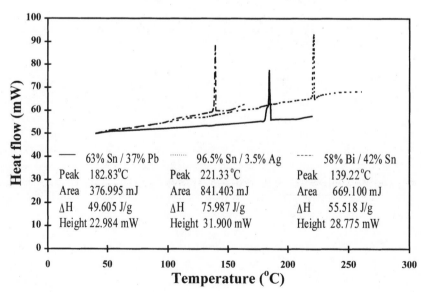

Figure 3.2 Heat flow–versus–temperature curves for melting point of solders from Vendor A.

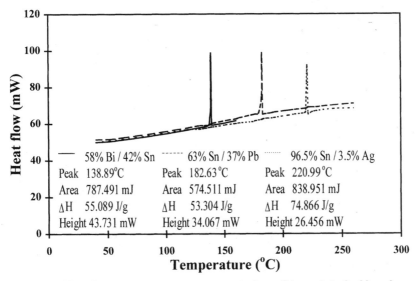

Figure 3.3 Heat flow–versus–temperature curves for melting point of solders from Vendor B.

literature.[3, 11] Also, the melting temperatures of the solders from Vendor A and Vendor B are almost the same.

3.3.2   Determining TCE with TMA[42]

The objective of thermal mechanical analysis (TMA) is to measure the change in dimension of a sample (expansion or contraction) as the sample is heated, cooled, or held at a constant (isothermal) temperature. The instrument consists of a platinum-wound furnace system that can be operated from –170 to +1000°C (Fig. 3.4). The scan rate is from 0.1

TABLE 3.4   Physical and Mechanical Properties of Solders

Type of solder	Melting point (°C)	TCE (ppm)	Storage modulus (GPa)			Moisture absorption (%)	
			30°C	70°C	125°C	Dry	20 hours of steam aging
63Sn-37Pb							
Vendor A	182.8	20.5	31	28.6	18.3	0.015	0.017
Vendor B	182.6	20	30	27.3	21.2	0.014	0.017
96.5Sn-3.5Ag							
Vendor A	221.3	22.1	34	32.2	18.7	0.022	0.023
Vendor B	221	23.2	33	32.7	29.2	0.012	0.015
58Bi-42Sn							
Vendor A	139.2	15.1	35	31.3	10	0.024	0.025
Vendor B	138.8	15.5	35.4	30.5	4.28	0.013	0.018

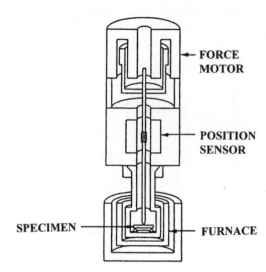

Figure 3.4   Schematic diagram of a TMA and DMA. TMA force motor applies static load, while DMA force motor applies dynamic load on specimen.

to 100°C/min. The specimen is mounted between a quartz platform and probe, and then a static load (F_s) is applied to the specimen while the dimensions of the specimen are monitored by the linear variable differential transducer (LVDT) throughout the analysis (Figs. 3.4 and 3.5). The TCE of the material is obtained by the slope of the dimension change–versus–temperature curve, and the glass transition of the material is obtained from the onset of the slope of the curve.

In this study, the samples are 6.4 ± 0.2 mm in diameter and 1.6 ± 0.1 mm in height, and they are measured in an expansion quartz system (from 50 to 160°C, except 48Sn-52Bi solder, which is measured from 50 to 120°C). The heating rate is 5°C/min. The typical expansion-versus-temperature curves for solders from Vendor A are shown in Fig. 3.6, and

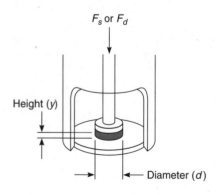

Figure 3.5   Quartz tube, probe, and specimen of TMA (with F_s) for TCE measurement.

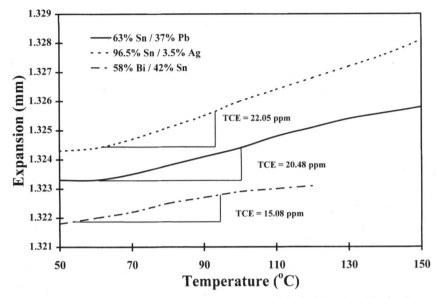

Figure 3.6 Expansion-versus-temperature curves for TCE of solders from Vendor A.

those for solders from Vendor B are shown in Fig. 3.7. It can be seen that for 63Sn-37Pb solder the TCE is about $21 \times 10^{-6}/°C$, for 96.5Sn-3.5Ag solder the TCE is about $22.5 \times 10^{-6}/°C$, and for 42Sn-58Bi solder the TCE is about $15.5 \times 10^{-6}/°C$. Again, these values (also shown in Table 3.4) are very similar to those reported in the literature and there is no significant difference between the solders from Vendor A and Vendor B.

3.3.3   Measuring storage modulus with DMA[42]

The dynamic (storage) modulus of the solders is measured with a three-point bending specimen (3.0 ± 0.3 mm $\times 2.9 \pm 0.3$ mm $\times 19 \pm 3$ mm) in a dynamic mechanical analysis (DMA) unit (30 to 130°C) at a heating rate of 5°C/min (Figs. 3.5 and 3.8). The objective of DMA is to measure mechanical properties such as modulus as a function of time, temperature, frequency, stress, or combinations of these parameters. The instrument consists of a force motor that can be programmed to apply constant stress, dynamic stress, or combinations of both (Figs. 3.5 and 3.8). The core rod applies stress to the sample and is held in place using an electromagnetic suspension. The ceramic furnace with platinum furnace element is capable of heating and cooling at a very high rate and can also reach temperatures of 1000°C. For electronics packaging materials, the flexural properties such as flexural modulus and dynamic mechanical properties such as storage modulus, loss modulus, and tan delta (tan δ) can be obtained with DMA.

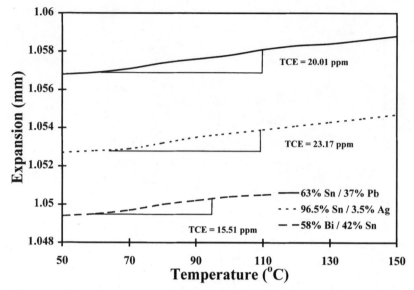

Figure 3.7 Expansion-versus-temperature curves for TCE of solders from Vendor B.

The storage (dynamic) modulus E_s is a measure of the energy stored per cycle of deformation and can be expressed as:

$$E_s = \frac{F_d\, x^3 \cos \delta}{4y^3 z \Delta}$$

where F_d is the dynamic load, δ is the phase angle, x is the clear span between the supports, y and z are the dimensions of the specimen, and Δ is the maximum dynamic deflection of the three-point bending solder specimen. Figure 3.9 shows the typical dynamic flexural storage modulus E_s as a function of temperature for the solders from Vendor A, and Fig. 3.10 shows these values for solders from Vendor B. It can be seen

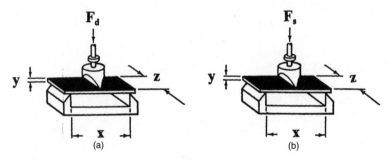

Figure 3.8 Three-point bending specimen setup for (a) DMA and (b) TMA.

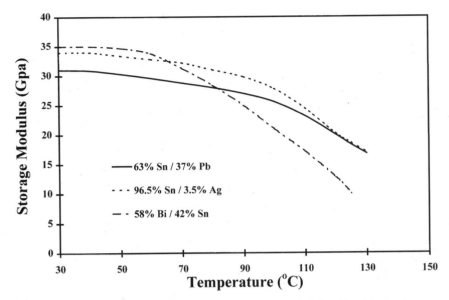

Figure 3.9 Storage modulus–versus–temperature curves of solders from Vendor A.

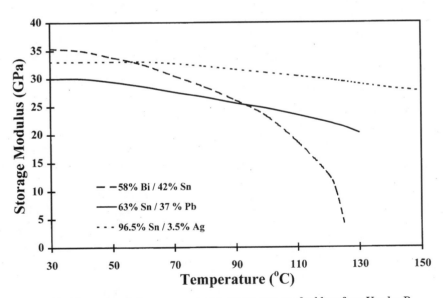

Figure 3.10 Storage modulus–versus–temperature curves of solders from Vendor B.

from Figs. 3.9 and 3.10 that the storage modulus is a function of temperature—the higher the temperature, the lower the storage modulus. Table 3.4 shows the average E_s at various temperatures. It can be seen that near room temperature (30°C), the storage modulus of 63Sn-37Pb solder is about 30.5 GPa, that for 96.5Sn-3.5Ag solder is about 33.5 GPa, and that for 42Sn-58Bi solder is about 35.4 GPa. Thus, near room temperature, the 42Sn-58Bi solder has the largest storage modulus.

At higher temperatures (60 to 70°C), however, the storage modulus of 42Sn-58Bi solder is smaller than that of 96.5Sn-3.5Ag solder (although still larger than that of 63Sn-37Pb solder). This could be because these temperatures exceed half the melting temperature of 42Sn-58Bi solder. The storage modulus of the 42Sn-58Bi solder starts to become smaller than that of the 63Sn-37Pb solder at about 83 to 93°C, and drops rapidly to much smaller values as the temperature approaches the melting point (138.89°C).

Modulus represents the stiffness of the material. The higher the stiffness, the stronger the material. Thus, near room temperature, the strength of the 42Sn-58Bi solder is higher than that of the 63Sn-37Pb and 96.5Sn-3.5Ag solders. However, at higher temperatures (especially beyond 85°C), the 42Sn-58Bi solder becomes very soft and should not be used as a joining material for electronic packaging and interconnections operated in these conditions.

Another drawback of 42Sn-58Bi solder is that it is not compatible with components and substrates having lead-coated finishing. When 42Sn-58Bi solder is soldered on surfaces containing Pb, the joint microstructure is completely different from the normal lamellar eutectic Sn-Bi, as shown in Fig. 3.11a. Even a very small amount (~1 wt%) of lead in the surface coating may significantly reduce the thermal fatigue life of the 42Sn-58Bi solder joint.[32] This is because the tiny amount of lead diffuses into the 42Sn-58Bi solder and forms a small amount of 16Sn-32Pb-52Bi ternary eutectic whose melting point is 95°C (Fig. 3.11b). The ternary eutectic is shown as point E in the liquidus temperature profiles of the Sn-Bi-Pb system.[33]

The microstructures of 42Sn-58Bi solder with (left side) or without (right side) dissolved Pb, after thermal cycling at either 18 to 75°C or −45 to +100°C, are shown in Fig. 3.11c. It can be seen that the larger thermal cycling range causes dramatic microstructural coarsening in the 42Sn-58Bi solder with dissolved Pb after 400 cycles. This microstructure coarsening ultimately causes premature solder joint failure. On the other hand, the smaller thermal cycling range has little effect on microstructural coarsening in the Sn-Bi solder both with and without dissolved Pb after 7000 cycles.[39]

Comparing the storage modulus between the 63Sn-37Pb and 96.5Sn-3.5Ag solders, it is found that the stiffness (strength) of the 96.5Sn-

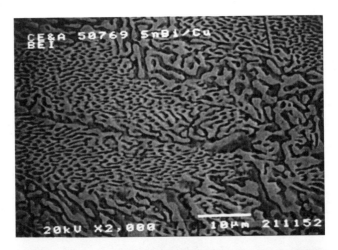

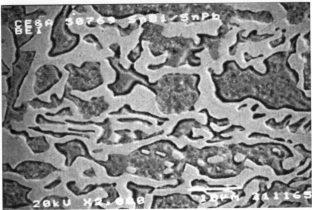

Figure 3.11 (*a*) Microstructures of eutectic Sn-Bi on organic-coated Cu (top) and on HAL Sn-Pb (bottom) finished surfaces.

3.5Ag solder is higher than that of the 63Sn-37Pb solder. Also, it is noted that at higher temperatures (over 100°C), the storage modulus of the 63Sn-37Pb and 96.5Sn-3.5Ag solders from Vendor B is higher than that for solders from Vendor A. This could be due to the purity, contamination, or preparation methods used by Vendor A and Vendor B.

3.3.4 Measuring moisture absorption with TGA[42]

Two sets of tests are carried out to determine the moisture content of the solders. One is for dry specimens and the other is for steam-aged specimens. The steam-aged specimen is prepared under steam evaporation for 20 h in a closed hot-water bath. All the specimen weights are about 20 to 40 mg. Weight loss of solders is measured with the TGA equipment (Fig. 3.12) under 102°C for 2 h.

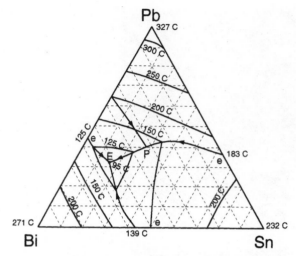

Figure 3.11   (*b*) Liquidus temperatures of Sn-Pb-Bi system. E, ternary eutectic; P, ternary peritectic; e, binary eutectic.

The objective of thermogravimetric analysis (TGA) is to measure the change in mass of a sample as the sample is heated, cooled, or held at a constant (isothermal) temperature. The instrument consists of a microbalance that allows the sensitive measurement of weight changes as small as a few micrograms; a furnace; and a sample holder area. The vertical analytical design also serves to isolate the balance mechanism from the furnace, eliminating temperature fluctuations that cause drifts and nonlinearity. This system can measure curie point, decomposition temperature, moisture uptake, and component separation.

The change in mass during thermal scan can be expressed as:

$$\frac{W_f - W_i}{W_i} \times 100\%$$

where W_f is the final weight after thermal scan and W_i is the initial weight before thermal scan. Figures 3.13 and 3.14 show, respectively, the typical percent weight change (moisture content) of solders from Vendor A before and after 20 h of steam aging. Figures 3.15 and 3.16 show these values for solders from Vendor B. The average moisture contents of the solders are shown in Table 3.4. It can be seen and expected that the moisture absorption of these solders is too small (less than 0.023 percent) to be significant.

3.3.5   Steady-state creep of lead-free solders[39, 40]

The steady-state creep data of 63Sn-37Pb and 42Sn-58Bi solders at 25, 65, and 90°C are shown in Figures 3.17 and 3.18, respectively. It can be seen that the creep rate of 42Sn-58Bi solder is slower than that of 63Sn-

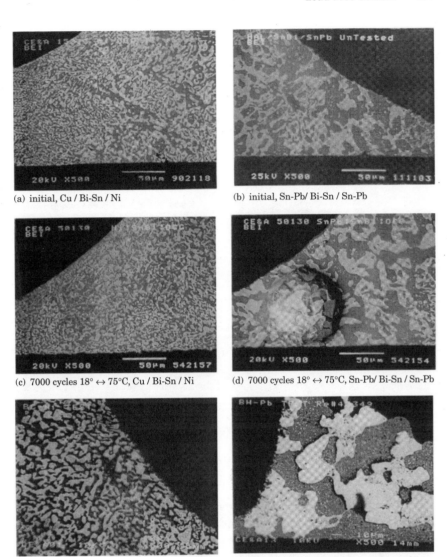

(a) initial, Cu / Bi-Sn / Ni

(b) initial, Sn-Pb/ Bi-Sn / Sn-Pb

(c) 7000 cycles 18° ↔ 75°C, Cu / Bi-Sn / Ni

(d) 7000 cycles 18° ↔ 75°C, Sn-Pb/ Bi-Sn / Sn-Pb

(e) 400 cycles −45° ↔ +100°C, Cu / Bi-Sn / Ni

(f) 400 cycles −45° ↔ +100°C, Sn-Pb/ Bi-Sn / Sn-Pb

Figure 3.11   (c) The solidified microstructures of 58Bi-4Sn solder joints between an OCC pad and a Ni-Pb-coated component lead (left side) and between a HAL 63Sn-37Pb pad and an 80Sn-20Pb-coated component lead (right side) after two different temperature cycling tests, namely, 7000 cycles between 18 and 75°C over 5 mo, and 400 cycles between −45 and 100°C over 16 days.

37Pb solder. Also, the creep rupture strain of 42Sn-58Bi is smaller than that of 63Sn-37Pb solder under the same loading conditions. Furthermore, the shear strength of 42Sn-58Bi solder is greater than that of 63Sn-37Pb solder at 25°C, and 42Sn-58Bi solder becomes much weaker as the temperature increases to 110°C, while the 63Sn-37Pb solder is still strong.[39]

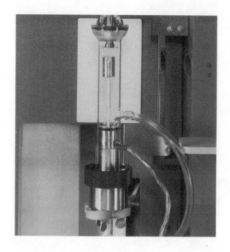

Figure 3.12 High-sensitivity weight measurements with TGA (thermal gravimetric analyzer).

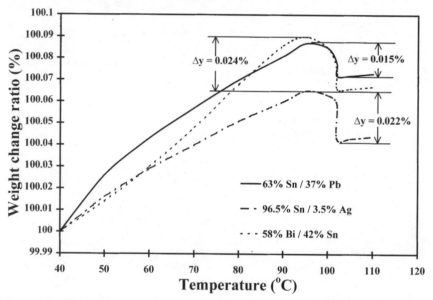

Figure 3.13 Weight change ratio–versus–temperature curves of solders from Vendor A before steam aging.

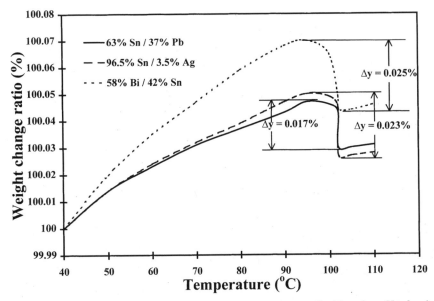

Figure 3.14 Weight change ratio–versus–temperature curves of solders from Vendor A after 20 hours of steam aging.

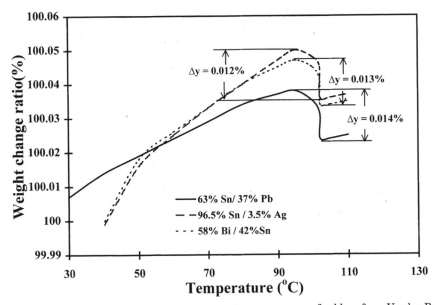

Figure 3.15 Weight change ratio–versus–temperature curves of solders from Vendor B before steam aging.

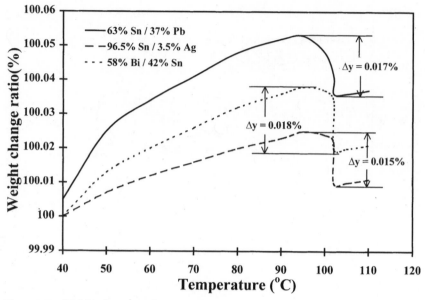

Figure 3.16   Weight change ratio–versus–temperature curves of solders from Vendor B after 20 hours of steam aging.

3.3.6   Isothermal fatigue of lead-free solders[39, 40]

The isothermal fatigue data for 63Sn-37Pb and 42Sn-58Bi solders at 25 and 70°C are shown in Fig. 3.19[39] The cycling frequency is about 10 min per cycle. It can be seen that at larger plastic shear strain range (~10 percent), the isothermal fatigue life of 42Sn-58Bi solder is poorer than that of 63Sn-37Pb. The lamellar structure of 42Sn-58Bi solder could be the reason for its poor isothermal fatigue life. However, isothermal fatigue life for both solders is very comparable at smaller plastic shear strain range.

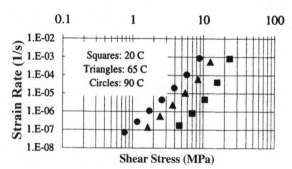

Figure 3.17   Steady-state creep of 63Sn-37Pb at room temperature, 65°C, and 110°C.

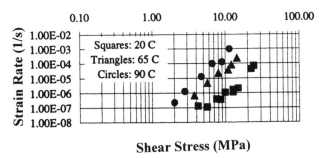

Shear Stress (MPa)

Figure 3.18 Steady-state creep of 58Bi-42Sn at room temperature, 65°C, and 110°C.

3.3.7 Thermal fatigue of lead-free solders[39, 40]

The thermal fatigue data for 63Sn-37Pb, 42Sn-58Bi, and 55Bi-41-Sn-2Ag solders are shown in Fig. 3.20.[39] The thermal cycling condition is –20 ↔ +110°C, with 1 h per cycle and a dwell time of 10 min at the low and high temperatures. The test samples are made of ceramic plates 4, 2, and 1 in. square and are soldered on an FR-4 PCB. The solder joints are 10 mils thick and 50 mils in diameter for all the solders. It can be seen from Fig. 3.20 that the thermal fatigue life of 63Sn-37Pb solder is better than that of 42Sn-58Bi solder. Also, a small amount of added silver dramatically increase the thermal fatigue life.

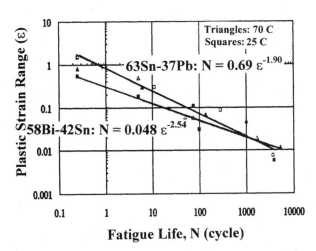

Fatigue Life, N (cycle)

Figure 3.19 Isothermal fatigue life of 58Bi-42Sn versus 63Sn-37Pb.

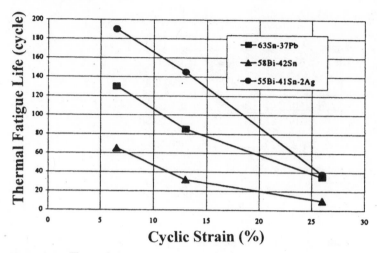

Figure 3.20   Thermal fatigue life of 58Bi-42Sn versus 63Sn-37Pb versus 55Bi-41Sn-2Ag.

3.4   Lead-Free Solders for Flip Chip Applications[41]

In order to minimize the soft error rate due to alpha particle emission by lead-bearing solder alloys, 19 different lead-free solder alloys have been studied by IBM[41] as possible replacements for the lead-based solders currently used in flip chip interconnections. These 19 solder alloys are based on Sn-Bi, Sn-Ag, and Sn-Sb combinations, as shown in Table 3.5.

3.4.1   Melting characteristics

The DSC is used to characterize the melting behavior of the selected solder alloys. The measurement results in terms of the extrapolated onset melting temperature (T_{ex}), peak melting temperature (T_p), and observed onset melting temperature (T_{ob}) are shown in Table 3.5.[41] The definitions of T_{ex}, T_p, and T_{ob} are shown in Fig. 3.21. It can be seen that solders S-11 and S-12 are not good candidates for replacing the lead-tin solder alloys because of their wide melting ranges.

3.4.2   Electrical resistivity

The bulk resistivity of the selected lead-free solder alloys is shown in Table 3.6.[41] In order to check the current measurement, a pure Sn foil is prepared like the other 19 solder alloys. The resistivity is measured to be 11.2 $\mu\Omega$/cm, which is in good agreement with the published value of 11.5 $\mu\Omega$/cm.

TABLE 3.5   Melting Characteristics of Lead-Free Solder Alloys

Solder ID	Composition (wt %)	T_{ex} (C)	T_p (C)	T_{ob} (C)
S-01	Sn95-Bi5	219.4	228.7	196.0
S-02	Sn90-Bi10	209.1	221.2	165.0
S-03	Sn96.5-Ag3.5	221.5	226.5	220.0
S-04	Sn95-Sb5	236.8	244.9	231.6
S-05	Sn90-Bi5-In5	205.5	218.4	177.0
S-06	Sn92-Bi5-Ag3	214.4	218.1	191.6
S-07	Sn89-Bi5-In5-Cu1	204.9	211.0	167.5
S-08	Sn94-Bi5-Cu1	216.7	226.7	184.8
S-09	Sn96-Ag3-Zn1	215.9	220.9	210.0
S-10	Sn94-Ag3-Zn3	217.8	222.5	206.8
S-11	Sn87-Ag3-Bi10	201.5	214.3	164.4
S-12	Sn87-Bi5-In5-Ag3	195.7	209.0	158.9
S-13	Sn92-Ag3-Bi2-Cu3	211.1	222.2	195.7
S-14	Sn93-Ag3-Sb2-Cu2	218.6	230.1	212.8
S-15	Sn93.5-Sb3-Bi2-Cu1.5	224.1	234.5	207.0
S-16	Sn92-Sb5-In3	226.7	240.3	211.7
S-17	Sn92-Ag3-Sb5	225.0	235.9	216.3
S-18	Sn94-Ag3-In3	210.7	214.7	193.3
S-19	Sn95-In5	219.7	223.4	211.5

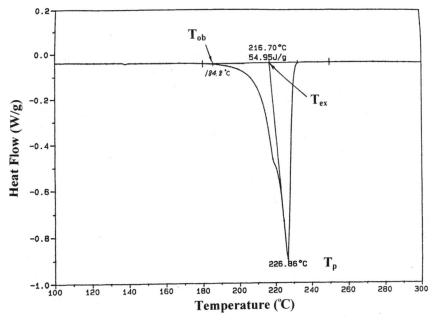

Figure 3.21 A typical heating profile of differential scanning calorimetry with a Pb-free solder alloy, illustrating the determination of T_{ex}, T_p, and T_{ob}.

TABLE 3.6   Electrical Resistivity, Wetability, and Hardness of Lead-Free Solder Alloys

Solder ID	Composition (wt %)	Resistivity ($\mu\Omega$/cm)	Wetability	Hardness (VHN)
S-01	Sn95-Bi5	11.0	97, 97	25.6
S-02	Sn90-Bi10	17.7	98, 97	33.0
S-03	Sn96.5-Ag3.5	7.7	97, 95	17.9
S-04	Sn95-Sb5	17.1	96, 96	17.2
S-05	Sn90-Bi5-In5	19.1	98, 95	28.0
S-06	Sn92-Bi5-Ag3	11.6	98, 98	29.9
S-07	Sn89-Bi5-In5-Cu1	14.1	96, 95	40.6
S-08	Sn94-Bi5-Cu1	10.5	96, 95	31.9
S-09	Sn96-Ag3-Zn1	10.4	98, 97	18.2
S-10	Sn94-Ag3-Zn3	4.8	97, 97	21.9
S-11	Sn87-Ag3-Bi10	8.8	—	—
S-12	Sn87-Bi5-In5-Ag3	12.1	—	—
S-13	Sn92-Ag3-Bi2-Cu3	10.6	97, 96	34.5
S-14	Sn93-Ag3-Sb2-Cu2	11.2	98, 98	28.6
S-15	Sn93.5-Sb3-Bi2-Cu1.5	8.0	98, 96	32.9
S-16	Sn92-Sb5-In3	9.8	98, 97	31.9
S-17	Sn92-Ag3-Sb5	10.5	99, 99	25.4
S-18	Sn94-Ag3-In3	7.7	99, 99	21.3
S-19	Sn95-In5	8.3	99, 97	20.8
	Sn100	11.2	—	10.5
	Sn63-Pb37	17.0	95, 91	12.9

3.4.3 Wetability

The wetting behavior of the selected lead-free solder alloys is determined by a Pessel solderability test, which produces a wetability ratio. Usually, wetability ratios larger than 90 mean good solderability. The wetability ratios of these 19 solder alloys are shown in Table 3.6. It can be seen that all values are larger than 90. This good solderability is attributed to the presence of the thin layer of Au on top of the NiFe layer. The flux used for the experiments is water soluble (Hi Grade WSF-200).[41]

3.4.4 Microhardness

The microhardness of the selected lead-free solder alloys is measured by a Leco M-400 Microhardness Tester using 25 g-forces for 5 s of loading. The results are shown in Table 3.6.[41] It can be seen that (1) all the lead-free solder alloys are harder than 63Sn-37Pb eutectic solder, and (2) the addition of Bi or Cu to Sn-rich solders increases the hardness much more than adding Sb, Zn, or In.

Acknowledgments

The author would like to thank Z. Mei, J. Glazer, and F. Hua of HP; S. Kang, J. Horkans, P. Andricacos, R. Carruthers, J. Cotte, M. Datta,

P. Gruber, J. Harper, K. Kwietniak, C. Sambucetti, L. Shi, G. Brouillette, and D. Danovitch of IBM; and C. Chang of EPS for sharing their useful knowledge with the industry.

References

1. Lau, J. H., and W.-S. R. Lee, *Chip Scale Package: Design, Materials, Process, Reliability, and Applications,* McGraw-Hill, New York, 1999.
2. Lau, J. H., C. P. Wong, J. L. Prince, and W. Nakayama, *Electronics Packaging: Design, Materials, Process, and Reliability,* McGraw-Hill, New York, 1998.
3. Lau, J. H., and Y.-H. Pao, *Solder Joint Reliability of BGA, CSP, Flip Chip, and Fine Pitch SMT Assemblies,* McGraw-Hill, New York, 1997.
4. Lau, J. H., *Flip Chip Technologies,* McGraw-Hill, New York, 1996.
5. Lau, J. H., *Ball Grid Technology,* McGraw-Hill, New York, 1995.
6. Lau, J. H., *Chip On Board Technologies for Multichip Modules,* Van Nostrand Reinhold, New York, 1994.
7. Lau, J. H., *Handbook of Fine Pitch Surface Mount Technology,* Van Nostrand Reinhold, New York, 1994.
8. Frear, D., H. Morgan, S. Burchett, and J. Lau, *The Mechanics of Solder Alloy Interconnects,* Van Nostrand Reinhold, New York, 1994.
9. Lau, J. H., *Thermal Stress and Strain in Microelectronics Packaging,* Van Nostrand Reinhold, New York, 1993.
10. Lau, J. H., *Handbook of Tape Automated Bonding,* Van Nostrand Reinhold, New York, November 1992.
11. Lau, J. H., *Solder Joint Reliability, Theory and Applications,* Van Nostrand Reinhold, New York, 1991.
12. Glazer, J., "Microstructure and Mechanical Properties of Pb-Free Solder Alloy for Low-Cost Electronic Assembly: A Review," *J. Electronic Material,* **23** (8): 693–700, 1994.
13. Kang, S. K., "Lead (Pb)-Free Solders for Electronic Packaging," *J. Electronic Material,* **23** (8): 701–707, 1994.
14. Shangguan, D., A. Achari, and W. Green, "Application of Lead-Free Eutectic Sn-Ag Solder in No-Clean Thick Film Electronic Modules," *IEEE Trans. Components, Packaging and Manufacturing Technology,* part B, **17** (4): 603–611, 1994.
15. Shangguan, D. A., and G. Gao, "Lead-Free and No-Clean Soldering for Automotive Electronics," *Soldering and Surface Mount Technology,* **9** (2): 5–8, July 1997.
16. Hwang, J. S., *Solder Paste in Electronics Packaging,* Van Nostrand Reinhold, New York, 1989.
17. Hwang, J. S., *Modern Solder Technology for Competitive Electronics Manufacturing,* McGraw-Hill, New York, 1996.
18. Frear, D. R., "The Mechanical Behavior of Interconnect Materials for Electronic Packaging," *J. Metals,* 49–53, May 1996.
19. Vianco, P., J. Rejent, I. Artaki, U. Ray, D. Finley, and A. Jackson, "Compatibility of Lead-Free Solders with Lead Containing Surface Finishes as a Reliability Issue in Electronic Assemblies," *Proceedings of the IEEE Electronic Components and Technology Conference,* pp. 1172–1183, May 1996.
20. Iida, A., Y. Kizaki, Y. Fukuda, and M. Mori, "The Development of Repairable Au-Al Solid Phase Diffusion Flip-Chip Bonding," *Proceedings of the IEEE Electronic Components and Technology Conference,* pp. 101–107, May 1997.
21. Vianco, P. T., K. L. Erickson, and P. L. Hopkins, *Solid State Intermetallic Compound Growth Between Copper and High Temperature, Tin-Rich Solder—Part 1: Experimental Analysis,* Sandia National Labs (Contract Number DE-AC04-94AL85000) Report, 1994.
22. Vianco, P. T., and D. R., "Issues in the Replacement of Lead-Bearing Solders," *J. Metals,* **45** (7): 36–40, July 1993.
23. Kang, S. K., and A. K. Sarkhel, "Lead (Pb)-Free Solders for Electronic Packaging," *J. Electronic Materials,* **23** (8): 701–708, August 1994.

24. Yang, W., L. E. Felton, and R. W. Messler, "The Effect of Soldering Process Variables on the Microstructure and Mechanical Properties of Eutectic Sn-Ag/Cu Solder Joints," *J. Electronic Materials,* **24** (10): 1465–1472, 1995.
25. Darveaus, R., and K. Banerji, "Constitutive Relations for Tin-Based Solder Joints," *Proceedings of the IEEE Electronic Components and Technology Conference,* pp. 538–551, May 1992.
26. Stromswold, E. I., *Characterization of Eutectic Tin-Silver Solder Joints,* Ph.D. dissertation, The University of Rochester, Rochester, NY, 1993.
27. McCormack, M., and S. Jin, "Progress in the Design of New Lead-Free Solder Alloys," *J. Metals,* **45** (7): 14–19, July 1993.
28. Felton, L. E., C. H. Taeder, and D. B. Knorr, "The Properties of Tin-Bismuth Alloys," *J. Metals,* **45** (7): 14–19, July 1993.
29. Grusd, A., "Lead Free Solders in Electronics," *Proceedings of Surface Mount International Conference,* pp. 648–661, August 1998.
30. Biocca, P., "Global Update on Lead-Free Solders," *Proceedings of Surface Mount International Conference,* pp. 705–709, August 1998.
31. Wege, S., G. Habenicht, and R. Bergmann, "Manufacture and Reliability of Alternate Solder Alloys," *Proceedings of Surface Mount International Conference,* pp. 699–704, August 1998.
32. Mei, Z., and H. Holder, "A Thermal Fatigue Failure Mechanism of 58Bi-42Sn Solder Joints," *ASME Trans., J. Electronic Packaging,* **118:** 62–66, June 1996.
33. Humpston, G., and D. Jacobson, *Principles of Soldering and Brazing,* ASM International, Materials Park, OH, 1993.
34. Lau, J. H., "Creep of 96.5Sn-3.5Ag Solder Interconnects," *Soldering and Surface Mount Technology,* **15:** 45–49, September 1993.
35. Ren, W., M. Lu, S. Liu, and D. Shangguan, "Thermal Mechanical Property Testing of New Lead-Free Solder Joints," *Soldering and Surface Mount Technology,* **9** (3): 37–40, October 1997.
36. Trumble, B. "Get the Lead Out," *IEEE SPECTRUM,* 55–60, May 1998.
37. NCMS, *Lead-Free Solder Project Final Report,* NCMS Report 040IRE96, August 1997.
38. Smith, E., III, and L. Swanger, "Lead Free Solders—A Push in the Wrong Directions?" *Proceedings of the IPC Technical Conference,* pp. F-10-1–F-10-6, March 1999.
39. Hua, F., Z. Mei, and J. Glazer, "Eutectic Sn-Bi as an Alternative to Pb-Free Solder," *Proceedings of Electronic Components and Technology Conference,* pp. 277–283, May 1998.
40. Mei, Z., F. Hua, J. Glazer, and C. Key, "Low Temperature Soldering," *Proceedings of IEMTS,* 463–476, October 1997.
41. Kang, S., J. Horkans, P. Andricacos, R. Carruthers, J. Cotte, M. Datta, P. Gruber, J. Harper, K. Kwietniak, C. Sambucetti, L. Shi, G. Brouillette, and D. Danovitch, "Pb-Free Solder Alloys for Flip Chip Applications," *Proceedings of Electronic Components and Technology Conference,* pp. 283–288, June 1999.
42. Lau, J. H., and C. Chang, "TMA, DMA, DSC, and TGA of Lead Free Solders," *Soldering and Surface Mount Technology,* **11** (2): 17–24, 1999.

4

High-Density PCB and Substrates

4.1 Introduction

In order to make high-density and cost-effective PCB and substrates, the sequential buildup (SBU) fabrication technique is a must. This is done by adding a minimum of one layer of dielectric to the double-sided or multilayer core. Non-through-hole microvias are selectively formed to reclaim "real estate" and are used to accommodate the fine I/O pitch and redistribute circuits from the chip to the internal layers of the PCB, to reduce PCB layer count and size, and to enhance electrical performance. By IPC's definition, holes 6 mils (0.15 mm) or less in size on PCB are called microvias.

There are many advantages of microvias: (1) much smaller pads can be used, saving on board size and weight; (2) more chips can be placed in less space or on a smaller PCB, which results in a low cost; and (3) electrical performance improves, because parasitic capacitance is increased due to the smaller via length and diameter and inductance is reduced due to the shorter pathway created by the microvia compared to the plated-through-hole (PTH).[1-7]

According to a recent study by TechSearch International, the market for microvia substrates will reach $1.6 billion this year. That figure is expected to grow to $8.6 billion by 2005. Currently, 75 companies around the world are producing microvia substrates. Japan is the world leader, with more than 50 percent of the world's microvia production, with Europe in second place and the Asia/Pacific region in third. Most of the products that use microvia technology are mobile phones (Fig. 4.1), ASICs, notebook computers, and other handheld products. The U.S. will not attain high-volume production until after the year 2000, because most U.S. companies are currently focusing on high-value applications like workstations, servers, and network systems.

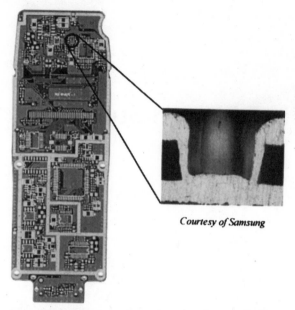

Courtesy of Samsung

Figure 4.1   Personal communication system with microvias.
(Source: Samsung)

Microvias are the focus of this chapter. The design, materials, processes, and reliability of microvias are presented. Also, microvia production in Japan is briefly reported. Finally, some useful charts for designing high-speed PCB are provided.

4.2   Categories of Vias

Typical via hole diameters range from 50 to 200 μm. These vias are divided into three categories, schematically shown in Fig. 4.2.

1. *Blind vias* are located on the outer layer of the top and bottom of the circuit board and are formed to such a depth as to make contact with the first inner layer. The depth of these holes usually does not exceed one aspect ratio (hole diameter).

2. *Buried vias* are plated within the core of the circuit board without access to the surface on either side of the board. These holes are formed before the board is laminated. The inner-layer material has the holes created by a through-hole processing method. The inner layers may be stacked several layers high during this hole-formation process.

3. *Through-hole vias* are formed through the entire thickness of the board. These vias are used as interconnects or as mounting locations for components.

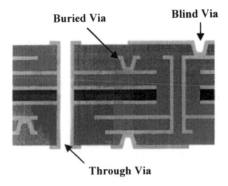

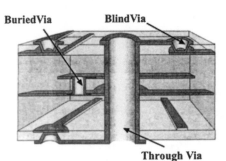

Figure 4.2   Via definitions.

IPC defines the typical via profiles and the processes that produce them as shown in Fig. 4.3. Basically, there are five major processes for microvia formation: (1) NC (numerical controlled) drilling; (2) laser via fabrication including CO_2 laser, UV-yttrium aluminum garnet (YAG) laser, and Excimer; (3) photo-defining vias, wet/dry; (4) etch via fabri-

Representative Appearence of Various Via Technologies as Produced

Mechanically Drilled - 1   Wet Etched - 2, 3, 6   Insulation Displacement - 8
Mechanically Punched -1   Dry Etched (Plasma) - 2   Conductive Paste Via - 9
Laser Drilled - 1, 4, 5   Abrasive Blast - 3   Conductive Bonding Sheets - 10
Photo Formed - 5, 6, 7   Post Pierced -8

Figure 4.3   Typical via profiles and the processes that produce them. *(Source: IPC)*

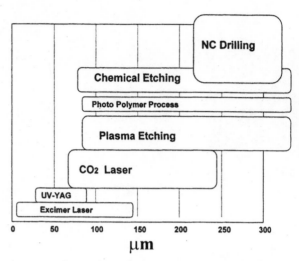

Figure 4.4   Comparison of microvia processes with via diameter.

cation, including chemical (wet) etching and plasma (dry) etching; and (5) conductive ink formation of vias, wet or dry. A comparison of microvia hole process versus via diameter is shown in Fig. 4.4. It can be seen that most of the NC drilling cannot make microvias. Also, even though photo-defined and chemical and plasma etching processes can make microvias, laser drills can make finer vias. Figure 4.5 shows a seven-layer cross section with blind vias and a buried via. Figure 4.6

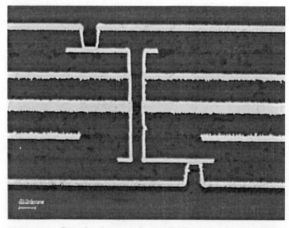

Figure 4.5   Standard seven-layer PCB with blind-buried-blind via net (total thickness = 460 μm; via diameter = 50 μm).

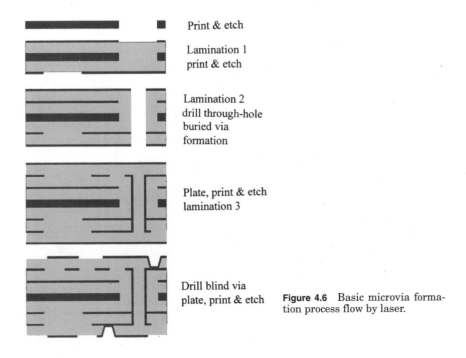

Print & etch

Lamination 1
print & etch

Lamination 2
drill through-hole
buried via
formation

Plate, print & etch
lamination 3

Drill blind via
plate, print & etch

Figure 4.6 Basic microvia formation process flow by laser.

shows the basic process flow of laser drilling and Fig. 4.7 shows a typical process flow of the photodefined via. It can be seen that the laser
drill process is much simpler. The materials and processes used in
forming these microvias are discussed in the following text.

4.3   Forming Microvias by Conventional Mechanical NC Drilling

Currently, NC drilling is the most common process for generating holes
in the PCB. However, NC drilling is technically limited to hole sizes of
200 µm and up (not microvias). Although smaller holes may be possible, these would come at the cost of significant productivity (minimal
stack height). In addition, creation of blind vias is virtually impossible
where typical dielectric thickness is 50 µm or less.[8]

Mechanical drilling for non-microvias can drill all kinds of PCB
materials such as FR-4, Mitsubishi Gas and Chemical's BT, General
Electric's GETEK, Hitachi Chemical's MegTran (FR-4.5), and Ma-
tsushita's MCL-679. There are many equipment vendors supplying
various mechanical drilling systems, such as Advanced Controls; ATI;
Ching Huei; Datron; Electro Scientific Industries, Inc.; Excellon;
Hitachi; OZO; Panasonic; Pluritec; Posalux; Sieb & Meyer; Technic; and
Yaskawa.

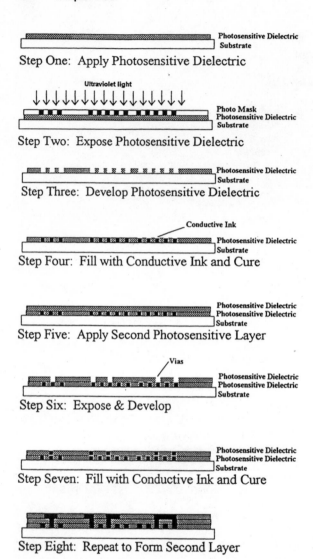

Figure 4.7 Cross-sectional view of a step-by-step process for creating the multilayer PCB using photodefinable dielectric and conductive ink technology.

4.4 Forming Microvias by Laser Drilling

This can be a single- or multiple-via generation technology that replaces the existing mechanical drilling process with a laser drilling process. Laser-drilled microvias have been reported by several authors,[9-21] and a typical PCB including blind vias and a buried via is shown in Fig. 4.5.[9] The laser drilling process is shown in Fig. 4.6. Laser drilling differs from mechanical drilling in that the focused beam used to create the holes can produce smaller holes than conventional drilling. One of the most important advantages of laser drilling is that it is compatible with many copper-clad or unclad dielectrics and reinforced or nonreinforced PCBs. Lasing can be used to create both blind vias and holes. It follows the standard multilayer process and is capable of resolving smaller features. Low productivity is one of the weaknesses of laser drilling, i.e., one beam produces one blind via at a time.[1] Several laser processes (Excimer, CO_2, and UY-YAG) have been developed to generate small via holes; these are categorized in the following text. The advantages and disadvantages of these processes are also discussed. The materials choices with laser drilling will be discussed first.

4.4.1 Materials choice with laser drilling

There is a wide range of materials available today for microvia formation. Some of the materials known to be qualified in the PCB industry today for laser drilling are:[20]

- Allied Signal RCC® and Polyclad RCF®
- Gore Speedboard® N and C
- Dupont Thermount® E210, E220, and E230
- FR4 and FR5 with glass fabrics: 1×104, 1×106, 1×1080, 1×1065, 1×2113, 1×3070, 2×104, 2×106, 3×104
- Novaclad® Polyimide
- Materials from other laser dielectric suppliers including BF Goodrich, Ciba, Specialty Chemical, Enthone-OMI, Hitachi Chemical, MacDermid, Mitsubishi Gas & Chemical, Park, Shipley, and Taiyo Ink.

For telecommunications applications, two types of resins that are typically used for microvia formation are Resin Coated Copper Foil® (RCC or RCF) for subtractive PCB processes and Thermal-Curing Resin (TCR) for additive PCB processes. RCC is the material of choice in the fabrication of the cellular phones. The reasons for choosing RCC and not any other dielectric in this application are:

- RCC is readily available.
- The cost of RCC has dropped over the last couple of years, making it more attractive.
- From a manufacturing perspective, it is easy to form microvias using the UV-YAG laser compared with photovia technology, and there is no need for any of the special chemical processes associated with other technologies.

- The dielectric has a laminated thickness from 40 to 60 µm, which allows for good impedance control.
- It is possible to make blind vias through multiple layers.

4.4.2 CO₂ laser

The CO_2 lasers have significantly higher productivity for holes larger than 70 µm in diameter. They can ablate more than 15,000 (100-µm diameter) holes through 50 µm thick dielectric in one minute. In general, carbon dioxide laser is used to drill on the dielectric layer and not used to drill holes on copper, because the visible wavelength (1060 nm) of CO_2 is too large to have enough energy to penetrate the copper at high speed. Therefore additional processing is required prior to drilling. For instance, windows must be selectively pre-etched through the copper foil at positions where holes in the dielectric are to be formed. (This is called the conformal method, and it presents the problem of higher costs and limitation of resolution in both via and circuits.) Recently, Hitachi has shown that with its CO_2 laser drilling machine, it can drill through copper foil (<9 µm) by coating a layer of black oxide (called *multibond* in Japan) or Cobra® onto the copper. Today, the companies that supply the CO_2 laser drilling machines are Hitachi, Lumonics, Panasonic (Matsushita), Mitsubishi, and Sumitomo.

4.4.3 UV-YAG laser

The UV-YAG laser, with smaller wavelengths (365 to 255 nm) and much higher energy, is used to drill holes on the copper and dielectric (the so-called imaging method) at a much lower speed. UV-YAG lasers have demonstrated great productivity through both dielectric and copper layers, but for both processes ablation time is proportional to hole diameter and neither process is therefore competitive above 100 µm diameters. Thus, the YAG laser is the preferred laser for drilling when the hole diameter becomes very small—50 µm, for example. Because YAG lasers will also ablate copper, it is more difficult to control the formation of blind vias using a copper stop-pad. Although the speed of the YAG is only ¹⁄₁₀ that of the CO_2 laser, when copper is to be drilled through, some makers of high-end multilayer boards (MLBs) prefer to use the YAG because it does not require that the copper be etched to create windows. Figure 4.8a shows a cross section of the microvia made by UV-YAG laser. Today, ESI, Excellon, Exotech, Hitachi, and Sumitomo supply the UV/Nd:YAG laser machines. Also, Hitachi and Lumonics of Canada now offer a combination of Nd:YAG/ CO_2 laser machines.

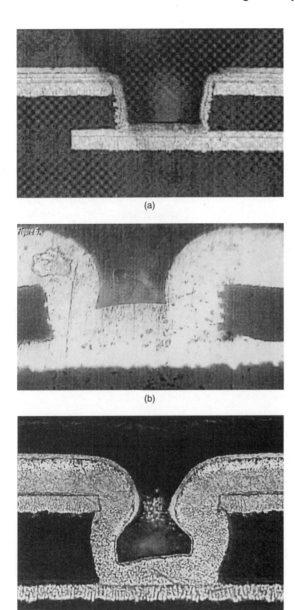

Figure 4.8 Microvias made by (*a*) laser drill, (*b*) photo-definition, and (*c*) plasma.

4.4.4 Excimer laser

The excimer laser has a wavelength of 248 nm with krypton fluoride (KrF) and one of 193 nm with argon fluoride (ArF). Excimer lasers can generate holes smaller than 50 µm in diameter through dielectric or copper layers. Controlled-depth drilling is also possible, thus allowing blind vias to be created. Their slow etch rate, however, makes them impractical for microvia formation. Nevertheless, the Excimer laser still can drill 10,000 holes per minute through a conformal mask, which has a predetermined hole pattern. However, an effective mask can only be made of quartz material and the cost of this mask is prohibitively expensive for small panel runs. Today, Litel, JPSA, and Tamarack supply UV:excimer machines, to which JPSA can add either T-CO_2 or diamond CO_2 laser heads.

4.4.5 Comparison of Excimer, UV-YAG, and CO_2 lasers[8]

Table 4.1 shows the comparison between the three laser processes for microvia formation. It can be seen that there are some important benefits of using the UV-YAG lasers, as shown in the following list.

■ The capability to drill through copper, which eliminates the printing and chemical etch step for via openings in the outer layer.

■ The capability to form microvias as small as 25 µm.

■ The capability to drill multilayer vias, providing opportunities to reduce sequential buildups.

Table 4.2[22] shows that microvias formed with the UV-YAG laser technology pass all qualification tests. Table 4.3 shows the test results of another case using the UV-YAG laser technology. It can be seen that the samples pass all of the qualification tests.[20] This indicates that UV-YAG laser technology is a robust process—which is why 70 percent of the world's microvia boards are made by laser ablation. Laser-ablated microvias are thermally and electrically more reliable than traditional PTH.[23, 24]

TABLE 4.1   Comparison of Laser Methods

	Excimer	UV-YAG	CO_2
Diameter (µm)	10–100	25–100	70–250 (50–250)
For copper	Very slow	Slow	No
For resin (epoxy and PI)	Slow	Slow	Fast
For FR-4	Very slow	Slow	Fair
Hole sharpness	Very good	Good	Slope
Process cost	High	High	Low

TABLE 4.2   Summary of Qualification Tests and Results

Test	Specification	Description	Substrates Tested	Results/ Comments
Preconditioning	JESD22-A113-A Level 3	30°C, 60%RH, 192 hours, followed by 3 reflows at 215°C	98	Pass
Thermal shock	MIL-STD-883 Method 1011 Condition B	500 cycles, −55 to 125°C, liquid to liquid	98	Pass
Thermal cycle	MIL-STD-883 Method 1011 Condition B	1000 cycles, −55 to 125°C, air to air	98	Pass
Pressure cooker	JESD22-A102-B	96 hours, 15 psig, 121°C	16	Pass
High-temperature storage	MIL-STD-883 Method 1008	150°C, 1000 hours	55	Pass
Temperature/ humidity bias	JESD22-A101-A	1000 hours, 85°C, 85%RH, 5-V bias	16	Pass

TABLE 4.3   Qualification Test and Results

Test	Condition	Result	Comments
Thermal cycle	−65 to 125°C 30 min/cycle 100 cycles 200 cycles 300 cycles	No corner crack Resistance change <10%	Pass
Wet proof	40 ± 2°C, 90–95%RH for 240 h	Resistance change < 10%	Pass
Voltage	20°C/65%RH for 96 h and 500 V DC for 60 s	No crack of copper in vias	Pass
Thermal shock	260°C (10 s) to 25°C (20 s) for 30 cycles	No corner crack No copper lift No S/R lift	Pass
Solder	260°C 10-s solder dip	No delamination No corner crack No S/R lift	Pass
	288°C solder float, 10 s	No delamination No corner crack No S/R lift	Pass
	288°C solder float, 5 s	No delamination No corner crack No S/R lift	Pass

It should be noted that laser drilling of SBU PCB and substrate is still in its infancy; therefore, further research in micromachining technology and reliability engineering is necessary. Specifically, the maintenance of laser pointing stability, power stability, and depth of field, as well as system positional tolerances, are critical issues to be resolved. In addition, lasers are not very energy efficient. Thus, putting lasers into around-the-clock manufacturing tools for making microvia SBU PCB and substrate poses a great challenge.

4.5 Photo-Defined Microvias

The first commercially used photoimageable dielectric was a modified liquid solder mask produced in 1990 by IBM at Yasu, Japan. Today, modern photoimageable dielectrics (PIDs) are in the form of either liquid or dry film. For photovias with liquid photodielectric, the dielectric is curtain-coated and cured, microvias are exposed, and the dielectric is developed. Then the panel plating follows with patterning to create signal traces. For photovias with dry-film photodielectric, the dielectric is laminated, microvias are exposed, and the dielectric is developed. The panel plating follows with patterning to create signal traces.[25]

4.5.1 Process for photo-defined vias

Microvias can be formed in mass production by photoimaging technology. The dielectric is applied over the base substrate, and the microvias are imaged and developed. Then this via layer is cured, adhesion-promoted, and copper-plated to connect the microvias. A final primary resist is applied to complete the outer-layer imaging; then the circuits are etched or pattern-plated. Conductive paste can also be used to fill the microvias and circuit pattern. A step-by-step process flow is illustrated in Fig. 4.7.[20]

- *Step 1:* The photosensitive polymer is applied to the base substrate.

- *Step 2:* The coated substrate is exposed to UV light through a right-reading mask containing the desired circuit pattern to selectively cross-link the material.

- *Step 3:* The dielectric is developed, removing the uncross-linked polymer, leaving the final photoexposed, cross-linked polymer with the desired conductor pattern.

- *Step 4:* The conductive ink is applied to fill the photoexposed conductor grooves.

- *Step 5:* A second layer of photosensitive dielectric is applied and exposed to a mask containing the pattern of the vias.

- *Step 6:* Repeat step 3.

- *Step 7:* Repeat step 4. The vias are filled with conductive ink and cured.

- *Step 8:* A third layer of photosensitive dielectric is applied. This layer is then exposed, developed, filled, and cured with the circuitry for the second conductor layer.

Figure 4.8*b* shows a cross section of a photo-defined via.

4.5.2 Notes on photo-defined vias

The benefit of using photo-defined via technology is that it can form thousands of vias at once (i.e., it is a mass-production process). However, there still are several drawbacks to the use of photo-defined via technology[20] that should be noted.

- Only photosensitive dielectric can be used, instead of the wide range of dielectrics that can be used with laser drilling technology.

- Making a dielectric photosensitive can degrade its electrical and reliability performance.

- Photovia technology is limited to blind vias with small aspect ratios (less than 2:1); vias must be formed and plated at every layer.

- This approach cannot be used with laminated metal foils.

- Minimum via size in production is limited to about 75 μm with most technologies.

- This technology generally requires starting with a rigid PCB core and building outward; this core has generally poor flatness (nonuniform dielectric thickness) and dimensional performance.

- Photovias cannot go through unpatterned metal layers.

4.5.3 Materials choice with photo-defined vias

There are many photoimageable dielectric dry films available in the market, such as:

- Ciba Probelec™

- DuPont ViaLux™ PDDF

- Enthone-Omi ENVISION® PDD-9015

- Photoimageable resins from Dow, MacDermid, Morton, Shipley, Taiyo Ink, and others

The deciding factor in materials selection is what characteristics are required. These can be categorized by dielectric type, e.g., epoxy or polyimide; the form the dielectric takes, e.g., liquid or film; and general physical characteristics of the dielectric, such as T_g, dielectric constant, or moisture absorption. In selecting a dielectric material on the basis of these criteria, both manufacturing choices and end product design requirements must be taken into consideration.

4.5.4 Design guidelines and equipment with photo-defined vias

Table 4.4 illustrates some design guidelines for dry-film or liquid-type photosensitive dielectrics.[8] Via diameter can be as small as 50 μm and trace width can be less than 25 μm. In comparison, via diameter can be 25 μm for laser drilling.

The equipment needed for photo-defined via technology includes exposure unit, developer, and wet processor. The suppliers of exposure units are Bacher, Byers, Colight, Csun, Dupont, Dynachem, Gyrex, Hi-Tech, Mirmir, Morton, Olec, Optical Radiation, ORC, Peak Measuring, Tamarac, and Theimer. The suppliers of developer are Advanced Chemill Systems, ASI, Chemcut, Ciba-Geigy, Circuit Services, Danippon Screen, Glenbrook, James River, Lantronic, Microplate, Quantum, Rexham Graphics, and Technifax. The suppliers of wet processors (develop-etch-strip) are ASI, Chemcut, Hollmuller, Lantronic, and Schmid.

4.5.5 Reliability data with photo-defined vias

Some qualification tests from Refs. 1, 6, and 26 show promising results for photo-defined via technology. Also, photovia processes are currently being used. Ibiden, IBM at Yasu, and MicroVia, Inc. are already running volume production.

TABLE 4.4   Comparison of Dry-Film and Liquid Technologies (Photo-Defined Vias)

	Dry film		Liquid	
	Epoxy Acrylic	Polyimide	Epoxy	Polyimide
Applicator	(Vacuum) laminator	Roll laminator	Screen print Spray coat Curtain coat	Screen print Spray coat Roll coat
Thickness range (μm)	25–50	25–50	10–25	10–20
Minimum opening (mm)	70	70	70	50
Trace width (μm)	50	50	50	25
Electrical performance	Good	Good	Good	Excellent
Chemical resistance	Acceptable	Good	Good	Very good
Handling	Easy	Difficult	Fair	Difficult
Cost	Fair	High	Low	High

4.6   Chemical (Wet)- and Plasma (Dry)-Etched Microvias

Microvias can be formed by various etching techniques. The most common is the use of a microwave gas plasma, which is a dry etching process. Wet etching by hot KOH has been used for polyimide films. Both methods are isotropic such that they etch inward while they etch downward. On the positive side, these formation techniques are capable of mass via generation in that they form all vias at the same time without regard to number or diameter.

4.6.1   Process for etched vias

Plasma etching and chemical etching can be extremely cost effective for generating high volumes of small holes in dielectric layers. The principle in this case is to create a mask that defines the positions and sizes of the holes. This may be achieved by using dry film to image, then etch a hole pattern in a copper layer, or simply by using the dry film as the etch mask by imaging and developing.

In either case, the process cost is derived from the number of holes in a given working area. All the holes are generated simultaneously, and the process time is dependent on how long it takes to erode or dissolve the unmasked dielectric. Plasma etching has the added benefit of removing organic contaminants, and with careful conditioning the amount of slope (undercut) created in the hole can be minimized. Chemical etching is the least expensive process for generating small holes on dielectrics. Both plasma and chemical etching processes can create blind vias, usually using the target pad as a means of defining the bottom of a blind via.[27, 28]

The dry etching process needs extra processes like plasma etching and copper thinning (etch back). Microvia hole (MVH) formation by gas-microwave plasma (GMP) has been promoted by Dyconex of Switzerland and its licensee, Hewlett-Packard. Dyconex has 16 licensees worldwide because it sells a basic plasma drill for $55,000, compared to a laser drill that costs $500,000 and up or a photoimaging facility that requires around $300,000. The plasma equipment can be obtained from Advanced Plasma Systems, Inc. or Plasma Etch Inc. A step-by-step process flow is illustrated as below.

- *Step 1:* Fabricate core by standard double-sided rigid or multilayer board methods.

- *Step 2:* Laminate adhesive-coated copper foil to core.

- *Step 3:* Print and develop via pattern.

- *Step 4:* Etch via pattern and strip etch resist.

- *Step 5:* Plasma-ablate dielectric and etch copper overhang.
- *Step 6:* Electroless-copper-plate panel.
- *Step 7:* Laminate photo resist; print and develop outer-layer image.
- *Step 8:* Pattern-electroplate copper.
- *Step 9:* Strip resist and etch.

There are some plasma-etchable materials like adhesiveless polyimide foil, FR-4 buttered Cu-foil, Aramid-paper reinforced FR-X, Liquid Crystal Polymer (LCP), and Cu foil buttered with FR-5 or Polyimide Glass or Cyanate Ester (CE) resin.

4.6.2 Notes on plasma-etched vias

Even though plasma technology can form many vias at one time, there are several drawbacks, shown in the following list, that limit popular acceptance of plasma techniques.

- Plasma technology batch processes are suitable only for small volumes or prototypes.
- Plasma techniques have a via size limit of 75 μm.
- They require isotropic etching to minimize undercutting.
- They require removal of the copper-clad overhang from blind via holes prior to copper plating.
- They rely on a special resin-coated copper foil (nonreinforced dielectric layer).

Figure 4.8c shows a via generated by the plasma formation process. The vias created by both laser drilling (Fig. 4.8a) and photoimaging (Fig. 4.8b) look very nice. However, the via generated by plasma shows undercut. This is why the plasma-etching process cannot surpass the laser-drilling and photoimaging processes.

A comparison of these three key SBU technologies is listed in Table 4.5.[29] It can be seen at a glance that photovia looks more promising than laser drilling. So why are 70 percent of the world's microvia boards made by laser ablation? Because laser drilling is a simple, straightforward process that, unlike photovia, does not require a photosensitive dielectric, tedious photolithography processing, and very clean rooms, all in addition to mechanical drilling.

4.7   Conductive-Ink-Formed Microvias

Conductive ink describes a single-layer dielectric with microvias formed by photoimaging, laser, or insulation displacement. A conduc-

TABLE 4.5 Comparison of Key SBU Technology

Items	Photovia (Bare Resin)	Laser Drill (Cladded Laminate)	Laser Drill (Coated Foil)	Laser Drill (Coated Foil)	Plasma Etch (Coated Foil)
Productivity	High (5)	Low (1)	Low (1)	Low (1)	Low-medium (2)
Initial investment	(3) Coater (LPI only) Printer Developer	(2) Coater (liquid only) Laser system (≈$500K)	(3) Laser system (≈$500K)	(3) Laser system (≈$500K)	(3) Plasma unit ($300–400K)
Dielectric cost	Medium-high (2)	Low-medium (4)	Medium (3)	High (1)	High (1)
Material used	LPI/dryfilm	Liquid film	Single-cladded laminates	Coated foil	Coated foil
Consumables	Phototool Developer	Phototool Photoresist Gases	Phototool Photoresist Gases	Phototool Photoresist Gases	Phototool Photoresist Gases
Process control	Average (3)	Low-average (4)	Average-high (2)	Average-high (2)	Average-high (2)
Technical challenge	Yields/cleanliness Material development	Productivity Process development	Cost Productivity	Cost Productivity	Cost Productivity
Resolution	25–50 µm	25–50 µm	50–80 µm	50–80 µm	75–100 µm

Relative rating: 1 = lowest, 5 = highest.

tive paste (or film) is used to fill the microvias and act as the conductive path between layers. Surface metallization may be accomplished either by laminating copper foil onto the dielectric surface or by chemical deposition. Three key advantages over standard PTH technology are: (1) no plated through-hole for inner layers; (2) no possible flow of prepreg during press lamination; and (3) no voiding in through-holes.

4.7.1 Materials choice

Usually, the conductive materials used are either DuPont's CB100 (wet) or Matsushita's (dry) conductive polymeric thick film (PTF). CB100 is a screen-printable paste made of silver, copper, and epoxy. Figure 4.9a shows a stacked via over filled PTH, and Fig. 4.9b shows an integrated IC-PCB assembly with the CB100 conductive via plug.[1] Matsushita's PTF is used to plug buried vias for any layer inner via hole (ALIVH™),[30–33] as shown in Fig. 4.10.

4.7.2 Fabrication process of CB100

CB100 is usually stencil-printed in the plated or unplated through-holes. After drying and curing, the plugged through-holes are planarized and plated to make them solderable. The design rules for CB100 are: aspect ratio = 1:1 to 6:1 with vacuum assist; via size = 6 to 25 mil (152 to 635 μm); and core thickness = 6 to 85 mil (152 to 2159 μm). The TCE is less than 35 ppm/°C. The CB100 assembly process is as follows:

- Drill through-holes on the double-sided copper clad substrate.
- Screen/stencil print CB100 into the through-holes.
- Dry and cure the via-plugged CB100.
- Scrub and planarize the surfaces.
- Plate the filled board with about 0.3 to 0.5 mil to enhance electrical conductivity and solderability. (With a multilayer core, the through-hole is first flash-copper-plated with a minimum of 0.1 mil copper to ensure interplane connection.)
- External flash-electroplate copper.
- Form wiring patterns of both sides of laminate by printing and etching.

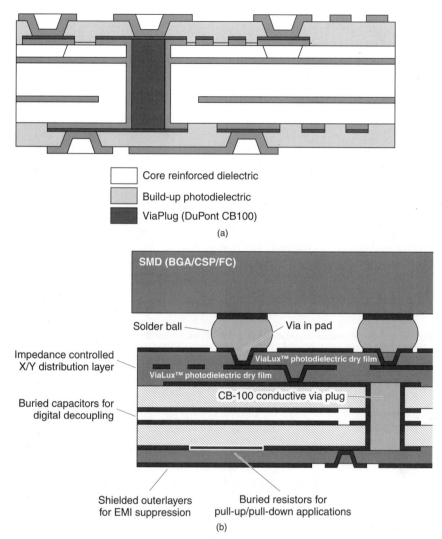

Figure 4.9 (a) The real estate lost to the PTH is recovered by filling the PTH with DuPont CB100 and then placing or stacking a microvia over the PTH. (b) Fully integrated IC-PCB with redistribution layer, controlled impedance, thermal dissipater, buried capacitor, buried resistor, and EMI shielding.

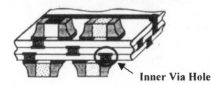

Figure 4.10   Schematic diagram of ALIVH substrate.

4.7.3   Fabrication process of ALIVH™

The fabrication process flow chart of a four-conductive-layer ALIVH substrate is shown in Figure 4.11.[33] The procedures are illustrated as follows.[33]

- The nonwoven aramid-epoxy prepreg is used as a substrate material. The B-staged aramid-epoxy sheets are drilled out to form via holes using a CO_2 laser for the electrical connection between conductive layers.

- The via-holes are filled with the via conductor paste. The via conductor paste consists of copper particles and a thermosetting resin. Two copper foils and the via-filled aramid-epoxy prepreg are stacked to be laminated, sandwiching the aramid-epoxy prepreg between two copper foils with no gap. The lamination and curing of the aramid-epoxy prepreg are performed under a pressure and a temperature. Wiring patterns are formed by etching the copper foil layers.

- The other via filled aramid-epoxy prepregs are stacked on both sides of the wiring-patterned aramid-epoxy sheet. Furthermore, two copper foils are stacked on both sides of the stacking laminates. Lamination is performed under a pressure and a temperature. Finally the wiring patterns of both sides of laminate are formed by etching.

4.7.4   Reliability of conductive-ink-formed vias

There are some qualification data for the CB100 ViaPlug, as shown in Ref. 1. It can be seen that the material can pass 1000 temperature

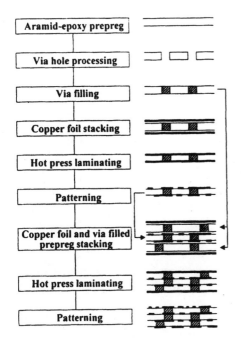

| Aramid-epoxy prepreg |
| Via hole processing |
| Via filling |
| Copper foil stacking |
| Hot press laminating |
| Patterning |
| Copper foil and via filled prepreg stacking |
| Hot press laminating |
| Patterning |

Figure 4.11 Fabrication process chart of ALIVH substrate containing four conductive layers.

cycles (−65 to 125°C) and pass 5 solder dips at 286°C. From Ref. 33, it can be seen that the ALIVH substrate can also pass 1000 temperature cycles (−55 to 125°C). The ALIVH technology is very popular in Japan.

4.8   Microvia Production in Japan

Since more than half of the world's microvia PCBs are made in Japan, it is interesting to note the status of some key manufacturers in this area. They are reported as follows.

4.8.1   Fujitsu Limited[34]

Fujitsu's FLD (Fujitsu Laminate and Deposit) buildup process is currently in production and utilizing a photoimaging system. The Fujitsu Laminate with Laser via and Deposit (FLLD) process is under development utilizing a laser drilling system. The technology is capable of producing 50-μm copper lines with 50-μm spaces and vias 100 μm in diameter for packaging substrates. Fujitsu is changing from photovias to laser vias for its next-generation packaging substrates. Highly Accelerated Stress Test (HAST) and reliability test results have been performed and are presented by Miyazawa et al.[34]

4.8.2   Hitachi Chemical Co.[35]

Hitachi has developed two types of PCBs, namely HITAVIA Type 1 and Type 3, using metal clad B stage insulation resin film (MCF). Type 1 is used for motherboards and liquid crystal display drivers; Type 3 is used for semiconductor packages such as PBGA. The interstitial via holes (IVHs) are formed using mechanically predrilled metal clad film. A CO_2 laser is used to etch the microvias. The technology is capable of producing 100-μm copper lines with 100-μm spaces for Type 1 and 50-μm lines with 50-μm spaces for Type 3. The size of the vias can be 250 μm on 500-μm pads for Type 1 and 100 μm on 250-μm pads for Type 3. Reliability tests have been done for both types. Both passed the tests.[35]

4.8.3   Ibiden[36]

Ibiden has developed a double-sided four-layer buildup board using fully additive copper plating and photosensitive epoxy. This technology is capable of producing 75-μm lines with 75-μm spaces for motherboards and 35-μm copper lines with 35-μm spaces for packages. Using the photovia process, it is possible to make 150-μm-diameter vias on 250-μm lands for motherboards and 80-μm-diameter via holes on 125-μm lands for packages. Reliability data[36] are available on this process and show that they pass all the tests.

4.8.4   IBM at Yasu[37–40]

The IBM at Yasu's Surface Laminar Circuit (SLC) also uses liquid dispensed photoimaged epoxy in the buildup layers on FR4 or BT substrates. The photosensitive epoxy resin is exposed and developed to form the via hole. The via diameter can be 100 μm and the land diameter can also be 100 μm. The line width and space can be 38 μm. The flip chip attached onto SLC board has passed reliability tests with 100°C delta temperature cycling and temperature/humidity/bias test.[37–40]

4.8.5   JVC[41]

The Victor Company of Japan (JVC) has produced the Variously Interconnected Layers (VIL) type of buildup PCB by applying thermosetting material to insulation and by the laser processing of via holes through the mask imaging method. With this technology, 100-μm lines and spaces have been produced. This process has succeeded in making a skip via hole as small as 280 μm in diameter. No reliability data are available for the process.[41]

4.8.6 Matsushita[33]

Matsushita has developed a unique stacked-type substrate technology called ALIVH, which has already been discussed in Sec. 4.7.[30–33] Laser ablation (currently CO_2 laser) and Cu paste are used for buried holes. It is reported that Matsushita's share of the Japanese cellular phone market has risen to 60 percent with the introduction of ALIVH substrates. Its target for both conductor line and space resolution is 50 μm in 1999. Reliability data for ALIVH-CSP can be found in Ref. 33.

4.8.7 NEC[42]

NEC Toyama has developed a high-density microvia (μV) PCB for advanced microelectronic packaging application. The more advanced technology can produce 50-μm copper lines with 50-μm spaces. The via diameter can be 50 μm and land diameter can be 150 μm. The via is formed by laser processing. Most of the samples pass reliability tests, except that there are fewer than 3 percent fails on interconnection reliability test, MIL-STD-107G (–65°C for 30 minutes to +125°C for 30 minutes), after 500 cycles.[42]

4.8.8 Toshiba[43]

Toshiba has introduced a unique Buried Bump Interconnection Technology (B²it™). Silver paste is used to form conductive bumps on copper foil. Conical bumps can penetrate the insulator and form a conductive path through the prepreg. The conductive lines and spaces, both currently 75 μm wide, are formed by subtractive processing using photoresist. The "via" (bump) diameter is 200 μm and land diameter is 400 μm. This technology is now under mass production.[43]

4.8.9 Summary

A summary of the line width and space width capabilities of these 8 Japanese microvia manufacturers is shown in Fig. 4.12. It can be seen that Ibiden can make the finest lines and spaces. Figure 4.13 is a summary of the microvia and pad diameter capabilities of six of the Japanese microvia manufacturers (no data for Matsushita). The smallest via and pad can be manufactured by NEC Toyama Corporation.

It should be pointed out that five of the eight Japanese companies (Hitachi, Ibiden, JVC, Matsushita, and Toshiba) have developed their own new materials for the microvia process and formation. Four (Hitachi, JVC, Matsushita, and NEC) use laser drilling. Two (Ibiden and IBM-Yasu) use photoimaging, and Fujitsu is changing from photoimaging to laser drilling. Toshiba has changed from the conventional (drilling and plating) process to a new process.

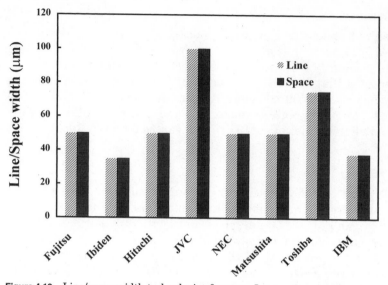

Figure 4.12   Line/space width technologies for some Japanese companies.

In general, laser drilling can create the smallest holes. Photo-defined via technology has the highest productivity. Laser drilling technology controls 70 percent of the microvia market today because of its straightforward nature and compatibility with all kinds of materials (including clad laminates, copper-coated foil, and dielectrics). The pho-

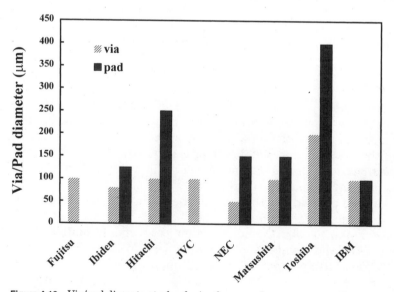

Figure 4.13   Via/pad diameter technologies for some Japanese companies.

toimaging method is the second most popular technology in making microvias.

Today, Japan is the leader in making microvias and the U.S. is very well back. However, some U.S. companies are catching up. For example, Johnson Matthey intends to be the largest North American supplier of microvias by the end of 1999. It has 10 laser drills (4 YAGs to ablate copper for RCC processing and 6 CO_2 units to remove dielectric). This electronic materials division was sold to Allied Signal's electronic materials unit in August 1999. Allied Signal sold its laminate systems business to Rutgers AG, based in Essen, Germany, in order to enter higher-margin specialty segments. Table 4.6 shows the specifications and capabilities of some microvia substrate suppliers.

4.9   Micro Via-in-Pad (VIP)

Microvias save PCB real estate. Combining them with via-in-pad (VIP) saves even more. Figures 4.14 through 4.17 show some of the applications. Figure 4.14 shows a microvia-pad C4 solder bump made by Motorola using a stencil printing process.[44] Figure 4.15 shows a 272-pin PBGA solder joint on an 0.075-mm blind microvia-in-pad.[45] It can been seen that there is a void in the solder joint; this is very common in these structures. Figure 4.16 shows the 272-pin solder joint on an 0.2-mm blind "micro" via-in-pad. Because of the large size of the VIP, there is no void in this case. Figure 14.17 shows a cross section of a solder-bumped flip chip on the microvia-in-pad on an organic substrate in a PBGA package. It is noted that the second-level solder joint is also on a larger VIP on the bottom side of the substrate. A novel and low-cost micro via-in-pad[46] will be presented in Chapter 11 of this book.

4.10   Useful Design Charts for High-Speed Circuits

In Ref. 47, IPC-2141 gives the equations to estimate the characteristic impedances for some very simple circuits. Since these are approximate formulas, very often they lead to great errors. For example, for the simplest case (surface microstrip), shown in Fig. 4.18, with $T = 35$ µm, $H = 794$ µm, $W = 3300$ µm, and $\varepsilon_r = 4.2$, IPC-2141's equation underpredicts the characteristic impedance by almost 30 percent. In this section, some useful design charts for the surface microstrip, differential edge-coupled surface microstrip, embedded microstrip, differential centered edge-coupled embedded microstrip, stripline, differential sysmmetrical broadside-coupled stripline, differential symmetrical edge-coupled stripline, and surface microstrips are shown in Figs. 4.18 through 4.29, respectively. It should be noted that all the curves are determined

TABLE 4.6   Capabilities of Microvia Substrate Suppliers

Company Name	Via Formation Method	Via/Pad Diameter (μm)	Line/Space (μm)	Maximum Number of Multilayers
ASTI	Laser	25/125	35/35	16
Canon	Laser	150/300	75/75	3+core+3
Compaq	Laser	75/195	75/30	10
Fujitsu	Photovia	80/130	40/40	6
	Laser	50/100	35/35	3+core+3
Hitachi Cable	Laser	50/80	25/25	2
Hitachi Chemical	Laser	100/125	50/50	10
Ibiden (IBSS)	Photovia	85/125	40/40	4-2-4
	Laser	150/250	75/75	10
Daisho Denshi	Laser	100/300	75/75	8
IBM (HPCC)	Laser	50/100	28/33	9
IBM (SLC)	Photovia	100/150	50/50	3+core+3
JCI	Laser	100/160	40/40	8
JVC	Laser	150/250	60/60	2+core+2
K&S (X-Lam)	Photovia	25/54	18/14	2-4
Kyocera/JME	Photovia	50/100	30/30	4-8-4
Matsushita	Laser	100/125	50/50	4+core+4
Mitsubishi	Laser	75/125	30/30	3+core+3
Multek	Photovia	75/150	75/75	8
NEC	Photovia	40/60	20/20	3+core+3
NTK	Photovia	90/130	35/35	3+core+3
	Laser	50/130	40/40	9
Samsung	Photovia with laser	150/200	50/50	3+core+3
Sheldahl	Laser	25/140	50/37.5	2
Shinko	Laser	50/110	45/45	3+core+3
SMI	Photovia	95/135	37/37	2+core+2
STP	Photovia	100/200	75/75	1+8+1
Toppan	Photovia	80/130	40/40	3+core+3
	Laser	100-125/300	75/75	—
Toshiba	B²it	200/400	75/75	4+core+4
W. L. Gore	Laser	50/87	25/25	9

Note: The number before and after the core is the number of build-up layer.

Figure 4.14   Solder bump on microvia-in-pad.

Figure 4.15   Solder joint on 0.075-mm microvia-in-pad with a large void.

Figure 4.16   Solder joint on 0.2-mm "micro" via-in-pad without voids.

Figure 4.17   Solder-bumped flip chip on microvia-in-pad on the substrate of a PBGA package. (The second-level solder joint is on a larger VIP on the bottom side of the substrate.)

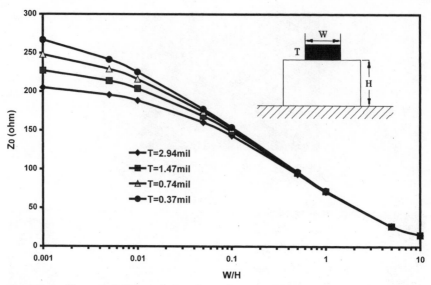

Figure 4.18 Characteristic impedance of a surface microstrip ($\varepsilon_r = 4.2$).

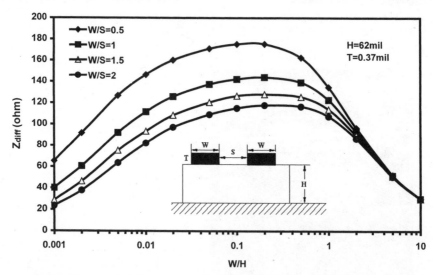

Figure 4.19 Characteristic impedance of a differential edge-coupled surface microstrip ($\varepsilon_r = 4.2$).

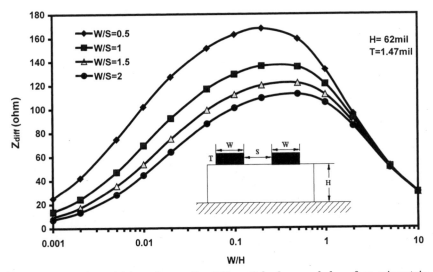

Figure 4.20 Differential impedance of a differential edge-coupled surface microstrip ($\varepsilon_r = 4.2$).

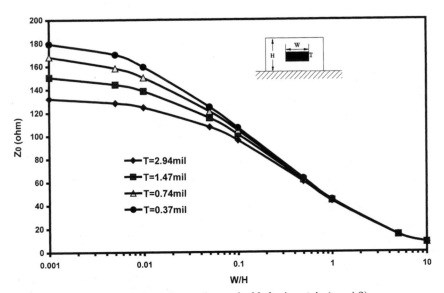

Figure 4.21 Characteristic impedance of an embedded microstrip ($\varepsilon_r = 4.2$).

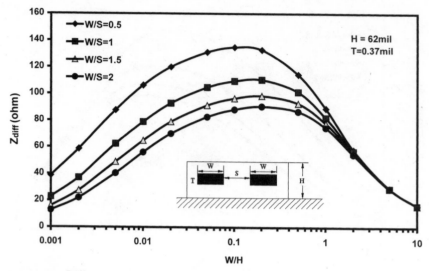

Figure 4.22 Differential impedance of a differential centered edge-coupled embedded microstrip ($\varepsilon_r = 4.2$).

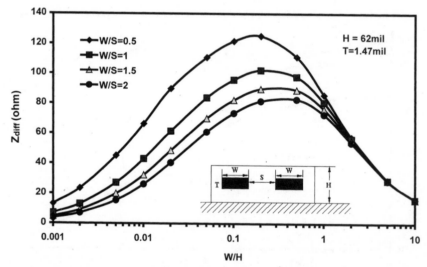

Figure 4.23 Differential impedance of a differential centered edge-coupled embedded microstrip ($\varepsilon_r = 4.2$).

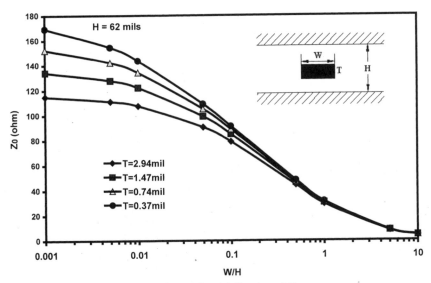

Figure 4.24 Characteristic impedance of a stripline ($\varepsilon_r = 4.2$).

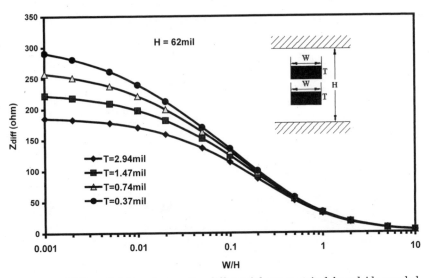

Figure 4.25 Differential impedance of a differential symmetrical broadside-coupled stripline ($\varepsilon_r = 4.2$).

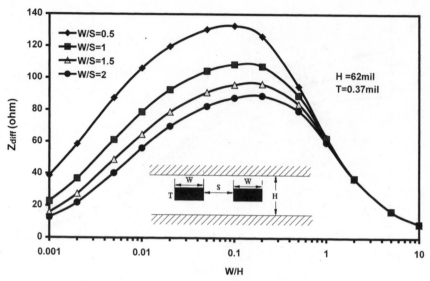

Figure 4.26 Differential impedance of a differential sysmmetrical edge-coupled stripline ($\varepsilon_r = 4.2$).

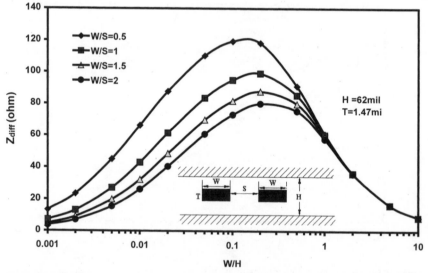

Figure 4.27 Differential impedance of a differential symmetrical edge-coupled stripline ($\varepsilon_r = 4.2$).

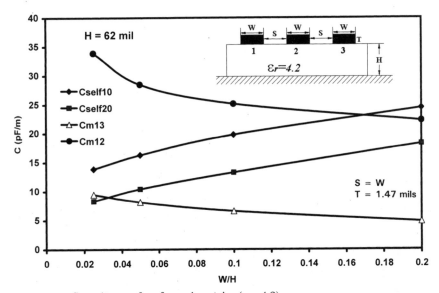

Figure 4.28 Capacitance of surface microstrips ($\varepsilon_r = 4.2$).

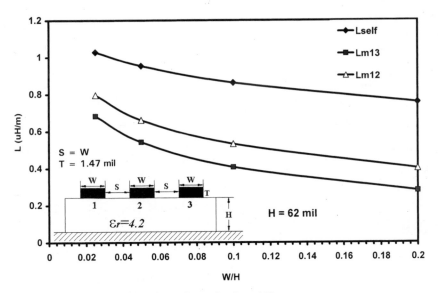

Figure 4.29 Inductance of surface microstrips ($\varepsilon_r = 4.2$).

(with $\varepsilon_r = 4.2$) from numerical simulations and checked randomly with some well-known solutions, such as those given in Ref. 48. In these charts, Z_0 is the characteristic impedance and $Z_{\mathrm{diff}} \cong 2 \times Z_{0o}$, where Z_{0o} is the odd-mode impedance.

Acknowledgments

The author would like to thank all the authors listed in the references of this chapter for sharing their useful know-how with the industry. I learned a lot from reading their papers.

References

1. Gonzalez, C. G., R. A. Wessel, and S. A. Padlewski, "Epoxy-Based Aqueous Processable Photo Dielectric Dry Film and Conductive Via Plug for PCB Build-Up and IC Packaging," *IEEE Trans. Advanced Packaging,* **22** (3): 385–390, August 1999.
2. Singer, A. T., "Microvia Cost Modeling," *Proceedings from IPC Works,* p. S-14-2, Washington, DC, October 1997.
3. Burgess, L. W., and P. D. Madden, "Blind Vias in SMD Pads," *Printed Circuit Fabrication,* **21** (1): 28–29, January 1998.
4. Castro, A., "Chip Carrier Package Constructions Made Easier with Dry Film Photo Dielectric," *Proceedings from IPC Works,* p. S01-5-1, Washington, DC, October 1997.
5. Thorne, J., "Using New Interconnection Technologies to Reduce Substrate Cost," *IPC EXPO '98,* p. S-10, Long Beach, CA, April 1998.
6. Nargi-Toth, K., and P. Gandhi, "Manufacturing Methodologies for High Density Interconnect Structures (HDIS)," *CSI Technical Symposium,* pp. 63–70, San Jose, CA, September 1998.
7. Nargi-Toth, K., "ITRI Microvia Technology Roadshow," *IPC EXPO '99,* p. S17-1, Long Beach, CA, March 1999.
8. Numakura, D. K., S. E. Dean, D. J. McKenney, and J. A. DiPalermo, "Micro Hole Generation Processes for HDI Flex Circuit," *HDI EXPO '99,* pp. 443–450.
9. Noddin, D. B., E. Swenson, and Y. Sun, "Solid State UV-LASER Technology for the Manufacture of High Performance Organic Modules," *Proceedings of 48th Electronic Components and Technology Conference,* pp. 822–827, Seattle, WA, May 1998.
10. Cable, A., "Improvements in High Speed Microvia Formation Using Solid State Nd:YAG UV Lasers," *IPC EXPO '97,* p. S7-2, San Jose, CA, March 1997.
11. Owen, M., "Production Experiences with CO_2 and UV YAG Drilling," *IPC EXPO '97,* p. S7-3, San Jose, CA, March 1997.
12. Tessier, T. G., and J. Aday, "Casting Light on Recent Advancements in Laser Based MCM-L Processing," *Proceedings 1995 International Conference on Multichip Modules,* pp. 6–13, 1995.
13. Illyefalvi-Vitez, Z., M. Ruszinko, and J. Pinkola, "Recent Advancements in MCM-L Imaging and Via Generation by Laser Direct Writing," *Proceedings of 48th Electronic Components and Technology Conference,* pp. 144–150, Seattle, WA, May 1998.
14. Moser, D., "Sights Set on Small Holes? How to Get There with Lasers," *Printed Circuit Fabrication,* pp. 20–22, February 1997.
15. Owen, M., E. Roelants, and J. Van Puymbroeck, "Laser Drilling of Blind Holes in FR4/Glass," *Circuit World,* **24** (1): 45–49, 1997.
16. Contini, H. S., "Machining Lasers Find Niches by Solving Very Small Problems," *Photonics Spectra,* 116–118, November 1997.
17. Illyefalvi-Vitez, Z. and J. Pinkola, "Application of Laser Engraving for the Fabrication of Fine Resolution Printed Wiring Laminates for MCM-Ls," *Proceedings of 47th Electronic Components and Technology Conference,* pp. 502–510, San Jose, CA, May 1997.
18. Kobayashi, K., N. Katagiri, and S. Koyama, "Development of a Build Up Package with High Density of Circuits for High Pin Count Flip Chip Application," *IPC Expo 99,* p. S01-4.

19. Burgess, L. W. and F. Pauri, "Optimizing BGA to PCB Interconnections Using Multi-depth Laser Drilled Blind Vias-in-Pad," *Circuit World,* **25** (2): 31–34, 1999.
20. Raman, S., J. H. Jeong, S. J. Kim, B. Sun, and K. Park, "Laser (UV) Microvia Application in Cellular Technology," *IPC EXPO '99,* p. S17-6, Long Beach, CA, March 1999.
21. Schaeffer, R. D., "Laser Microvia Drilling: Recent Advances," *CircuiTree,* **12:** 38–44, 1998.
22. Petefish, W. G., D. B. Noddin, and D. A. Hanson, "High Density Organic Flip Chip Package Substrate Technology," *Proceedings of 48th Electronic Components and Technology Conference,* pp. 1089–1097, Seattle, WA, May 1998.
23. Young, T. and F. Polakovic, "Thermal Reliability of Laser Ablated Microvias and Standard Through-Hole Technologies," *IPC Expo 99,* p. S17-2.
24. Gaku, M., H. Kimbara, N. Ikeguchi, and Y. Kato, "CO_2 Laser Drilling Technology for Glass Fabrics Base Copper Clad Laminate," *IPC Expo 99,* p. S17-3.
25. Estes, W. E., T. R. Overcash, S. Padlewski, B. D. Neve, E. B. Murray, R. E. Anderson, R. C. Mason, J. P. Lonneville, W. L. Hamilton, and M. Periyasamy, "Photodielectric Dry Films for Ultra High Density Packaging," *SMI Proceedings,* pp. 47–53, San Jose, CA, September 1997.
26. McDermott, B. J., and S. Tryzbiak, "The Practical Application of Photo-Defined Micro-Via Technology," *SMI Proceedings,* pp. 199–207, San Jose, CA, September 1997.
27. Reboredo, L., "Microvias: A Challenge for Wet Processes," *IPC Expo 99,* p. S12-1.
28. Schmidt, W., "High Performance Microvia PWB and MCM Applications," *IPC Expo 99,* p. S17-5.
29. Ho, I., "What's Up with SBU Technology?" *Printed Circuit Fabrication,* 64–68, March 1997.
30. Felten, J. J., and S. A. Padlewski, "Electrically Conductive Via Plug Material for PWB Applications," *IPC EXPO '97,* p. S6-6, San Jose, CA, March 1997.
31. Wessel, R. A., J. F. Henderson, J. J. Felten, S. Padlewski, M. A. Saltzberg, P. Charest, J. L. Parker, and P. T. Miscikowski, "A New Approach . . . to Filled, Conductive Vias in PCBs," *Printed Circuit Fabrication,* 42–45, November 1997.
32. Nishii, T., S. Nakamura, T. Takenaka, and S. Nakatani, "Performance of Any Layer IVH Structure Multi-Layered Printed Wiring Board," *Proceedings of 1995 Japan IEMT Symposium,* pp. 93–96, 1995.
33. Itagaki, M., K. Amami, Y. Tomura, S. Yuhaku, Y. Ishimaru, Y. Bessho, K. Eda, and T. Ishida, "Packaging Properties of ALIVH-CSP using SBB Flip-Chip Bonding Technology," *IEEE Trans. Advanced Packaging,* **22** (3): 366–371, August 1999.
34. Miyazawa, Y., T. Shirotsuki, H. Sugai, and Y. Yoneda, "Highly Accelerated Stress Test and Reliability Analysis for Build-up Circuits," *1998 International Symposium on Microelectronics,* pp. 430–434, November 1998.
35. Arike, S., K. Otsuka, N. Urasaki, A. Nakaso, K. Shibata, K. Kobayashi, K. Tsuyama, K. Suzuki, and H. Nakayama, "PWB using Mechanically Pre-drilled Metal Clad Film for IVH and Build-up PWB with Laser Via Hole for Semiconductor Package Substrate," *1998 International Symposium on Microelectronics,* pp. 425–429, San Diego, CA, November 1998.
36. Enomoto, R., M. Asai, and N. Hirose, "High Density MLB using Additive and Build-up Process," *1998 International Symposium on Microelectronics,* pp. 399–404, San Diego, CA, November 1998.
37. Tsukada, Y., and S. Tsuchida, "Surface Laminar Circuit, a Low Cost High Density Printed Circuit Board," *Proceedings of Surface Mount International,* vol. 1, pp. 537–542, August 1992.
38. Tsukada, Y., Y. Mashimoto, T. Nishio, and N. Mii, "Reliability and Stress Analysis of Encapsulated Flip Chip Joint on Epoxy Base Printed Circuit Board," *Proceedings of ASME/JSME Joint Conference on Electronic Packaging,* vol. 2, pp. 827–835, September 1992.
39. Tsukada, Y., Y. Maeda, and K. Yamanaka, "A Novel Solution for MCM-L Utilizing Surface Laminar Circuit and Flip Chip Attach Technology," *Proceedings of 2nd International Conference on Multichip Modules,* pp. 252–259, April 1993.
40. Tsukada, Y., "Solder Bumped Flip Chip Attach on SLC Board and Multichip Module," *Chip on Board,* Lau, J. H., ed., Van Nostrand Reinhold, New York, pp. 410–443, 1994.

41. Segawa, K., "Build-up PWB with Laser-Processed Via Holes 'VIL,'" *1998 International Symposium on Microelectronics*, pp. 419–424, San Diego, CA, November 1998.
42. Maniwa, R., "Finer Micro-via PWB by Laser and Additive Process," *1998 International Symposium on Microelectronics*, pp. 413–418, San Diego, CA, November 1998.
43. Fukuoka, Y., T. Oguma, and Y. Tahara, "New High Density Substrates with Buried Bump Interconnection Technology (B²it™)—Design Features of Electrical and Thermal Performances with the Actual Applications," *1998 International Symposium on Microelectronics*, pp. 405–412, San Diego, CA, November 1998.
44. Jimarez, M., L. Li, C. Tytran, C. Loveland, and J. Obrzut, "Technical Evaluation of a Near Chip Scale Size Flip Chip/Plastic Ball Grid Array Package," *Proceeding of IEEE ECTC*, pp. 495–502, June 1998.
45. Mawer, A., K. Simmons, T. Burnette, and B. Oyler, "Assembly and Interconnect Reliability of BGA Assembled onto Blind Micro and Through-Hole Drilled Via in Pad," *Proceedings of SMI Conference*, pp. 21–28, August 1998.
46. Lau, J., C. Chang, C.-C. Chen, C. Ouyang, R. Lee, T.-Y. Chen, D. Cheng, T. J. Tseng, and D. Lin, "Via-In-Pad (VIP) Substrates for Solder Bumped Flip Chip Applications," *Proceedings of SMI Conference*, pp. 128–136, September 1999.
47. IPC-2141-Controlled Impedance Circuit Boards and High-Speed Logic Design, April 1996.
48. Wadel, B., *Transmission Line Design Handbook*, Artech House, 1991.

Flip Chip on Board with Solderless Materials

5.1 Introduction

Recently, for the sake of a green environment, solderless materials such as conductive adhesives[1–23] have been evaluated for use in assembling flip chips on substrates. There are two different kinds of conductive adhesives: isotropic conductive adhesives and anisotropic conductive materials. Isotropic conductive adhesives, such as the Ag-Pd paste used in the stud bump bonding (SBB) flip chip on board technology given in Chapter 12 of Lau,[4] electrically conduct in all directions and will not be discussed in this book.

Anisotropic conductive materials, also called z-axis conductive materials, electrically conduct only in the vertical direction. There are two groups of anisotropic conductive materials: anisotropic conductive film (ACF) and anisotropic conductive adhesive (ACA). ACF looks like paper and consists of thermosetting adhesive, conductive particles, and release film. ACA looks like paste and consists of thermosetting adhesive and conductive particles. In this chapter the design, materials, process, and reliability of flip chip assemblies with ACF and ACA will be discussed.

5.2 FCOB Assemblies with ACF[2]

In this section, DCA with solderless flip chip on PCB with ACF will be considered. The chip is bumped with three different metallurgies: Cu, Au, and Ni-Au. The Cu pads on the FR-4 epoxy PCB are with electroless Ni-Au and are organic coated. Hitachi's ACF[7, 8] is used for this study. The design, materials, and assembly process flow are shown in

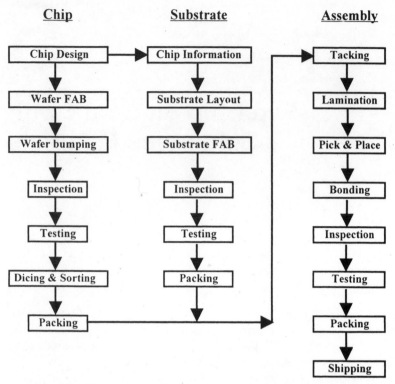

Figure 5.1   Flip chip on board with anisotropic conductive film (ACF) process.

Fig. 5.1. In the following sections some of the major steps will be discussed. Also, some thermal cycling and surface insulation resistance (SIR) test results are presented.

5.2.1 The wafer

The chip size is 0.5 in. (12.7 mm) square and 25 mils (0.64 mm) thick. The street width between all the chips is 6 mils (0.15 mm). The chip has 8-mil (0.2-mm) square pads and 14-mil (0.36-mm) pitch. All of the pads are arranged symmetrically around the perimeter of the chip and interconnected via traces on the chip in an alternating pattern so as to provide daisy-chained connections when the chip is attached to the FR-4 PCB.

The silicon wafer consists of a patterned aluminum layer on a layer of silicon dioxide that is covered with a patterned silicon nitride passivation layer. The wafer fabrication process flow starts with deposition of an 0.25-μm layer of silicon dioxide by Plasma Enhanced Chemical Vapor Deposition (PECVD) on a <111> silicon substrate. For the metal layer, an 0.85-μm layer of Al-1%Si-0.1%Ti alloy is sputtered over the silicon dioxide. The metal pattern is then defined by coating with positive resist, exposed by projection alignment, developed, and wet-chemical-etched. After the photoresist is removed by plasma stripping, the metal pattern is sintered at 450°C to remove film stresses. For the passivation layer, a 0.75-μm layer of silicon nitride is deposited over the entire surface of the wafer by PECVD. The pad opening in the passivation layer is then defined by coating with positive resist, exposed by projection alignment, developed, and plasma-etched. Finally, the resist is removed by plasma stripping, leaving the silicon nitride passivation layer to overlap the perimeter of the Al pads by 10 μm.

5.2.2 Wafer bumping with Au, Cu, and Ni-Au

In this study, three different bumps (Au, Cu, and Ni-Au) are put on the wafers (Fig. 5.2). The materials and process for each kind of bump are briefly discussed.

Wafer bumping with Au. The wafers are Au-bumped by the electroplating method (Fig. 5.3). The under-bump metallurgies (UBMs) of the wafers are titanium (Ti) and tungsten (W). They are sputtered on the entire surface of the wafer, 0.1 to 0.2 μm of Ti first, followed by 0.3 to 0.5 μm of W. A 20-μm layer of resist is then overlaid on the Ti-W and a bump mask is used to define the bump pattern. The openings in the resist are 7 to 10 μm wider than the pad openings in the passivation layer. A 20-μm-thick layer of Au is then plated over the Ti-W. The resist is then removed and the Ti-W stripped off with a hydrogen peroxide etch. Figure 5.4 shows a cross section of the electroplated Au bump on the Si wafer. The average microhardness of the Au bumps is 120 on the Vickers scale (HVN).

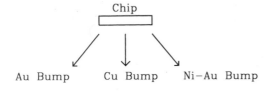

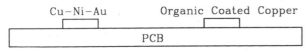

Figure 5.2 Flip chip with three different kinds of bumps on two different kinds of PCBs.

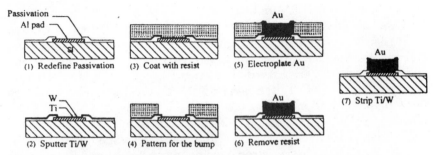

Figure 5.3 Electroplated Au wafer-bumping process.

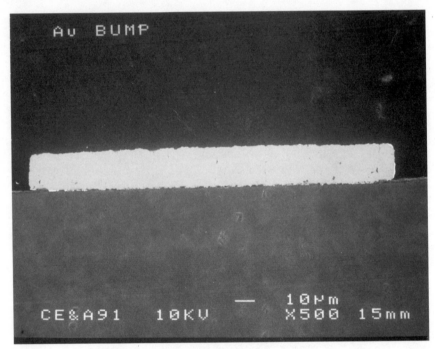

Figure 5.4 A typical Si chip with Au bump.

Wafer bumping with Cu. The wafer-bumping process with Cu is almost the same as that with Au except for the UBM, which is Ti (0.1 to 0.2 μm) and Cu (0.5 to 0.8 μm). On top of the UBM, the plated Cu thickness is 20 μm. The average microhardness of the Cu bumps is 100 (HVN). Since the copper surface oxidizes and corrodes very easily, the Cu bumps are electroless-plated with a very thin layer of Sn (i.e., immersed or simply dipped). Figure 5.5 shows a cross section of the electroplated Cu bump on the Si wafer.

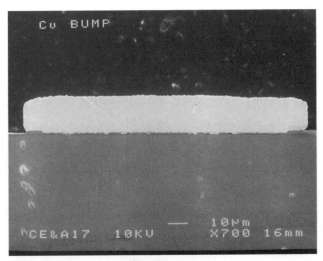

Figure 5.5 A typical Si chip with Cu bump.

Wafer bumping with Ni-Au. The Ni-Au bumps on the wafers were made by PICOPAK in Finland in 1995. The average bump height is 24 μm and the average shear strength is 80 g. PICOPAK's Ni-Au process in 1995 is shown in the following list.

1. Visual inspection of wafer

2. Test runs with diced wafer samples to find optimal process conditions

3. Application of photoresist to cover possible ink dots, undesired openings in dicing lanes, etc., using a standard photolithography process

4. Application of photoresist on back side of wafer

5. Plasma cleaning of exposed surfaces (<5 min at very low wafer temperature)

6. First zincate [~2 min at room temperature (RT)], rinse in DI water

7. Zinc strip (~1 min at RT), rinse in DI water

8. Second zincate (~2 min at RT), rinse in DI water

9. Nickel plating (~1 h at <100°C), rinse in DI water

10. Visual inspection of wafer

11. Bump height measurements

12. Immersion in gold (~15 min at <100°C), rinse in DI water

13. Removal of photoresist in hot (~100°C) acidic solution

14. Visual inspection of wafer

15. Bump shear test

16. Bump wetting test (not necessary for ACF applications)

17. Documentation and shipping of goods

Figure 5.6 shows a cross section of the electroless Ni-Au bump on the Si wafer. It should be noted that for such a large Ni-Au bump (24 μm), the Ni creates a large amount of stress that could crack the passivation (Figs. 5.7 through 5.11). Great care in monitoring of all process steps and

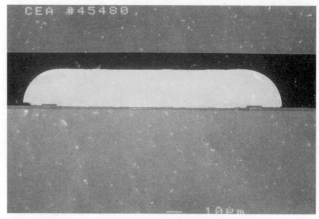

Figure 5.6   A typical Si chip with Ni-Au bump.

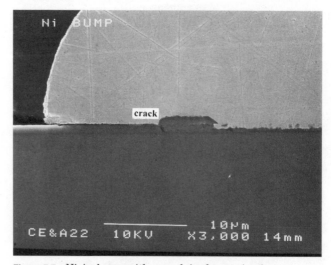

Figure 5.7   Ni-Au bump with a crack in the passivation.

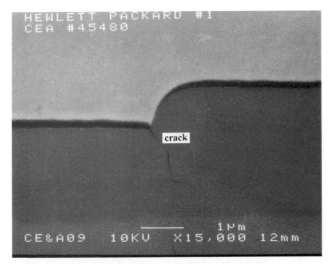

Figure 5.8 Ni-Au bump with a crack in the passivation.

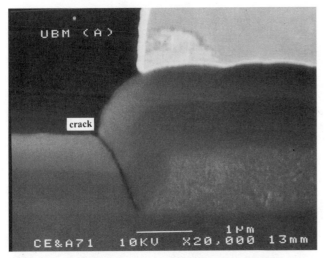

Figure 5.9 Ni-Au bump with a crack in the passivation.

tight contamination control of the Zn, Ni, and Au solution tanks must be taken in order to obtain the Ni-Au bump shown in Figs. 5.6 and 5.12.

5.2.3 PCB

A matching PCB is designed along with the chip. The Cu pads are round (8 mils or 0.2 mm in diameter) and in a 14-mil (0.36-mm) pitch. In this study, two Cu pad finishings, electroless Ni-Au and organic, are

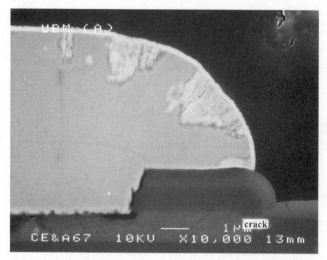

Figure 5.10   Ni-Au bump with a crack in the passivation.

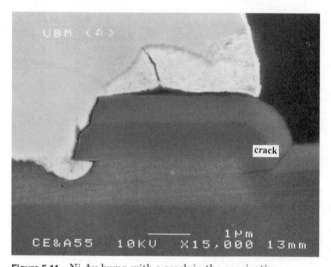

Figure 5.11   Ni-Au bump with a crack in the passivation.

considered (Fig. 5.2). Most of the pads are interconnected via traces on the PCB in an alternating pattern so as to provide daisy-chained connections with the chip.

5.2.4   ACF[7, 8]

The ACF used for this study is Hitachi Chemical's double-layer ACF (Fig. 5.13). It consists of nonfilled thermal-setting adhesive and thermal-

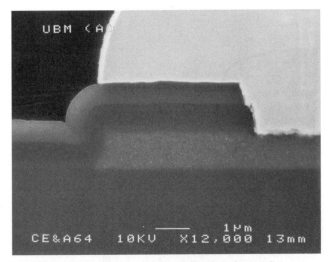

Figure 5.12 Ni-Au bump with no cracks (a perfect one!).

setting adhesive layers filled with Ni (2 to 5-μm) conducting particles. Each layer is about 30 μm thick. It has been shown by Hitachi Chemical that there are more conductive particles between the bumps of the chip and the pads of the substrate if the particle-filled thermal-setting adhesive layer in the ACF faces the pad of the substrate (Fig. 5.14).[7,8] For this study, the ACF is sandwiched between two layers of release paper.

5.2.5 FCOB assembly with ACF

More than 100 chips have been bonded on the Cu-Ni-Au and organic-coated copper (OCC) FR-4 PCB. These chips have three different kinds of bumps: 20-μm Cu studs, 20-μm Au studs, and 24-μm Ni-Au bumps.

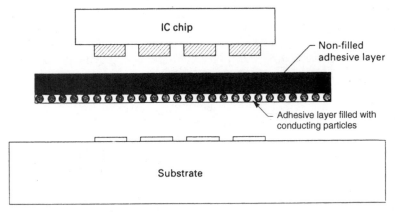

Figure 5.13 Hitachi's double-layer ACF.

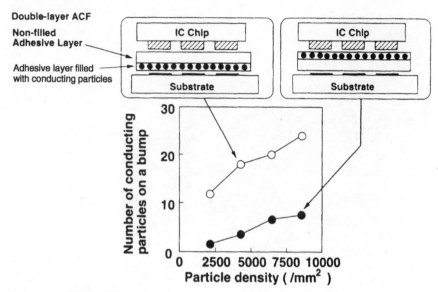

Figure 5.14   Effect of Hitachi's ACF direction on the number of conducting particles per bump.

The assembly process of the ACF is very simple and clean (fluxless). First of all, the ACF is cut to a little larger than the size of the chip and one of the release papers is removed. The ACF is then placed on the FR-4 PCB with the other release paper facing upward. This is called *tacking*.

The next step is to prepress the ACF under conditions of 80°C, 10 kg/cm², and 5 s. This is called *lamination*. In the next step, a lookup camera is used to read in some of the bumps of the chip. The release paper is removed, and then a lookdown camera is used to read in the corresponding pad locations of the PCB. After the necessary adjustment, the chip is placed on top of the ACF on the PCB. This is called *pick and place* and is done on Hitachi's aligning machine. Finally, the chip on board assembly is transported to the Hitachi bonder, which performs the *bonding* under conditions of 180°C, 5 kg/μm², and 20 s. Figure 5.15a shows a typical flip chip on PCB with ACF assembly. A typical cross section of the assembly is shown in Fig. 5.15b. There are many voids.

Figures 5.16 through 5.19 show the scanning electron microscopy (SEM) of a cross section of the ACF-bonded flip chip with Au bumps on PCB with Cu-Ni-Au pads. It can be seen that there are a few Ni conducting particles. Since the electroplated Au bump on the chip is softer than the electroless Au-Ni on the PCB, these Ni conducting particles penetrate more into the Au bump on the chip. One of the disadvantages of anisotropic conductive materials is the waste of conductive particles, as shown in Fig. 5.19.

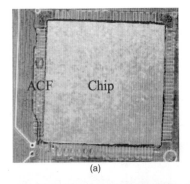

(a)

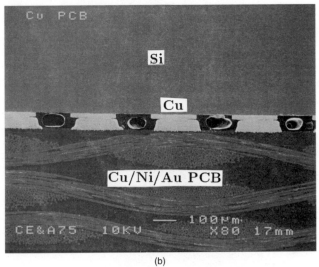

(b)

Figure 5.15   (*a*) An assembled FCOB with ACF. (*b*) A typical cross section of the ACF-bonded FCOB assembly.

Figure 5.20 shows the SEM of a cross section of the ACF-bonded flip chip with Cu bumps on PCB with Cu-Ni-Au pads. It can be seen that most of the Ni conducting particle penetrates into the electroplated Cu (with a coating of Sn) bump on the chip. Again, this is because the microhardness of the Au-Ni pads on the PCB is greater than that of the Sn-Cu bumps on the chip.

Figure 5.21 shows the SEM of a cross section of the ACF-bonded flip chip with Ni-Au bumps on PCB with Cu-Ni-Au pads. It can be seen that the penetration of the Ni conducting particles into both the Ni (with a coating of Au) bump on the chip and Au-Ni pads on the PCB is small. As a matter of fact, some of the Ni conducting particles have been badly deformed.

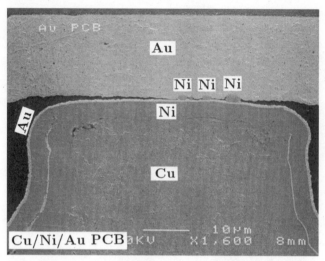

Figure 5.16   ACF-bonded flip chip with Au bump on Cu-Ni-Au PCB.

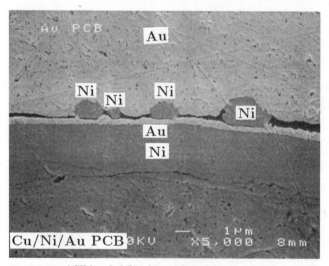

Figure 5.17   ACF-bonded flip chip with Au bump on Cu-Ni-Au PCB.

Figure 5.22 shows the SEM of a cross section of the ACF-bonded flip chip with Cu bumps on PCB with OCC pads. It can be seen that the Ni conducting particles penetrate into both the Cu (with a coating of Sn) bump on the chip and the OCC pad on the PCB.

In general, the ACF assembly yield is strongly affected by the kinds of bumps on the chip and the flatness of the PCBs. In our cases, chips

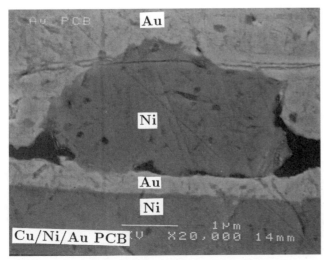

Figure 5.18 ACF-bonded flip chip with Au bump on Cu-Ni-Au PCB.

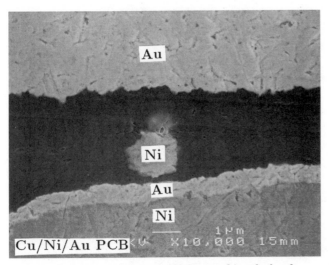

Figure 5.19 A Ni particle is wasted by not making the bond.

with Cu bumps have the highest yield and chips with Ni-Au bumps have the lowest yield. This could be due to the microhardness of the bumps, since the Cu bump is the smallest and the Ni-Au bump is the largest. Also, PCB with OCC pads leads to a better assembly yield than Cu-Ni-Au pads. This could be due to the less curved surface (greater flatness) of the OCC finishing.

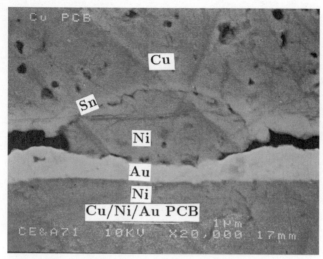

Figure 5.20   ACF-bonded flip chip with Cu bump on Cu-Ni-Au PCB.

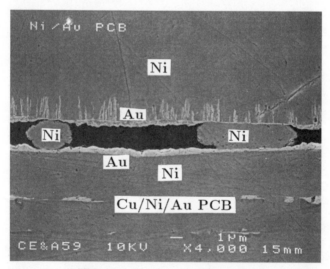

Figure 5.21   ACF-bonded flip chip with Ni-Au bump on Cu-Ni-Au PCB.

5.2.6   Thermal cycling test of FCOB assemblies with ACF[2]

Forty ACF-bonded flip chips (20 with Au bumps and 20 with Cu bumps) on Cu-Ni-Au PCB were subjected to thermal cycling. The temperature profile is shown in Fig. 5.23. Temperatures ranged from −20 to +110°C, and the cycle time was 1 h. The test results are shown in Fig. 5.24. It can be seen that after 1000 cycles, there was no opening in

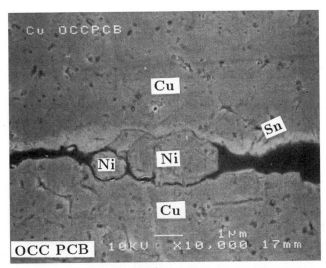

Figure 5.22   ACF-bonded flip chip with Cu bump on OCC PCB.

either case. However, the resistance of the FCOB assemblies with Cu bumps increased to about 29 percent. On the other hand, the increase in resistance of the FCOB assemblies with Au bumps was only 5.3 percent. This could be due to the greater degradation of the Cu-bumped FCOB than the Au-bumped FCOB assemblies after thermal cycling.

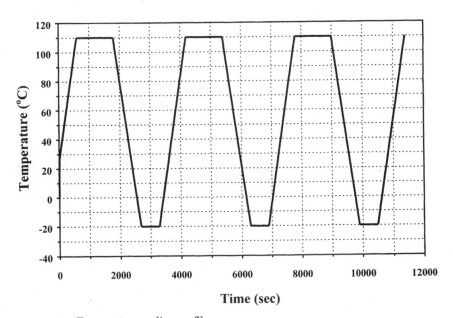

Figure 5.23   Temperature cycling profile.

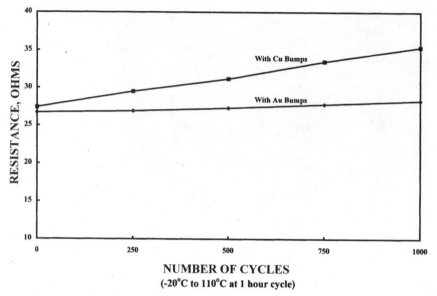

NUMBER OF CYCLES
(-20°C to 110°C at 1 hour cycle)

Figure 5.24   Thermal cycling results of ACF-bonded flip chip on Cu-Ni-Au PCB.

From the present results it is expected that the Au-bumped FCOB with ACF assemblies should have a better thermal fatigue life than the Cu-bumped FCOB with ACF assemblies.

5.2.7   SIR test results of FCOB assemblies with ACF

Surface insulation resistance testing is one of the most widely used techniques for assessing electrical performance reliability in electronic packaging. Leakage currents are monitored as a function of time at predetermined temperature, humidity, and bias voltage conditions. Leakage currents between closely spaced bumps/pads are sensitive indicators and are good signals of potential field risks. SIR values of 100 MΩ or higher are acceptable for commercial and industrial applications, and 500 MΩ is acceptable for military requirements. Figure 5.25 shows the SIR test (85°C/85%RH at 10-V bias) results for the ACF-bonded FCOB assemblies. It can be seen that ACF is acceptable for all the cases under consideration.

5.2.8   Summary

The Au-bumped, Cu-bumped, and Ni-Au-bumped flip chips have been assembled on Cu-Ni-Au and OCC PCBs with ACF. Important parameters and process steps such as the wafer; wafer bumping with Au, Cu, and Ni-Au; PCB; ACF; tacking; lamination; pick and place; and bonding have been discussed. Furthermore, the ACF-bonded FCOB assemblies

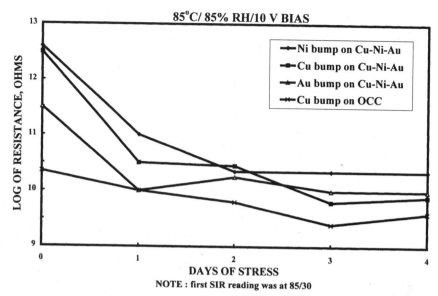

Figure 5.25   Surface insulation resistance (SIR) test results of ACF-bonded flip chip on PCBs.

have been subjected to thermal cycling and SIR tests. Some important results are summarized as follows.

1. The ACF-bonded FCOB assembly yield is strongly affected by the kind of bumps (Au, Cu, or Ni-Au) on the chip. Flip chip with Cu bumps has the highest assembly yield, and that with Ni-Au bumps has the lowest.

2. The ACF-bonded FCOB assembly yield is strongly affected by the flatness of the PCB. Also, in this study, PCB with OCC pads leads to a better assembly yield than PCB with Cu-Ni-Au pads.

3. There is no opening in the ACF-bonded flip chip on Cu-Ni-Au PCB assemblies after 1000 temperature cycles (−20 to +110°C). The resistance change due to 1000 temperature cycles of the Au-bumped flip chip assemblies (5.3 percent) is smaller than that of Cu-bumped flip chip assemblies (29 percent).

4. In this study, the ACF-bonded FCOB assemblies meet SIR test requirements for commercial, industrial, and military applications.

5.3   FCOB Assemblies with ACA[3]

In this section, materials, process, and reliability of flip chip on PCB assemblies with different ACAs are discussed. Most of the results presented herein are provided by the Technical University of Berlin, Center of Microperipherics Research, and Fraunhofer-Institute of Reliability and Microintegration.[3]

5.3.1 IC, PCB, and ACA materials

Five different ACA materials with different conductive particles are considered. Their actual curing conditions are determined by DSC and are shown in Table 5.1. Three different sizes of ICs with various daisy-chained I/Os are considered in Table 5.2. Two different bumps are considered, namely the electroless Ni-Au and Au (electroplated Au and Au stud), as shown in Table 5.2. The PCB is made of FR-4 and the Cu finishing is Ni-Au (3 μm-0.2 μm). The objective of the study is to compare the performance of different combinations of bumps and conductive particles as shown in Table 5.3.

Figures 5.26 and 5.27 show, respectively, the effect of prebaked PCB with adhesives A and B. The dark spots are voids. It can be seen that there is no void within the adhesive layer when prebaked substrates are used. (The gray treelike structures are particle sediments of filler particles used to enhance the mechanical properties of the cured adhesives.) As a matter of fact, increasing the preheat time leads to

TABLE 5.1 Properties of Processed Adhesives and Underfill Materials

Sample	Conductive filler particles	Recommended cure schedule	Investigated cure schedule
A	5.0 μm Ni	30 s @ 150°C	5 s @ 200°C
B	None	30 s @ 150°C	10 s @ 200°C
C	3.0 μm Au/Poly	30 s @ 150°C	10 s @ 200°C
D	3.7 μm Au/Poly	30 s @ 150°C	10 s @ 200°C
E	5.0 μm Ni	30 s @ 150°C	10 s @ 200°C

TABLE 5.2 Specifications of PCB and ICs

Substrate	
Material	FR-4
Size	2 × 2 in.
Electrode	Cu-Ni-Au: 17-3-0.2 μm
IC	
Size	5.0×5.0 mm^2
	7.5×7.5 mm^2
	10.0×10.0 mm^2
Pitch	200 μm, 300 μm
I/Os	136, 88, 184, 120
Bumps	Electroless Ni-Au:
	φ 75 μm, height 10 μm
	Electroplated Au:
	φ 75 μm, height 20 μm
	Stud Au

TABLE 5.3   Combinations of Bumps and Conductive
Particles Under Study

	A	B	C	D	E
Ni-Au		X	X	X	X
Stud Au	X	X			
Electroplated Au				X	X

decreased voiding. Thus, only preheated PCBs are used for further experiments.

5.3.2   FCOB with Ni-Au bumps and ACA

The flip chips with Ni-Au bumps are bonded on the PCB with forces between 50 and 150 g per bump. The contact resistance is measured and is shown in Table 5.4. Since there is no conductive particle with adhesive B, the bonds are all open for low pressure (50 g per bump). However, for greater pressures (100 and 150 g per bump), contact resistance of 4 mΩ as well as open contacts are observed. Figure 5.28 shows the cross sections of contacts for two different applied forces. It can be

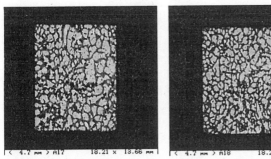

Figure 5.26   C-mode images of die bonded on the original PCB (*left*) and prebaked PCB (*right*) with adhesive A.

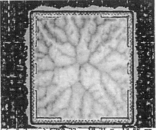

Figure 5.27   C-mode images of die bonded onto original PCB (*left*) and prebaked PCB (*right*) with adhesive B.

TABLE 5.4   Influence of Bond Force on Contact Resistance

Conductive particles	Bond force (g/bump)	Contact resistance Average (mΩ)	Range (mΩ)
B None	50	All open	—
	100	50% open	—
	150	Some open	—
C 3.0-μm Au-Poly	50	80	50
	70	60	45
	100	50	20
D 3.7-μm Au-Poly	50	20	10
	70	65	22
E 5.0-μm Ni	50	60	27
	100	20	5

seen on the left (50 g per bump) that because of some nonconductive fillers within the adhesive, it is difficult to have a good contact. On the other hand, the contact pads on the right (150 g per bump) are pushed downward into the substrate matrix by bending of the electrodes. The bending for this applied force is too large, which may lead to electrode cracking and die touching. Thus, bonding forces of no more than 100 g per bump will be used for the next experiments.

Figure 5.29a shows the cross section of a flip chip assembly with adhesive C (5.0-μm Au-poly) with a bond force of 50 g per bump. It can be seen that the particle deformation is not sufficient to ensure a good contact. However, with a bond force of 100 g per bump (Fig. 5.29b), it can be seen that the particles are partially crushed and the conductance is provided by direct contact between the bump and pad. For adhesive D, which has larger conductive Au/Poly particles (5.7 μm), there is a good contact resistance of 20 mΩ after bonding at 50 g per bump (Table 5.4). However, for a bond force of 70 g/bump, the contact

Figure 5.28   Bending of bond pads at bonding force of 50 g per bump (left) and 150 g per bump (right).

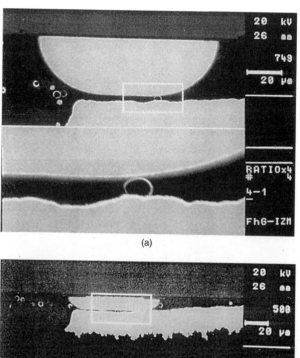

(a)

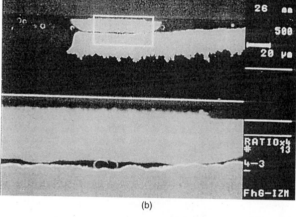

(b)

Figure 5.29 (*a*) Particle deformation at bonding force of 50 g per bump (adhesive C). (*b*) Particle deformation and direct contact at bonding force of 100 g per bump (adhesive C).

resistance is in the range of 65 mΩ. This means that the bond quality has been decreased by crushing the particles and forming a direct contact between the hard bump and the electrode.

As for the 5.0-μm Ni conductive particles in adhesive E, a bond force of 50 g per bump leads to a high contact resistance with large deviations. However, a bond force of 100 g per bump lowers the contact resistance to 20 mΩ with the smallest deviations of all. This is due to the large particle size, which ensures a good contact over the entire bump surface even with bent and sunk electrodes.

5.3.3   FCOB with Au bumps and ACA

Figure 5.30 shows the cross section of a Au-bumped flip chip on board with Au/polymer-filled adhesive D (bonding force of 50 g per bump). It can be seen that there are two current paths, via particles and directly between bump and electrode, because the electroplated gold bump itself is deformed and forms a direct contact with the PCB electrode. Also, the gold-coated polymer particles have not only been deformed but have penetrated into the Au bump.

Figure 5.31 shows the cross section of a Au-bumped flip chip on board with 5.0-μm Ni conductive particles of adhesive E (bonding force of 50 g per bump). It can be seen that the Ni particles penetrate into the Au bump. Also, the electrodes are bent where they are supported by the glass fibers.

5.3.4   Accelerated aging test and results

Table 5.5 shows the bonding forces and curing schedules for the samples subjected to the following accelerated aging tests: (1) high-temperature storage at 125°C, (2) temperature cycling at –55 to +125°C, (3) humidity storage at 85°C/85% RH, and (4) reflow test after humid-

Figure 5.30 Interconnection using Au-coated polymer-filled ACA (adhesive D).

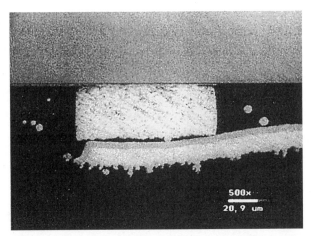

Figure 5.31   Interconnection using Ni-filled ACA (adhesive E).

ity storage. Table 5.6 shows the test results. A failed contact is defined as a complete opening or a 10-fold increase in initial contact resistance. Figure 5.32 (humidity storage test) shows that Ni-filled adhesive E performs the worst. This could be due to the corrosion of the Ni particles.

Figure 5.33 (temperature cycling test) shows that most of the failures during temperature cycling tests are observed with the Ni-filled adhesive and electroplated Au bumps. The adhesive with the best performance is filled with 5.7-μm Au-coated polymer (adhesive D) with Ni-Au bumps and electroplated Au bumps.

5.3.5   Summary[3]

1. Unlike on polyimide, ACA materials may form voids on PCB. Either prebaking of the PCB or preheating of the ACA materials could reduce voiding to zero.

TABLE 5.5   Sample Preparation for Reliability Tests

Adhesive	Bumps	Bond force	Cure schedule
A	Au stud	70 g/bump	5 s @ 200°C
B	Au stud	70 g/bump	10 s @ 200°C
D	Ni-Au	50 g/bump	10 s @ 200°C
D	Electroplated Au	50 g/bump	10 s @ 200°C
E	Ni-Au	100 g/bump	10 s @ 200°C
E	Electroplated Au	50 g/bump	10 s @ 200°C

TABLE 5.6   Results of Reliability Tests

Tested combination	Results after 1000 h high-temperature storage	Results after 1000 thermal cycles	Results after 1000 h humidity storage
A (stud Au bumps)	Not tested	15% failed	Not tested
B (stud Au bumps)	No increase in contact resistance	No increase in contact resistance	No increase in contact resistance
D (electro-plated Au bumps)	No increase in contact resistance	No increase in contact resistance	36% failed
D (Ni-Au bumps)	No increase in contact resistance	No increase in contact resistance	2× increase in contact resistance
E (electro-plated Au bumps)	No increase in contact resistance	22% failed	100% failed
E (Ni-Au bumps)	No increase in contact resistance	No increase in contact resistance	58% failed

2. Bond forces have to be adjusted carefully according to bump metallurgies and type of conductive particles. Excess bond forces may lead to badly deformed electrodes.

3. Reliability of flip chip on board strongly depends on the bump metallurgies and kind of conductive particles. The combination of hard Ni-Au bumps and soft Au-coated polymer particles performs the best through all tests.

4. The combination of soft electroplated Au bumps and large (5.0-μm) Ni-filled adhesive performs the worst.

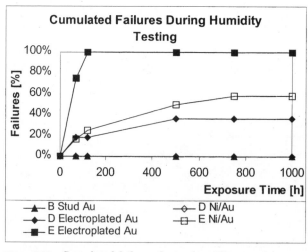

Figure 5.32   Cumulated failures during humidity storage test.

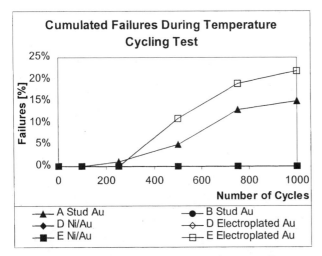

Figure 5.33   Cumulated failures during temperature cycling test.

Acknowledgments

The author would like to thank the management (especially Anita Danford) of Hewlett-Packard for their strong support of the low-cost flip chip program. Thanks are also due to R. Miebner, R. Aschenbrenner, and H. Reichl for providing the industry so much useful ACA information, and to the friendly people at PICOPAK for making the Ni-Au bumps. Finally, the author thanks the very intelligent and hardworking team (especially Dr. Itsuo Watanabe) of Hitachi Chemical for constructive contributions.

References

1. Lau, J. H., C. P. Wong, J. L. Prince, and W. Nakayama, *Electronic Packaging: Design, Materials, Process, and Reliability,* McGraw-Hill, New York, 1998.
2. Lau, J. H., "Flip Chip on PCBs with Anisotropic Conductive Film," *Adv. Packaging,* 44–48, July/August 1998.
3. Miebner, R., R. Aschenbrenner, and H. Reichl, "Reliability Study of Flip Chip on FR4 Interconnections with ACA," *Proceedings of IEEE Electronic Components and Technology Conference,* pp. 595–601, June 1999.
4. Lau, J. H., *Flip Chip Technologies,* McGraw-Hill, New York, 1996.
5. Lau, J. H., *Chip on Board Technologies for Multichip Modules,* Van Nostrand Reinhold, New York, 1994.
6. Gustafsson, K., S. Mannan, J. Liu, Z. Lai, D. Whalley, and D. Williams, "The Effect of Temperature Ramp Rate on Flip-Chip Joint Quality and Reliability Using Anisotropically Conductive Adhesive on FR-4 Substrate," *IEEE/ECTC Proceedings,* pp. 561–566, May 1997.
7. Watanabe, I., K. Takemura, N. Shiozawa, O. Watanabe, K. Kojima, A. Nagai, and T. Tanaka, "Anisotropic Conductive Adhesive Films for Flip-Chip Interconnection," *Proceedings of the 9th International Microelectronics Conference,* pp. 328–332, Omiya, Japan, 1996.

8. Watanabe, I., N. Shiozawa, K. Takemura, and T. Ohta, "Flip Chip Interconnection Technology Using Anisotropic Conductive Adhesive Films," in *Flip Chip Technologies,* Lau, J. H., ed., McGraw-Hill, New York, pp. 301–315, 1996.
9. Aschenbrenner, R., R. Miebner, and H. Reichl, "Adhesive Flip Chip Bonding on Flexible Substrates," *Proceedings of the IEEE Polymeric Electronics Packaging,* pp. 86–94, October 1997.
10. Wong, C. P., D. Lu, L. Meyers, S. Vona, and Q. Tong, "Fundamental Study of Electrically Conductive Adhesives (ECAs)," *Proceedings of the IEEE Polymeric Electronics Packaging,* pp. 80–85, October 1997.
11. Lu, D., C. P. Wong, and Q. Tong, "Mechanism Underlying the Unstable Contact Resistance of Conductive Adhesives," *Proceedings of IEEE Electronic Components and Technology Conference,* pp. 342–346, June 1999.
12. Nguyen, G., J. Williams, F. Gibson, and T. Winster, "Electrical Reliability of Conductive Adhesives for Surface Mount Applications," *Proceedings of International Electronic Packaging Conference,* pp. 479–486, 1993.
13. Nguyen, G., J. Williams, and F. Gibson, "Conductive Adhesives: Reliable and Economical Alternatives to Solder Paste for Electrical Applications," *Proceedings of ISHM Symposium,* pp. 510–517, 1992.
14. Li, L., and J. Morris, "Reliability and Failure Mechanism of Isotropically Conductive Adhesive Joints," *Proceedings of IEEE Electronic Components and Technology Conference,* pp. 114–120, May 1995.
15. Chung, K., T. Devereaux, C. Monti, and M. Yan, "Z-Axis Conductive Adhesives as Solder Replacement," *Proceedings of International SAMPE Electronic Conference,* vol. 7, pp. 473–481, 1994.
16. Yamaguchi, M., F. Asai, F. Eriguchi, and Y. Hotta, "Development of Novel Anisotropic Conductive Film (ACF)," *Proceedings of IEEE Electronic Components and Technology Conference,* pp. 360–364, June 1999.
17. Hotta, Y., "Development of 0.025 mm Pitch Anisotropic Conductive Film," *Proceedings of IEEE Electronic Components and Technology Conference,* pp. 1042–1046, June 1998.
18. Dernevik, M., R. Sihlbom, K. Axelsson, Z. Lai, J. Liu, and P. Starski, "Electrically Conductive Adhesives at Microwave Frequencies," *Proceedings of IEEE Electronic Components and Technology Conference,* pp. 1026–1030, June 1998.
19. Kang, S. K., and S. Purushothaman, "Development of Low Cost, Low Temperature Conductive Adhesives," *Proceedings of IEEE Electronic Components and Technology Conference,* pp. 1031–1035, June 1998.
20. Yim, M., K. Paik, T. Kim, and Y. Kim, "Anisotropic Conductive Film (ACF) Interconnection for Display Packaging Applications," *Proceedings of IEEE Electronic Components and Technology Conference,* pp. 1036–1041, June 1998.
21. Wei, Y., and E. Sancaktar, "A Pressure Dependent Conduction Model for Electrically Conductive Adhesives," *Proceedings of International Symposium on Microelectronics,* pp. 231–236, 1955.
22. Liu, J., and R. Rorgren, "Joining of Displays Using Thermosetting Anisotropically Conductive Adhesives," *J. Electronics Manufacturing,* **3:** 205–214, 1993.
23. Ito, S., M. Mizutani, H. Noro, M. Kuwamura, and A. Prabhu, "A Novel Flip Chip Technology Using Non-Conductive Resin Sheet," *Proceedings of IEEE Electronic Components and Technology Conference,* pp. 1047–1051, June 1998.

Flip Chip on Board with Conventional Underfills

6.1 Introduction

Flip chip is defined as a chip mounted on a substrate with various interconnect materials and methods such as TAB, wire interconnects, isotropic and anisotropic conductive adhesives, metal bumps, compliant bumps, fluxless solder bumps, and pressure contacts, as long as the chip surface (active area or I/O side) is facing the substrate.[1] Compared to the face-up chip mounting configurations, flip chip provides the shortest possible leads, the lowest inductance, the highest frequencies, the best noise control, the highest density, the greatest number of I/Os, the smallest device footprints, and the lowest profile.[1–239] In this chapter, the focus is on flip chips with solder-bump materials.

The solder-bumped flip chip was introduced by IBM in the early 1960s for its solid logic technology (SLT), which became the logical foundation of the IBM 360/System computer line. The so-called C4 (controlled-collapse chip connection) technology utilizes solder bumps deposited on wetable metal terminals on the chip and a matching footprint of solder wetable terminals on the substrate. The solder-bumped flip chip is aligned to the substrate, and the solder joints are made simultaneously by reflowing the solder bumps (Fig. 6.1).

In this book, the solders on the chip before it is joined to the substrate are called *solder bumps*. After the solder bumps have been reflowed on the substrate, they are called *solder joints*.

For its C4 bumps, IBM used the 5wt%Sn-95wt%Pb solder alloy, which has solid and liquid temperatures of 308 and 312°C, respectively. Sometimes, IBM also used the 3wt%Sn-97wt%Pb solder alloy, which

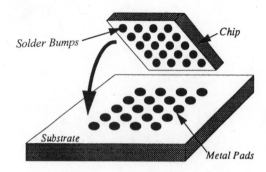

Figure 6.1   Solder-bumped flip chip technology.

has solid and liquid temperatures of 314 and 320°C, respectively. The substrate materials IBM used were usually ceramics, which have a thermal expansion coefficient of about 4 to 6 × 10⁻⁶/°C. During reflow, because of the surface tension of the molten solder, the chip will self-align to the substrate; this leads to a very high-yield manufacturing process (Fig. 6.2). For large chips, sometimes underfill encapsulant is used to enhance solder joint reliability.

6.2   FCOB with High-Temperature Solder Bumps

IBM at Yasu, Japan, has been assembling solder-bumped flip chip on organic PCB since 1990[33–35] (Fig. 6.3). Applications include personal computers, personal computer memory card international association (PCMCIA) cards, and Token Ring local area network (LAN) adapter cards. IBM's results show that solder-bumped flip chip assembly technology is applicable for low-cost PCB as well as high-cost substrates such as ceramics.

The DCA technology of IBM at Yasu is based on IBM's C4 technology (Figs. 6.3 and 6.4). That means the melting temperatures of the C4 solder bumps are higher than 310°C. In order to be SMT compatible (at a maximum reflow temperature less than 240°C), the copper pads on the

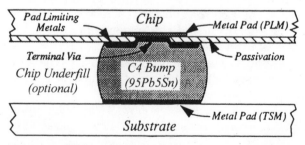

Figure 6.2   IBM's controlled-collapse chip connection (C4) technology.

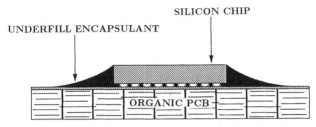

Figure 6.3 Low-cost solder-bumped flip chip on board.

PCB are coated with 63wt%Sn-37wt%Pb solder alloy, like those shown in Fig. 2.72, using conventional electroplating method or IBM at Yasu's molten solder injection machine, as shown in Fig. 2.45 (replacing the wafer with PCB). A typical cross section of the FCOB assembly is shown in Fig. 6.5. It should be noted that the high-temperature C4 solder bump is not reflowed.

In order to reduce the cost of substrate even further, Motorola[72] developed E3 solder bumps (Sec. 2.3.1; Chap. 6 of Lau[6]) for FCOB assembly

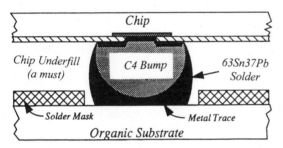

Figure 6.4 Direct chip attach with IBM's C4 technology.

Figure 6.5 Cross section of IBM's low-cost FCOB.

applications. In that process, the unreflowed E3 solder bumps are placed on the OCC PCB and mass-reflowed with other SMT components.

6.3   FCOB with Low-Temperature Solder Bumps

Unlike IBM and Motorola, which use composite solder interconnects (high-lead solder bump with eutectic solder) and the evaporation method, Oki, Sharp, Hewlett-Packard, Toshiba, and many other companies use the 63Sn-37Pb eutectic solder bumps as shown in Fig. 6.6. The substrates for such non-C4 solder-bumped flip chips can be standard FR-4 or bismaleimide triazine (BT) with either Cu-Ni-Au, Sn-Pb, or OCC PCBs. These companies do not use the E-3 process or coat the PCB with eutectic solder; rather, they treat the eutectic solder-bumped chips like other SMT components, which can be mass-reflowed on the low-cost PCBs with all the other SMT components.

The non-C4 solder-bumped flip chip assemblies perform very well under thermal cycling tests. For example, Sharp's wafers are bumped with 60Sn-40Pb solder (Fig. 6.7) by electroplating. That company's substrates for the solder-bumped flip chips are standard FR-4 PCBs with a Cu-Ni-Au metal finish (Fig. 6.8). Sharp conducts thermal cycling tests (–45 to +100°C and 1 cycle per hour) of five different square flip chip assemblies (3, 6, 9, 12, and 15 mm) with underfill; no failures are observed at 1000 cycles. For the smaller flip chip assemblies (3, 6, and 9 mm) with underfill, no failures are observed even after 9000 cycles, as shown in Fig. 6.9.[65, 66] However, without underfill, even the smaller chips failed before 300 cycles.

Another set of thermal cycling tests is performed on the chip shown in Fig. 6.10. It can be seen that the chip dimensions are 6.7 × 6.5 mm for a single and ~14 × 14 mm for a quad. The chips are bumped with 63Sn-37Pb solder using the electroplating method. The bump height is about 100 μm. The chips are assembled on the OCC PCB with a thickness of 1.57 mm. The underfill materials used are the CNB11 and 4511, provided by HYSOL.

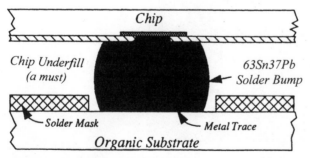

Figure 6.6   Low-cost solder-bumped flip chip on board with 63Sn-37Pb solder alloy.

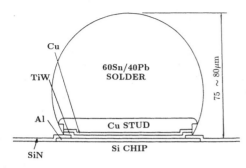

Figure 6.7   Sharp's eutectic solder bump.

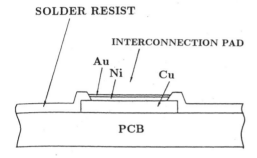

Figure 6.8   Sharp's low-cost PCB.

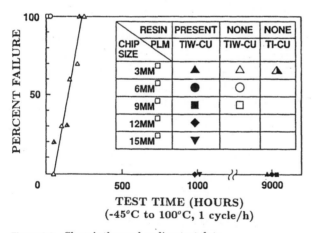

TEST TIME (HOURS)
(-45°C to 100°C, 1 cycle/h)

Figure 6.9   Sharp's thermal cycling test data.

The temperature loading imposed on the flip chip assemblies is shown in Fig. 6.11. It can be seen that for each cycle (60 min) the temperature is between −20 and +110°C, with 15 minutes ramp, 20 minutes hold at hot, and 10 minutes hold at cold. There are two reasons for choosing this temperature profile: (1) the glass transition temperature of the PCB is 120°C and we don't want to introduce additional failure mechanisms of the solder joint due to the degradation of the PCB; and

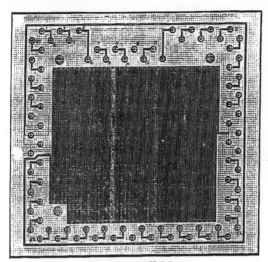

Chip size = 6.7 X 6.5 mm
Number of bumps = 84
Bump diameter = 150 μm
Bump pitch = 250 μm

Figure 6.10   A very useful test chip.

(2) the behavior of solder below –20°C is not very well understood. The test is run nonstop for more than 6 months.

The thermal cycling test results represented by the Weibull life distribution are shown in Figs. 6.12 and 6.13, respectively, for the single and quad flip chip assemblies. It can be seen that: (1) as expected, the smaller chip performs better than the larger chip, but not by very much, and (2) the solder joints in both cases are reliable for use in most operating conditions.

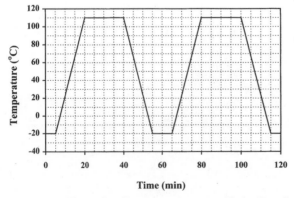

Figure 6.11   Thermal cycling temperature profile (on board).

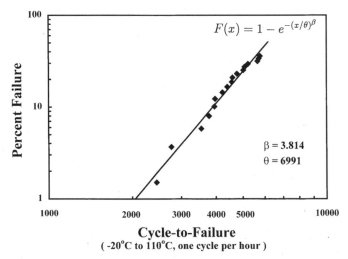

Figure 6.12   Solder joint Weibull life distribution of the single test chip.

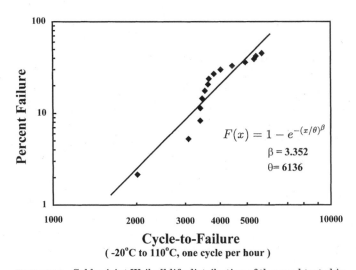

Figure 6.13   Solder joint Weibull life distribution of the quad test chip.

6.4   Most Desirable Features of Underfills[187]

One of the major reasons why solder-bumped flip chip on low-cost PCB works is because of the underfill encapsulant. It reduces the effect of the global thermal expansion mismatch between the silicon chip ($2.5 \times 10^{-6}/°C$) and the organic FR-4 PCB ($18.5 \times 10^{-6}/°C$). (Since the chip, underfill, and PCB are deformed together as a unit—i.e., the relative deformation between the chip and the PCB is very small—the shear deformation of the solder joint is very small.) The other advantages of underfill encapsulant are that it protects the chip from moisture, ionic contaminants, radiation, and hostile operating environments such as thermal and mechanical conditions, shock, and vibration. Also, without heat sink, underfill is usually the major heat path from the chip to the PCB.

The most desired features of underfill materials are:[187]

- Low viscosity (fast flow), which can increase throughput
- Low curing temperature/fast curing time, which can reduce cost and be less harmful to other components
- Low TCE, which can reduce thermal expansion mismatch between the chip, solder bumps, and PCB
- High modulus, which leads to good mechanical properties
- High glass transition temperature, which enables endurance of higher temperature environments
- Low moisture absorption, which can extend shelf life
- Good adhesion, which can improve product life time

Some companies that provide various underfill materials are Ablestik, Alpha Metal, Dexter, Emersion, Epoxy Technology, HYSOL, Johnson Matthey, Kester, Loctite, Matsushita, Thermoset, and Zymet. It should be emphasized that the curing conditions and material properties of the underfill materials provided by the vendors are for reference purposes only. For good engineering, the end users should verify the data through measurements.

6.5 Handling and Application of Underfills

Usually, underfill materials are premixed at the supplier sites, packed in plastic syringes (5 to 10 cc), and then frozen packed at $-40°C$ to prevent curing. Shipping of these underfill materials requires special handling to maintain the low temperature continuously. When the package is received, it should be unwrapped and the syringes should be quickly removed and stored in an uninterrupted temperature-controlled ($-40°C$) freezer. In this case, most of the underfill encapsulants would have approximately 1 year of storage life.

Before the underfill materials are applied, they are removed from the freezer and thawed at room temperature. It takes about 1 h to thaw a 5- to 10-cc syringe. Once thawed, most of the underfill materials have a pot life of 14 to 16 h.

Dispensing the underfill encapsulants through a syringe is accomplished with a version system (to locate the edges of chip) and a pumping system (to control the amount of underfill materials). The temperature of the heating plate is around 90°C. After the underfill materials are dispensed on either one side or two adjacent sides of the chip, the chip and substrate should remain on the heating plate for about 30 to 90 s (depending on the chip size and the gap between the chip and the substrate) to enhance the flow rate of the underfill. After the underfill has flowed out from the other sides of the chip (sometimes touch-up is necessary), the chip on board will be put into a curing oven for 10 to 60 min at 130 to 170°C depending on the curing conditions of the underfill materials.

6.6 Curing Conditions of Underfills

Ten different underfill materials with different sizes and filler contents (underfills A through J) are under consideration, as shown in Table 6.1. For all these materials, the filler is silica. Most components of the underfills are bis-phenol-type epoxies and trade secret resins.

To determine the curing conditions of underfill materials, they are put into an aluminum pan (which will form a disc sample 6.4 ± 0.2 mm in diameter and 1.6 ± 0.1 mm in height), weighed, and then put into DSC equipment. Since the system is always directly measuring energy flow to or from the sample, the DSC can directly measure temperature at onset of curing. The kinetic software enables analysis of a DSC peak to obtain specific kinetic parameters that characterize a reaction process.

Figure 6.14 shows the typical DSC thermal scan curve. It can be seen that the peak curing temperatures for underfills B and F are 135.2 and 141.1°C, respectively. The exothermic peak of curve B is wider than that of curve F, which is very sharp. This means that underfill F has to be cured within a narrow temperature range while underfill B can be cured along a wide temperature range. Also, underfill B needs more time to be fully cured than underfill F (10 min at 150°C for underfill B, and 5 min at 150°C for underfill F). In general, lower temperatures and

TABLE 6.1   Average Curing Conditions for Underfills A Through J

Underfill	Filler Content and Size	Curing Temperature (°C)	Curing Conditions [temperature (°C)/time (min)]				Temperature (°C) at 15 minutes of Curing
A	40% < 10 μm	136.2	150 / 27	160 / 10	165 / 7	170 / 4	155.8
B	60% < 10 μm	135.2	150 / 10	160 / 4	165 / 2	170 / 1.2	146.3
C	40% < 10 μm	117	150 / 8	160 / 5	165 / 4	170 / 3	139.9
D	60% < 10 μm	143.3	165 / 83	170 / 38	180 / 9	185 / 4	175.4
E	60% < 10 μm	142.9	165 / 693	170 / 268	180 / 43	185 / 18	188.3
F	50% < 20 μm	141.1	150 / 5	152 / 3	154 / 2	156 / 1	146.4
G	60% < 20 μm	137.1	150 / 297	160 / 54	165 / 24	170 / 11	167.9
H	55% < 20 μm	141.9	150 / 186	160 / 60	165 / 35	170 / 21	172.9
I	60% < 20 μm	153.6	165 / 182	170 / 112	180 / 44	185 / 28	191.7
J	60% < 10 μm	139.7	150 / 169	160 / 60	165 / 36	170 / 22	173.3

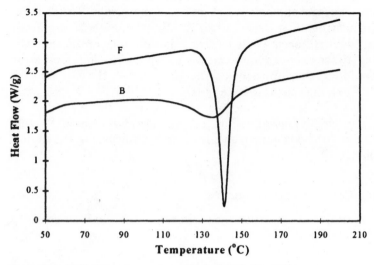

Figure 6.14   DSC thermal scan curve for underfills B and F.

longer times for full curing of underfill materials will lead to lower thermal stresses and better physical and mechanical properties. This could be due to the fact that the molecular chains have more time to reorganize and pack well.

Figure 6.15 shows the curing temperature–versus–time curves. It can be seen that underfill D cured faster than underfill E. Table 6.1 summarizes the average curing conditions for underfills A through J. It can be seen that the curing conditions varied with different underfill materials and that materials cured faster at higher temperatures.

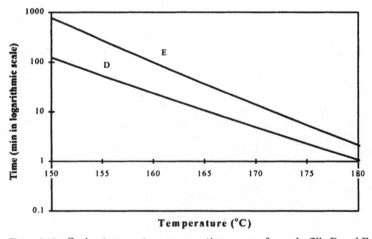

Figure 6.15   Curing temperature–versus–time curves for underfills D and E.

Also, underfills B, C, and F cured faster than the others. Furthermore, the temperatures of these three materials (less than 156°C) at 15 min of curing are lower than those of the others. For most of the manufacturing environments, underfill materials that can be fully cured at 150°C within 15 min are acceptable.

6.7 Material Properties of Underfills

The values for TCE, storage modulus, tangent delta, glass transition temperature, moisture content, and flow rate of underfills A through J are obtained in this section. Also, the Young's modulus, stress-strain relations, and creep curves are provided. In addition, the fracture toughness of underfills, of underfill-FR-4 interfaces, and of underfill-passivation interfaces are presented.

6.7.1 TCE

The TCE values for underfills A through J (with sample dimensions of 6.4 ± 0.2 mm in diameter and 1.6 ± 0.1 mm in height) are determined by TMA in an expansion quartz system (50 to 200°C) at a heating rate of 5°C/min. The TCE of the material is obtained by the slope of the dimension change–versus–temperature curve, and T_g can be obtained from the onsets of the two different slopes of the curve.

Figure 6.16 shows typical expansion curves for underfills C and H. Table 6.2 summarizes the average results for underfills A through J. It

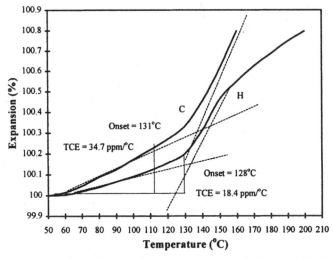

Figure 6.16 Typical expansion curves for underfills C and H for determining TCE.

TABLE 6.2   Average Material Properties of Underfills A Through J

Underfill	Filler Content and Size	TCE (ppm/ °C)	T_g (°C)	Width of T_g peak (°C)	Storage Modulus (GPa) 25°C	55°C	110°C	Percent Drop from 55 to 110°C
A	40% < 10 μm	37	149	18	4.62	4.56	4.22	7.5
B	60% < 10 μm	29	162	19	4.88	4.87	4.74	2.7
C	40% < 10 μm	34.7	140	15.8	3.4	3.3	2.93	11.2
D	60% < 10 μm	29.8	170	24.7	3.73	3.69	3.44	6.8
E	60% < 10 μm	32.5	124	19.5	3.78	3.76	2.7	28.2
F	50% < 20 μm	32.5	134	58	4.4	4.34	2.45	44
G	60% < 20 μm	17.7	155	26	4.5	4.5	4.12	8.4
H	55% < 20 μm	18.4	142	43.2	4.5	4.45	3.69	17.1
I	60% < 20 μm	16.7	159	36.9	2.96	2.96	2.67	9.8
J	60% < 10 μm	14.6	127	37	4.0	3.96	2.5	36.9

can be seen that the TCE of underfill A (40 percent filler content) is the largest (37×10^{-6}/°C) and the TCE of underfill J (60 percent filler content) is the smallest (14.6×10^{-6}/°C). The T_g value obtained using the TMA method is usually about 10°C lower than that obtained with the DMA method, which will be discussed in Secs. 6.7.2 and 6.7.3.

In general, the higher the filler content, the lower the TCE value. This is because of the fact that the filler (made from silica) expands less than the epoxy resin. From solder joint thermal fatigue reliability points of view, it is preferable to use underfill materials having lower TCEs ($<27 \times 10^{-6}$/°C) to reduce the thermal expansion mismatch between the chip, the solder joints, and the low-cost substrate.

6.7.2   Storage modulus

The modulus values for underfills A through J are measured with a three-point bending specimen (3.0 ± 0.3 mm $\times$ 2.9 ± 0.3 mm $\times$ 19 ± 3 mm) in a DMA unit (50 to 200°C) at a heating rate of 5°C/min. For electronics packaging materials, flexural properties such as the storage modulus, loss modulus, and tangent delta (tan δ) can be obtained with DMA. In material sciences, the flexural storage modulus E_s is a measure of the energy stored per cycle of deformation.

Figure 6.17 shows typical dynamic flexural storage modulus curves for underfills B and F as a function of temperature. The stiffness of underfill B at low temperatures (less than 100°C) is temperature independent, while the stiffness of underfill F is very poor, dropping quickly when the temperature increases. The average storage modulus results for underfills A through J are reported in Table 6.2. It can be seen that the storage modulus of all the underfill materials is temperature dependent, i.e., the higher the temperature, the lower the modulus. The

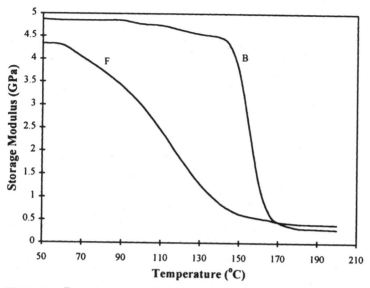

Figure 6.17   Typical storage modulus curves for underfills B and F.

percent drop in modulus from 55 to 110°C, shown in the last column of Table 6.2, can be as high as 44 percent for some underfill materials.

6.7.3   Tan δ and T_g

Mechanical energy supplied at the temperature and frequency corresponding to the transition will be absorbed by the material, then partially stored elastically and partially dissipated in the form of heat. Tangent delta (tan δ) (a measure of material-related damping properties) for underfills A and J is shown in Fig. 6.18. The temperature at the peak of a tangent delta curve is often reported in the literature as glass transition temperature (T_g).

The damping is the ratio of heat dissipation to energy storage, and is dependent on the temperature and frequency. Near the glass transition temperature, polymers display maximum values of loss modulus and tan δ, which are measures of energy dissipation—or, in other words, their "lossiness" is at a maximum. The wider range of tan δ may involve friction between fillers or between fillers and polymer and may provide additional means of dissipating energy. Contributions from the inhomogeneous distribution of fillers and from thermal stresses may also affect damping properties. In this study, it can be seen from Fig. 6.18 that curve A is sharper than curve J (that is, the transition temperature range is smaller), indicating that the stiffness of underfill A is better than that of underfill J. The underfills will become softer during a

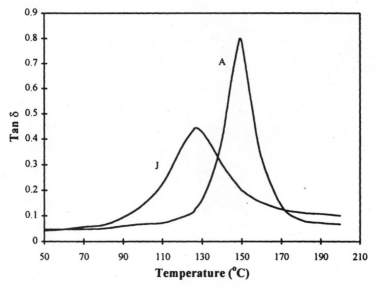

Figure 6.18   Typical tangent delta curves for underfills A and J.

wider transition temperature range. The average values of T_g for all the underfills are shown in Table 6.2.

6.7.4 Moisture content

Porosity exists in materials due to trapped air and outgasing, and is responsible for the absorption of moisture. The pores serve to concentrate stresses that may cause further growth of cracks when external loads or thermal stresses are applied. In general, the more porosity, the more moisture absorption. Three sets of tests are carried out to determine the moisture content. One is for dry specimens, the second is for moisture-soaked specimens, and the third is for steam-aged specimens. The moisture-soaked specimen is subjected to 85°C and 85 percent relative humidity for 168 h. The steam-aged specimen is prepared using steam evaporation for 20 h in a closed hot-water bath. All the specimens are 6.4 ± 0.2 mm in diameter and 1.6 ± 0.1 mm in height. Weight changes for underfills A through J are measured with the TGA equipment at 102°C for 2 h.

The change in mass during thermal scan can be expressed as $(W_f - W_i)/W_i$, where W_f is the final weight after thermal scan and W_i is the initial weight before thermal scan. Figure 6.19 shows typical percent weight change ratio (moisture content) for underfill E before and after moisture soak and after 20 h of steam aging. The average moisture contents of all the underfills are shown in Table 6.3. It can be seen that the average moisture content of all the underfills under moisture-soak con-

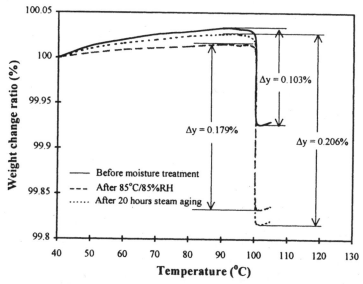

Figure 6.19 Typical weight change (moisture content) for underfill E before and after moisture soak and after 20 hours of steam aging.

ditions is very close to that after 20 h of steam aging. Based on this study, it is recommended that the acceptable moisture absorption for dry underfill materials is 0.1 percent and that for moisture-soaked underfill materials is 0.2 percent.

6.7.5 Young's modulus[189, 223]

The Young's modulus of the HYSOL FP4526 underfill has been determined by researchers at Wayne State University.[189, 223] The glass transi-

TABLE 6.3 Average Flow Rates for FCOB with Underfills A Through J and Moisture Content

Underfill	Filler Content and Size	Flow Rate (mm/s)	Moisture content (%)			Percent Increase After 85°C/85% RH for 168 Hours
			Dry	20 Hours of Steam Aging	85°C/85% RH for 168 Hours	
A	40% < 10 μm	0.42	0.115	0.279	0.302	162.6
B	60% < 10 μm	0.63	0.105	0.202	0.216	105.7
C	40% < 10 μm	0.42	0.153	0.855	0.686	348.4
D	60% < 10 μm	0.105	0.121	0.228	0.19	57
E	60% < 10 μm	0.21	0.103	0.206	0.179	73.8
F	50% < 20 μm	0.31	0.124	0.385	0.292	135.5
G	60% < 20 μm	0.63	0.088	0.195	0.177	101.1
H	55% < 20 μm	0.07	0.077	0.194	0.196	154.5
I	60% < 20 μm	0.07	0.075	0.142	0.15	100
J	60% < 10 μm	0.105	0.07	0.172	0.187	167.1

tion temperature of this underfill is 133°C as measured by Dexter. Figures 6.20 and 6.21 show the Young's modulus of the HYSOL FP4526 underfill at different strain rates and temperatures. It can be seen that the Young's modulus is temperature dependent; that is, the higher the temperature, the lower the Young's modulus, and the higher the strain rate, the higher the Young's modulus. It should be pointed out that, for a given underfill material, the Young's modulus is always larger than the storage modulus. The relationship between these two is not yet known.

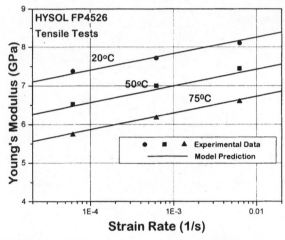

Figure 6.20 Strain rate and temperature dependence of Young's modulus of the HYSOL FP4526 underfill.

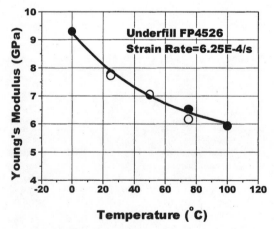

Figure 6.21 Temperature dependence of Young's modulus of the HYSOL FP4526 underfill at a certain strain rate.

6.7.6   Stress-strain relations[189, 223]

The stress-strain relations of the HYSOL FP4526 underfill at different strain rates and temperatures have been determined.[189, 223] Figures 6.22, 6.23, and 6.24 show the stress-strain relations of the underfill at different strain rates at 20, 50, and 75°C, respectively. For all the temperatures and strains, it can be seen that the higher the strain rates, the higher the stresses, and the higher the temperatures, the lower the stresses.

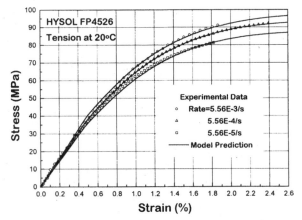

Figure 6.22 Stress-strain relations of the HYSOL FP4526 underfill at different strain rates at 20°C.

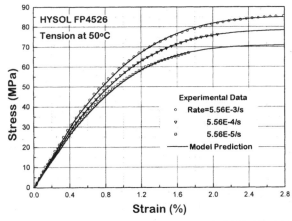

Figure 6.23 Stress-strain relations of the HYSOL FP4526 underfill at different strain rates at 50°C.

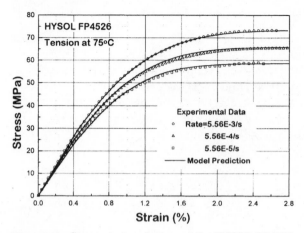

Figure 6.24 Stress-strain relations of the HYSOL FP4526 underfill at different strain rates at 75°C.

6.7.7   Creep curves[189, 223]

The creep responses of the HYSOL FP4526 underfill at different temperatures and pressures have been obtained.[189, 223] Figures 6.25 and 6.26 show the creep curves of the underfill at 20°C and 70 MPa, and at 75°C and 45 and 50 MPa. It can be seen that the creep responses of the HYSOL FP4526 underfill are dominated by transient (first-stage) creep. There is no tertiary-stage creep before the rupture of specimens. Also, for the same temperature, the higher the applied stresses, the higher the creep strains. Furthermore, a constitutive model for the underfill, which compares very well with the measurement data, has been developed.[189, 223]

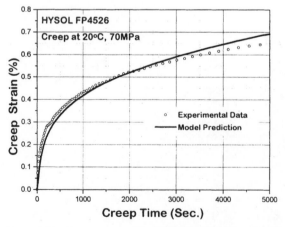

Figure 6.25 Creep curve of the HYSOL FP4526 underfill at 70 MPa and at 20°C.

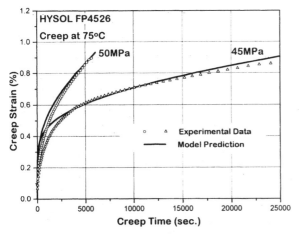

Figure 6.26   Creep curves of the HYSOL FP4526 underfill at 45 and 50 MPa and 75°C.

6.7.8   Fracture toughness of underfills[218]

Fracture toughness is a material property in the same sense that yield strength is a material property. It is independent of crack length, geometry, or loading system. Physically, fracture toughness can be considered a material property that describes the inherent resistance of the material to failure in the presence of a cracklike defect (flaw). The unit for fracture toughness is $MPa\sqrt{m}$ or $Psi\sqrt{in}$.

The opening mode I (pure tensile load) fracture toughness (K_{IC}) of three different underfill materials has been determined by IBM[218] through the plane strain specimen shown in Fig. 6.27. The K_{IC} of

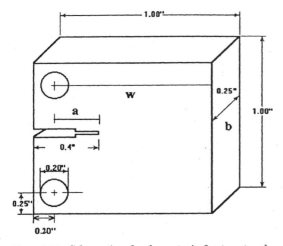

Figure 6.27   Schematics of a plane strain fracture toughness test specimen.

these underfill materials cured at temperatures between 100 and 200°C is shown in Fig. 6.28. It can be seen that, within the practice today (less than 170°C), the higher the curing temperature, the higher the K_{IC}.

Also, the fracture surfaces are different with different K_{IC}. For example, Fig. 6.29a (shown in macroscopic scale) is a typical fracture surface signified by the fanning-out river pattern for two different materials. However, on a smaller scale (Fig. 6.29b), the fracture surface of the tougher underfill material shows a consistent fish-scale pattern, which implies a localized plastic deformation as well as the directional change of crack front during propagation.[218]

Furthermore, Fig. 6.30 shows the opening mode I tensile fracture surface of underfill materials with fillers. It can be seen that the weakest interface is the filler-epoxy interface, which implies that more flexible filler (absorbing more energy) or better interfacial adhesion might further improve the fracture toughness.[218]

6.7.9   Fracture toughness of underfill-chip interfaces[214]

Interfacial fracture toughness is a material property and is defined as the interface's resistance to decohesion. Quantitatively, it equals to the amount of energy (per unit area) required to separate a bonded inter-

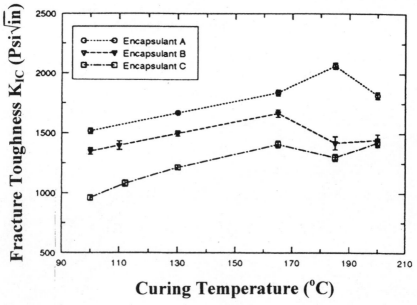

Figure 6.28 Fracture toughness under mode I loading of three underfills cured at various temperatures.

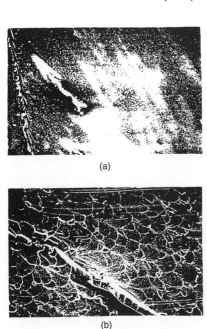

(a)

(b)

Figure 6.29 Tensile fracture surfaces on a macroscopic scale (*a*) and on a smaller scale (*b*).

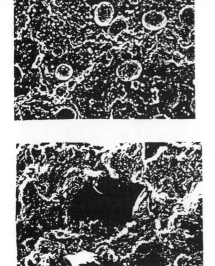

Figure 6.30 Tensile fracture surfaces of two underfills with fillers.

face between two different materials. The interfacial fracture toughness between different underfills and the chip surface and between different underfills and the PCB with solder mask surface has been determined by researchers at the Georgia Institute of Technology (GIT)[214] through four-point bending tests.

In Ref. 214, four different underfill materials are formulated by adding 1% (wt) of the selected silane coupling agents to a base resin (called Base) as shown in Table 6.4, where the storage modulus values for these materials are also given. The interfacial fracture toughness between these underfill materials and alumina, polyimide (PI), and benzocyclobutene (BCB) have been measured and are shown in Figs. 6.31 through 6.33, respectively.[214] It can be seen that, for all the cases considered (alumina, PI, and BCB), (1) underfill materials S1, S2, and S4 yield higher interfacial toughness than the base material, indicating that the silane additives in these materials enhance the interface toughness; (2) the epoxy functionalized silane has the maximum interfacial fracture toughness enhancement, and the methacrylate functionalized silane has no significant effect; and (3) the mode mixities are near 45°C (46.37 to 52.3°C), indicating that they are almost evenly mixed between open and shear modes of fractures. Also, for the interface between BCB and underfill, even the S3 material leads to higher interfacial toughness than the base material. Finally, it seems that PI yields the best results.

6.7.10   Fracture toughness of underfill-PCB interfaces[214]

The interfacial toughness between the underfill materials and the FR-4 PCB with solder mask has also been obtained[214] and is shown in Fig. 6.34. It can be seen that (1) all the underfill materials containing silane additives yield higher interfacial toughness than the base material on FR-4 PCB with solder mask, and (2) just as on the chip surface, the epoxy functionalized silane has the maximum interfacial toughness enhancement, and the methacrylate functionalized silane has the minimum effect on the interface toughness of FR-4 PCB with solder mask. Also, it is interesting to note that the interface fracture toughness between the underfills and the FR-4 PCB with solder mask is greater than that between the underfills and the alumina, PI, and BCB on the chip surface.

6.8   Flow Rate of FCOB with Underfills

The flow rates of underfills A through J, discussed in Secs. 6.6 and 6.7.1 through 6.7.4, are measured underneath a solder-bumped flip chip on

TABLE 6.4   Storage Modulus of the Base Underfill and Four Underfills

	Storage Moduli (GPa)			
	Silane Coupling Agents with			
Base	Amine (S1)	Epoxy (S2)	Methacrylate (S3)	Vinyl Organofunctional (S4)
3.03	3.18	3.17	3.43	2.99

Underfills formulated by adding 1% (wt) of selected silane agents to the base resin.

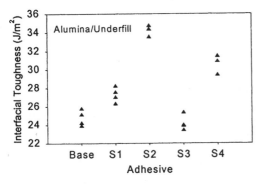

Figure 6.31 Interfacial fracture toughness of alumina/underfill interface.

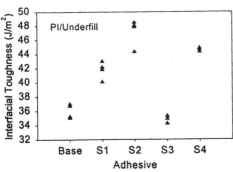

Figure 6.32 Interfacial fracture toughness of polyimide/underfill interface.

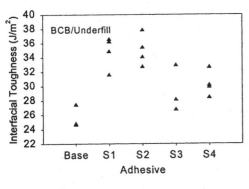

Figure 6.33 Interfacial fracture toughness of benzocyclobutene/underfill interface.

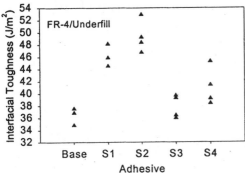

Figure 6.34 Interfacial fracture toughness of FR-4 with solder mask/underfill interface.

a BT substrate. The chips measure 6.3×3.6 mm and have 32 bumps. The BT substrate has two metal layers and 32 vias (0.254 mm in diameter). The standoff height of the solder bumps is 0.07 ± 0.02 mm.

The assembled flip chip is placed on a hot plate at 80 to 90°C. Approximately 0.025 cc of room-temperature underfill is placed around two chip sides in an L-shaped pattern. The time for the underfill to completely fill the gap is recorded. It should be pointed out that there is no flow of any underfill materials to the bottom side of the substrate. The average flow rates for underfills A through J are shown in Table 6.3. It can be seen that, in general, the smaller the filler size, the faster the flow rate.

6.9 Shear Test of FCOB with Underfills

Filler content, filler shape, filler size, and polymer type all affect the mechanical performance of underfill materials. Homogeneous mixing of fillers in epoxy is very difficult to achieve, since most of the underfill is particle-reinforced polymer. Furthermore, porosity or void absorbs moisture and serves to reduce the amount of load-bearing of the underfill materials. The Royce Instruments Model 550 is used to perform the mechanical shear tests. The shear wedge is placed very close to the substrate and against one edge of the solder-bumped flip chip with underfill on the BT substrate, which is clamped on the stage. A push of the wedge is applied to shear the chip/bumps/underfill away. The speed of the shear wedge is 0.004 in./s (0.1 mm/s).

6.9.1 Test results

Mechanical shear testing is performed on four sets of specimens: dry, steam-aged, 85°C/85 percent RH for 168 h, and level 1 precondition (reflow 3 times after 168 h at 85°C/85 percent RH). The average test results for underfills A through J are shown in Table 6.5, and shear force for different underfills in the solder-bumped flip chip on board is shown in Fig. 6.35. It can be seen from Table 6.5 that the average shear force varies with the different underfills under consideration. It can also be seen that, in most of the cases, the shear force of the solder-bumped flip chip with underfill is about the same for moisture soak and after 20 h of steam aging. Underfills E and I show better adhesion after the level 1 precondition, probably due to the reflow temperature (220°C maximum), which helps the underfills to be fully cured. (The curing temperature for underfills E and I is very high compared with the rest, Table 6.1.)

6.9.2 Failure modes

In general, there are four types of failures that occur under shear testing. The first occurs when the underfill has good adhesion with the chip

TABLE 6.5   Effects of Moisture on Mechanical Shear Test Performance of FCOB with Underfills A Through J

	Shear force (kgf)				
Underfill	Dry Conditions	85°C/85 % RH	20 Hours of Steam Aging	Level 1 Precondition	Percent Drop After 85°C/85% RH
A	52.2	31.6	30.1	34.3	39.5
B	53.3	35.4	25.4	32.2	33.6
C	51.2	45	46.7	34.4	12.1
D	44.8	33	32.7	34.2	26.3
E	47.6	34.5	31.2	54.8	27.5
F	46.3	33.2	32.2	34	28.3
G	68.6	33.7	32.7	32.1	50.9
H	72.7	35	29.5	30.8	51.9
I	73.4	28.2	27.6	57.5	61.6
J	40.4	23.3	30.2	29.5	42.3

and the PCB. For example, Fig. 6.36 shows the chip with underfill A and most of the solder mask from the PCB. Figure 6.37 shows the corresponding PCB surface with very little solder mask. This phenomenon occurs for most of the samples under dry conditions. Figure 6.38 shows another example of this type of failure, in which the chip is broken into pieces and some broken chips still remain on the substrate. This happens for most of the samples under moisture conditions.

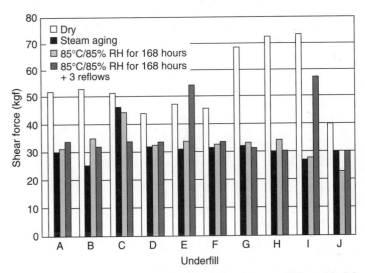

Figure 6.35   Shear force for solder-bumped flip chip assemblies with different underfills under dry conditions, steam aging, 85°C/85% RH for 168 hours, and 85°C/85% RH for 168 hours + 3 SMT reflows.

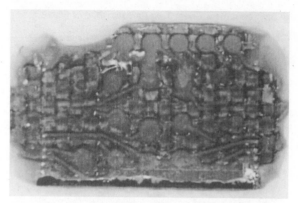

Figure 6.36   Die surface with solder mask and underfill.

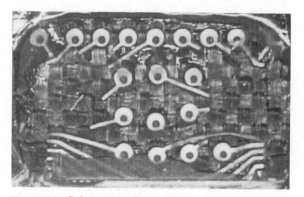

Figure 6.37   Substrate surface without solder mask.

Figure 6.38   Substrate surface with partial underfill and broken chips.

The second type of failure under shear testing occurs when the underfill has good adhesion with the chip and poor adhesion with the solder mask. Figure 6.39 shows the chip with most of the underfill and very little solder mask. Figure 6.40 shows the corresponding PCB surface with most of the solder mask and very little underfill remaining. This happens for most of the samples under moisture conditions.

The third type of failure under shear testing occurs when the underfill has good adhesion with the solder mask on the PCB and poor adhesion with the passivation of the chip. Figure 6.41 shows the PCB with most of the underfill. The corresponding chip surface is clean (not shown). This means that the adhesion between the underfill and the solder mask is stronger than that between the underfill and the passivation of the chip.

The fourth type of failure under shear testing occurs when the underfill has good adhesion with the solder mask on the PCB and poor

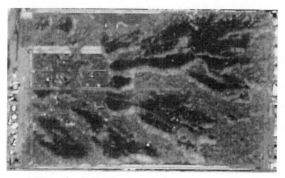

Figure 6.39 Die surface of flip chip with most of the underfill and very little solder mask.

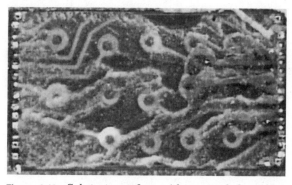

Figure 6.40 Substrate surface with most of the solder mask and very little underfill.

Figure 6.41 Substrate surface with most of the underfill.

adhesion with the chip due to passivation cracking. This type of failure on the PCB is very similar to that illustrated in Fig. 6.41.

It should be noted that the purpose of the destructive shear tests is to quantify the strength of these underfills. It does not mean that underfill is not reliable to use. How is an underfill material selected? The answer is not easy, since the performance of underfill depends on many factors such as, among others, the curing temperature and time, the TCE value, the storage modulus, the Young's modulus, the glass transition temperature, the moisture uptake, the flow rate, the shear strength, the fracture toughness, and the adhesion strength. However, based on the measurement data for the 10 different underfill materials under consideration, a simple method has been proposed to rank (select) the proper underfill. Interested readers are encouraged to read Ref. 170.

Acknowledgments

The author would like to thank C. P. Wong, Q. Yao, J. Qu, and J. Wu of Georgia Institute of Technology; Z. Qian, J. Yang, M. Lu, W. Ren, and S. Liu of Wayne State University; T. Wu, Y. Tsukada, and W. Chen of IBM; and C. Chang of EPS for sharing their important and useful information with the industry.

References

1. Lau, J. H., *Ball Grid Array Technology,* McGraw-Hill, New York, 1995.
2. Tummala, R., E. Rymaszewski, and A. Klopfenstein, *Microelectronics Packaging Handbook,* 2nd ed., Chapman-Hall, New York, 1997.
3. Lau, J. H., and R. Lee, *Chip Scale Package, Design, Materials, Process, Reliability, and Applications,* McGraw-Hill, New York, 1999.
4. Lau, J. H., C. Wong, J. Prince, and W. Nakayama, *Electronic Packaging: Design, Materials, Process, and Reliability,* McGraw-Hill, New York, 1998.
5. Lau, J. H., and Y. H. Pao, *Solder Joint Reliability of BGA, CSP, Flip Chip, and Fine Pitch SMT Assemblies,* McGraw-Hill, New York, 1997.

6. Lau, J. H., *Flip Chip Technologies*, McGraw-Hill, New York, 1996.
7. Lau, J. H., *Chip on Board Technologies for Multichip Modules*, Van Nostrand Reinhold, New York, 1994.
8. Lau, J. H., *Handbook of Tape Automated Bonding*, Van Nostrand Reinhold, New York, 1992.
9. Harman, G., *Wire Bonding in Microelectronics, Materials, Processes, Reliability, and Yield*, McGraw-Hill, New York, 1997.
10. Garrou, P. E., and I. Turlik, *Multichip Module Technology Handbook*, McGraw-Hill, New York, 1998.
11. Elshabini-Riad, A., and F. D. Barlow, III, *Thin Film Technology Handbook*, McGraw-Hill, New York, 1998.
12. Suwa, M., T. Miwa, Y. Tsutsumi, and Y. Shirai, "Development of a 1000-Pin Fine-Pitch BGA for High Performance LSI," *Proceedings of IEEE Electronic Components and Technology Conference*, pp. 430–434, June 1999.
13. Ouimet, S., and M. Paquet, "Overmold Technology Applied to Cavity Down Ultrafine Pitch PBGA Package," *Proceedings of IEEE Electronic Components and Technology Conference*, pp. 458–462, May 1998.
14. Dehkordi, P., K. Ramamurthi, D. Bouldin, H. Davidson, and P. Sandborn, "Impact of Packaging Technology on System Partitioning: A Case Study," *Proceedings of the IEEE Multichip Module Conference*, pp. 144–149, Santa Cruz, CA, February 1993.
15. Dehkordi, P., and D. Bouldin, "Design for Packageability: The Impact of Bonding Technology on the Size and Layout of VLSI Dies," *Proceedings of the IEEE Multichip Module Conference*, pp. 153–159, Santa Cruz, CA, February 1993.
16. Singh, P., and D. Landis, "Optimal Chip Sizing for Multi-Chip Modules," *IEEE Trans. Components, Packaging, and Manufacturing Technology*, part B, 17 (3): 369–375, August 1994.
17. Sandborn, P., M. Abadir, and C. Murphy, "The Tradeoff Between Peripheral and Area Array Bonding of Components in Multichip Modules," *IEEE Trans. Components, Packaging, and Manufacturing Technology*, part A, 17 (2): 249–255, June 1994.
18. Dehkordi, P., and D. Bouldin, "Design for Packageability, Early Consideration of Packaging from a VLSI Designer's Viewpoint," *IEEE Computer*, 76–81, April 1993.
19. Lau, J. H., "Cost Analysis: Solder Bumped Flip Chip versus Wire Bonding," *IEEE Trans. Electronics Manufacturing*, January 2000.
20. Davis, E., W. Harding, R. Schwartz, and J. Corning, "Solid Logic Technology: Versatile, High Performance Microelectronics," *IBM J. Res. Dev.*, 102–114, April 1964.
21. Suryanarayana, D., and D. S. Farquhar, "Underfill Encapsulation for Flip Chip Applications," in *Chip On Board Technologies for Multichip Modules*, Lau, J. H., ed., Van Nostrand Reinhold, New York, pp. 504–531, 1994.
22. Totta, P. A., and R. P. Sopher, "SLT Device Metallurgy and Its Monolithic Extension," *IBM J. Res. Dev.*, 226–238, May 1969.
23. Goldmann, L. S., and P. A. Totta, "Chip Level Interconnect: Solder Bumped Flip Chip," in *Chip On Board Technologies for Multichip Modules*, Lau, J. H., ed., Van Nostrand Reinhold, New York, pp. 228–250, 1994.
24. Goldmann, L. S., R. J. Herdzik, N. G. Koopman, and V. C. Marcotte, "Lead Indium for Controlled Collapse Chip Joining," *Proceedings of the IEEE Electronic Components Conference*, pp. 25–29, 1977.
25. Totta, P., "Flip Chip Solder Terminals," *Proceedings of the IEEE Electronic Components Conference*, pp. 275–284, 1971.
26. Goldmann, L. S., "Geometric Optimization of Controlled Collapse Interconnections," *IBM J. Res. Dev.*, 251–265, May 1969.
27. Goldmann, L. S., "Optimizing Cycle Fatigue Life of Controlled Collapse Chip Joints," *Proceedings of the 19th IEEE Electronic Components and Technology Conference*, pp. 404–423, 1969.
28. Goldmann, L. S., "Self Alignment Capability of Controlled Collapse Chip Joining," *Proceedings of the 22nd IEEE Electronic Components and Technology Conference*, pp. 332–339, 1972.
29. Shad, H. J., and J. H. Kelly, "Effect of Dwell Time on Thermal Cycling of the Flip Chip Joint," *Proceedings of the ISHM*, pp. 3.4.1–3.4.6, 1970.

30. Hymes, I., R. Sopher, and P. Totta, "Terminals for Microminiaturized Devices and Methods of Connecting Same to Circuit Panels," U.S. Patent 3,303,393, 1967.
31. Karan, C., J. Langdon, R. Pecararo, and P. Totta, "Vapor Depositing Solder," U.S. Patent 3,401,055, 1968.
32. Seraphim, D. P., and J. Feinberg, "Electronic Packaging Evolution," *IBM J. Res. Dev.,* 617–629, May 1981.
33. Tsukada, Y., Y. Mashimoto, and N. Watanuki, "A Novel Chip Replacement Method for Encapsulated Flip Chip Bonding," *Proceedings of the 43rd IEEE/EIA Electronic Components and Technology Conference,* pp. 199–204, June 1993.
34. Tsukada, Y., Y. Maeda, and K. Yamanaka, "A Novel Solution for MCM-L Utilizing Surface Laminar Circuit and Flip Chip Attach Technology," *Proceedings of the 2nd International Conference and Exhibition on Multichip Modules,* pp. 252–259, April 1993.
35. Tsukada, Y., S. Tsuchida, and Y. Mashimoto, "Surface Laminar Circuit Packaging," *Proceedings of the 42nd IEEE Electronic Components and Technology Conference,* pp. 22–27, May 1992.
36. Miller, L. F., "Controlled Collapse Reflow Chip Joining," *IBM J. Res. Dev.,* 239–250, May 1969.
37. Miller, L. F., "A Survey of Chip Joining Techniques," *Proceedings of the 19th IEEE Electronic Components and Technology Conference,* pp. 60–76, 1969.
38. Miller, L. F., "Joining Semiconductor Devices with Ductile Pads," *Proceedings of ISHM,* pp. 333–342, 1968.
39. Norris, K. C., and A. H. Landzberg, "Reliability of Controlled Collapse Interconnections," *IBM J. Res. Dev.,* 266–271, May 1969.
40. Oktay, S., "Parametric Study of Temperature Profiles in Chips Joined by Controlled Collapse Technique," *IBM J. Res. Dev.,* 272–285, May 1969.
41. Bendz, D. J., R. W. Gedney, and J. Rasile, "Cost/Performance Single Chip Module," *IBM J. Res. Dev.,* 278–285, 1982.
42. Blodgett, A. J., Jr., "A Multilayer Ceramic Multichip Module," *IEEE Trans. Components, Hybrids, and Manufacturing Technology,* 634–637, 1980.
43. Fried, L. J., J. Havas, J. Lechaton, J. Logan, G. Paal, and P. Totta, "A VLSI Bipolar Metallization Design with Three-Level Wiring and Area Array Solder Connections," *IBM J. Res. Dev.,* 362–371, 1982.
44. Clark, B. T., and Y. M. Hill, "IBM Multichip Multilayer Ceramic Modules for LSI Chips—Designed for Performance Density," *IEEE Trans. Components, Hybrids, and Manufacturing Technology,* 89–93, 1980.
45. Blodgett, A. J., and D. R. Barbour, "Thermal Conduction Module: A High-Performance Multilayer Ceramic Package," *IBM J. Res. Dev.,* 30–36, 1982.
46. Dansky, A. H., "Bipolar Circuit Design for a 5000-Circuit VLSI Gate Array," *IBM J. Res. Dev.,* 116–125, 1981.
47. Oktay, S., and H. C. Kammer, "A Conduction-Cooled Module for High-Performance LSI Devices," *IBM J. Res. Dev.,* 55–66, 1982.
48. Chu, R. C., U. P. Hwang, R. E. Simons, "Conduction Cooling for an LSI Package: A One-Dimensional Approach," *IBM J. Res. Dev.,* 45–54, 1982.
49. Howard, R. T., "Packaging Reliability and How to Define and Measure It," *Proceedings of the IEEE Electronic Components Conference,* pp. 376–384, 1982.
50. Howard, R. T., "Optimization of Indium-Lead Alloys for Controlled Collapse Chip Connection Application," *IBM J. Res. Dev.,* 372–389, 1982.
51. Satoh, R., K. Arakawa, M. Harada, and K. Matsui, "Thermal Fatigue Life of Pb-Sn Alloy Interconnections," *IEEE Trans. Components, Hybrids, and Manufacturing Technology,* **14** (1): 224–232, March 1991.
52. Logsdon, W. A., P. K. Liaw, and M. A. Burke, "Fracture Behavior of 63Sn-37Pb," *Eng. Fracture Mechanics,* **36** (2): 183–218, 1990.
53. Lau, J. H., "Thermal Fatigue Life Prediction of Flip Chip Solder Joints by Fracture Mechanics Method," *Int. J. Eng. Fracture Mechanics,* **45:** 643–654, July 1993.
54. Lau, J. H., "Thermomechanical Characterization of Flip Chip Solder Bumps for Multichip Module Applications," *Proceedings of IEEE International Electronics Manufacturing Technology Symposium,* pp. 293–299, September 1992.

55. Lau, J. H., "Thermal Fatigue Life Prediction of Encapsulated Flip Chip Solder Joints for Surface Laminar Circuit Packaging," ASME Paper No. 92W/EEP-34, 1992 ASME Winter Annual Meeting.
56. Lau, J. H., and D. Rice, "Thermal Fatigue Life Prediction of Flip Chip Solder Joints by Fracture Mechanics Methods," *Proceedings of the 1st ASME/JSME Electronic Packaging Conference*, pp. 385–392, April 1992.
57. Lau, J. H., M. Heydinger, J. Glazer, and D. Uno, "Design and Procurement of Eutectic Sn/Pb Solder-Bumped Flip Chip Test Die and Organic Substrates," *Circuit World*, **21:** 20–24, March 1995.
58. Wun, B., and J. H. Lau, "Characterization and Evaluation of the Underfill Encapsulants for Flip Chip Assembly," *Circuit World*, **21:** 25–32, March 1995.
59. Kelly, M., and J. H. Lau, "Low Cost Solder Bumped Flip Chip MCM-L Demonstration," *Circuit World*, **21:** 14–17, July 1995.
60. Lau, J. H., "A Brief Introduction to Flip Chip Technologies for Multichip Module Applications," in *Flip Chip Technologies*, Lau, J. H., ed., McGraw-Hill, New York, pp. 1–82, 1996.
61. Lau, J. H., and D. W. Rice, "Solder Joint Fatigue in Surface Mount Technology: State of the Art," *Solid State Technology*, 91–104, October 1985.
62. Lau, J. H., T. Krulevitch, W. Schar, M. Heydinger, S. Erasmus, and J. Gleason, "Experimental and Analytical Studies of Encapsulated Flip Chip Solder Bumps on Surface Laminar Circuit Boards," *Circuit World*, **19** (3): 18–24, March 1993.
63. Lau, J. H., "Solder Joint Reliability of Flip Chip and Plastic Ball Grid Array Assemblies Under Thermal, Mechanical, and Vibration Conditions," *Proceedings of IEEE Japan IEMT Symposium*, pp. 13–19, December 1995.
64. Wong, C. P., J. M. Segelken, and C. N. Robinson, "Chip On Board Encapsulation," in *Chip On Board Technologies for Multichip Modules*, Lau, J. H., ed., Van Nostrand Reinhold, New York, pp. 470–503, 1994.
65. Rai, A., Y. Dotta, H. Tsukamoto, T. Fujiwara, H. Ishii, T. Nukii, and H. Matsui, "COB (Chip On Board) Technology: Flip Chip Bonding Onto Ceramic Substrates and PWB (Printed Wiring Boards)," *ISHM Proceedings*, pp. 474–481, 1990.
66. Rai, A., Y. Dotta, T. Nukii, and T. Ohnishi, "Flip-Chip COB Technology on PWB," *Proceedings of IMC*, pp. 144–149, June 1992.
67. Lowe, H., "No-Clean Flip Chip Attach Process," *International TAB/Advance Packaging and Flip Chip Proceedings*, pp. 17–24, February 1994.
68. Giesler, J., S. Machuga, G. O'Malley, and M. Williams, "Reliability of Flip Chip on Board Assemblies," *International TAB/Advance Packaging and Flip Chip Proceedings*, pp. 127–135, February 1994.
69. Tsukada, Y., "Solder Bumped Flip Chip Attach on SLC Board and Multichip Module," in *Chip On Board Technologies for Multichip Modules*, Lau, J. H., ed., Van Nostrand Reinhold, New York, pp. 410–443, 1994.
70. Hu, K., C. Yeh, R. Doot, A. Skipor, and K. Wyatt, "Die Cracking in Flip-Chip-on-Board Assembly," *Proceedings of IEEE Electronic Components and Technology Conference*, pp. 293–299, May 1995.
71. Sweet, J., D. Peterson, J. Emerson, and R. Mitchell, "Liquid Encapsulant Stress Variations as Measured with the ATC04 Assembly Test Chip," *Proceedings of IEEE Electronic Components and Technology Conference*, pp. 300–304, May 1995.
72. Greer, S. "An Extended Eutectic Solder Bump for FCOB," *Proceedings of IEEE Electronic Components and Technology Conference*, pp. 546–551, May 1996.
73. McLaren, T., S. Kang, W. Zhang, D. Hellman, T. Ju, and Y. Lee, "Thermosonic Flip-Chip Bonding for an 8 × 8 VCSEL Array," *Proceedings of IEEE Electronic Components and Technology Conference*, pp. 393–400, May 1995.
74. Nishimori, T., H. Yanagihara, K. Murayama, Y. Kama, and M. Nakamura, "Characteristics and Potential Application of Polyimide-Core-Bump to Flip Chip," *Proceedings of IEEE Electronic Components and Technology Conference*, pp. 515–519, May 1995.
75. Suryanarayana, D., J. Varcoe, and J. Ellerson, "Repairability of Underfill Encapsulated Flip-Chip Packages," *Proceedings of IEEE Electronic Components and Technology Conference*, pp. 524–528, May 1995.

76. Shaukatullah, H., B. Hansen, W. Storr, and F. Andros, "Thermal Enhancement of Flip-Chip Packages with Radial-Finger-Contact Spring," *Proceedings of IEEE Electronic Components and Technology Conference,* pp. 865–871, May 1995.

77. Kim, D., H. Han, S. Park, G. Joo, M. Song, N. Hwang, S. Kang, H. Lee, and H. Park, "Application of the Flip-Chip Bonding Technique to the 10 Gbps Laser Diode Module," *Proceedings of IEEE Electronic Components and Technology Conference,* pp. 872–875, May 1995.

78. Perfecto, E., R. Shields, R. Master, S. Purushothaman, and C. Prasad, "A Low Cost MCM-D Process for Flip Chip and Wirebonding Applications," *Proceedings of IEEE Electronic Components and Technology Conference,* pp. 1081–1086, May 1995.

79. Howell, W., D. Brouillette, J. Korejwa, E. Sprogis, S. Yankee, and J. Wursthorn, "Area Array Solder Interconnection Technology for the Three-Dimensional Silicon Cube," *Proceedings of IEEE Electronic Components and Technology Conference,* pp. 1174–1178, May 1995.

80. Chen, W., J. Gentile, and L. Higgins, "FCOB Reliability Evaluation Simulating Multiple Rework/Reflow Processes," *Proceedings of IEEE Electronic Components and Technology Conference,* pp. 1184–1195, May 1996.

81. Clementi, J., G. Dearing, J. Zimmerman, and C. Bergeron, "Reliability and Analytical Evaluations of No-Clean Flip-Chip Assembly," *Proceedings of IEEE Electronic Components and Technology Conference,* pp. 1191–1196, May 1995.

82. Ogashiwa, T., T. Arikawa, H. Murai, A. Inoue, and T. Masumoto, "Reflowable Sn-Pb Bump Formation on Al Pad by a Solder Bumping Method," *Proceedings of IEEE Electronic Components and Technology Conference,* pp. 1203–1208, May 1995.

83. Eldring, J., K. Koeffers, H. Richter, A. Baumgartner, and H. Reichl, "Flip Chip Attachment of Silicon Devices Using Substrate Ball Bumping and the Technology Evaluation on Test Assemblies for 20 Gbit/s Transmission," *Proceedings of IEEE Electronic Components and Technology Conference,* pp. 1209–1216, May 1995.

84. Okuno, A., N. Oyama, K. Nagai, and T. Hashimoto, "Flip-Chip Packaging Using PES (Printing Encapsulation Systems) and PES Underfill Epoxy Resin," *Proceedings of IEEE Electronic Components and Technology Conference,* pp. 1240–1243, May 1995.

85. Goldstein, J., D. Tuckerman, P. Kim, and B. Fernandez, "A Novel Flip-Chip Process," *Proceedings of SMI Conference,* pp. 59–71, August 1995.

86. Magill, P., and G. Rinne, "Implementation of Flip Chip Technology," *Proceedings of SMI Conference,* pp. 72–79, August 1995.

87. Lowe, H., and R. Lyn, "Real World Flip Chip Assembly: A Manufacture's Experience," *Proceedings of SMI Conference,* pp. 80–87, August 1995.

88. Aschenbrenner, R., E. Zakel, G. Azdasht, A. Kloeser, and H. Reichl, "Fluxless Flip Chip Bonding on Flexible Substrates: A Comparison Between Adhesive Bonding and Soldering," *Proceedings of SMI Conference,* pp. 91–101, August 1995.

89. Liu, J., K. Boustedt, and Z. Lai, "Development of Flip-Chip Joining Technology on Flexible Circuitry Using Anisotropically Conductive Adhesives and Eutectic Solder," *Proceedings of SMI Conference,* pp. 102–109, August 1995.

90. Patterson, T., "A Practical Versatile Approach to Flip Chip on Flex," *Proceedings of SMI Conference,* pp. 110–114, August 1995.

91. Koh, W., and M. Edwards, "Performance Enhancement of Newly Developed Chip on Board (COB) Encapsulants," *Proceedings of SMI Conference,* pp. 138–145, August 1995.

92. Schiesser, T., E. Menard, T. Smith, and J. Akin, "Microdynamic Solder Pump vs. Alternatives Comparative Review of Solder Bumping Techniques for Flip Chip Attach," *Proceedings of SMI Conference,* pp. 171–178, August 1995.

93. McDermott, B., "Impact of Direct Chip Attach on the Printed Circuit Board," *Proceedings of SMI Conference,* pp. 1026–1030, August 1995.

94. Schrand, J., "Using Die in PC Card Applications," *Proceedings of SMI Conference,* pp. 1031–1038, August 1995.

95. Koopman, N., and S. Nangalia, "Fluxless Flip-Chip Solder Joining," in *Flip Chip Technologies,* Lau, J. H., ed., McGraw-Hill, New York, pp. 83–121, 1996.

96. Sharma, R., and R. Subrahmanyan, "Solder Bumped Flip Chip Interconnect Technologies: Materials, Processes, Performance and Reliability," in *Flip Chip Technologies,* Lau, J. H., ed., McGraw-Hill, New York, pp. 123–153, 1996.

97. Chung, T., T. Dolbear, and Dick Nelson, "Large High I/O Solder Bumped Flip-Chip Technology," in *Flip Chip Technologies*, Lau, J. H., ed., McGraw-Hill, New York, pp. 155–179, 1996.

98. Degani, Y., T. D. Dudderar, R. C. Frye, K. L. Tai, M. Lau, and B. Han, "Micro-Interconnect Technology: The Large Volume Fabrication of Cost Effective Flip-Chip MCMs," in *Flip Chip Technologies*, Lau, J. H., ed., McGraw-Hill, New York, pp. 181–221, 1996.

99. Estes, R., and F. Kulesza, "Conductive Adhesive Polymer Materials in Flip Chip Applications," in *Flip Chip Technologies*, Lau, J. H., ed., McGraw-Hill, New York, pp. 223–267, 1996.

100. Breen, M., D. Duane, R. German, K. Keswick, and R. Nolan, "Compliant Bumps for Adhesive Flip Chip Assembly," in *Flip Chip Technologies*, Lau, J. H., ed., McGraw-Hill, New York, pp. 269–287, 1996.

101. Date, H., Y. Hozumi, H. Tokuhira, M. Usui, E. Horikoshi, and T. Sato, "Anisotropic Conductive Adhesive for Fine Pitch Flip Chip Interconnections," in *Flip Chip Technologies*, Lau, J. H., ed., McGraw-Hill, New York, pp. 289–300, 1996.

102. Watanabe, I., N. Shiozawa, K. Takemura, and T. Ohta, "Flip-Chip Interconnection Technology Using Anisotropic Conductive Adhesive Films," in *Flip Chip Technologies*, Lau, J. H., ed., McGraw-Hill, New York, pp. 301–315, 1996.

103. Lee, C. H., "Anistropic Conductive Flip Chip-on-Glass Technology," in *Flip Chip Technologies*, Lau, J. H., ed., McGraw-Hill, New York, pp. 317–339, 1996.

104. Baba, S., and W. Carlomagno, "Wirebonding Flip Chip Technology for Multichip Modules," in *Flip Chip Technologies*, Lau, J. H., ed., McGraw-Hill, New York, pp. 341–356, 1996.

105. Tsunoi, K., T. Kusagaya, and H. Kira, "Flip Chip Mounting Using Stud-Bumps and Adhesive for Encapsulation," in *Flip Chip Technologies*, Lau, J. H., ed., McGraw-Hill, New York, pp. 357–366, 1996.

106. Moresco, L., D. Love, W. Chow, and V. Holalkere, "Wire Interconnect Technology: An Ultra High Density Flip Chip to Substrate Connection Method," in *Flip Chip Technologies*, Lau, J. H., ed., McGraw-Hill, New York, pp. 367–386, 1996.

107. DiStefano, T., and J. Fjelstad, "A Compliant Chip Size Packaging Technology," in *Flip Chip Technologies*, Lau, J. H., ed., McGraw-Hill, New York, pp. 387–413, 1996.

108. Zakel, E., and H. Reichl, "Flip Chip Assembly Using the Gold, Gold-Tin and Nickel-Gold Metallurgy," in *Flip Chip Technologies*, Lau, J. H., ed., McGraw-Hill, New York, pp. 415–490, 1996.

109. Gilg, L., "Assurance Technologies for Known Good Die," in *Flip Chip Technologies*, Lau, J. H., ed., McGraw-Hill, New York, pp. 491–521, 1996.

110. Rinne, G., "Burn-In and Test Substrate for Flip Chip ICs," in *Flip Chip Technologies*, Lau, J. H., ed., McGraw-Hill, New York, pp. 523–538, 1996.

111. Doi, K., N. Hirano, T. Okada, Y. Hiruta, T. Sudo, and M. Mukai, "Prediction of Thermal Fatigue Life for Encapsulated Flip Chip Interconnection," *Proceedings of International Symposium on Microelectronics*, pp. 247–252, October 1995.

112. Zoba, D., and M. Edwards, "Review of Underfill Encapsulant Development and Performance of Flip Chip Applications," *Proceedings of International Symposium on Microelectronics*, pp. 354–358, October 1995.

113. Bessho, Y., Y. Tomura, T. Shiraishi, M. Ono, T. Ishida, and K. Omoya, "Advanced Flip-Chip Bonding Technique to Roganic Substrates," *Proceedings of International Symposium on Microelectronics*, pp. 359–364, October 1995.

114. Yatsuda, H., and T. Eimura, "Flip-Chip Assembly Technique for SAW Devices," *Proceedings of International Symposium on Microelectronics*, pp. 365–370, October 1995.

115. Gupta, D., "Evaluation of Alternative Processes for Au-Sn-Au Flip Chip Bonding of Power Devices," *Proceedings of International Symposium on Microelectronics*, pp. 371–377, October 1995.

116. Chrusciel, R., P. Delivorias, and K. Rispoli, "Flip Chip Interconnect, a Versatile Known Good Die Technology," *Proceedings of International Symposium on Microelectronics*, pp. 384–389, October 1995.

117. Amano, T., M. Kohno, and Y. Obara, "Solder Bumping Through Super Solder," *Proceedings of IEEE International Electronics Manufacturing Technology Symposium*, pp. 1–4, October 1995.

118. Schwiebert, M., and W. Leong, "Underfill Flow as Viscous Flow Between Parallel Plates Driven by Capillary Action," *Proceedings of IEEE International Electronics Manufacturing Technology Symposium,* pp. 8–13, October 1995.
119. Goenka, L., and A. Achari, "Void Formation in Flip-Chip Solder Bumps—Part I," *Proceedings of IEEE International Electronics Manufacturing Technology Symposium,* pp. 14–19, October 1995.
120. Kallmayer, C., D. Lin, J. Kloeser, H. Oppermann, E. Zakel, and H. Reichl, "Fluxless Flip-Chip Attachment Techniques Using the Au/Sn Metallurgy," *Proceedings of IEEE International Electronics Manufacturing Technology Symposium,* pp. 20–28, October 1995.
121. Koopman, N., G. Adema, S. Nangalia, M. Schneider, and V. Saba, "Flip Chip Process Development Techniques Using a Modified Laboratory Aligner Bonder," *Proceedings of IEEE International Electronics Manufacturing Technology Symposium,* pp. 29–35, October 1995.
122. Imler, W., T. Hildebrandt, S. Paolini, K. Scholz, M. Cobarruviaz, and V. Nagesh, "Design and Fabrication of 600-dpi Light-Emitting Diode Printheads Using Precision Flip-Chip Solder Bump Technology," *Proceedings of IEEE International Electronics Manufacturing Technology Symposium,* pp. 154–159, September 1994.
123. Aintila, A., A. Bjorklof, E. Jarvinen, and S. Lalu, "Electroless Ni/Au Bumps for Flipchip-on-Flex and TAB Applications," *Proceedings of IEEE International Electronics Manufacturing Technology Symposium,* pp. 160–164, September 1994.
124. Zakel, E., J. Gwiasda, J. Kloeser, J. Eldring, G. Engelmann, and H. Reichl, "Fluxless Flip Chip Assembly on Rigid and Flexible Polymer Substrates Using the Au-Sn Metallurgy," *Proceedings of IEEE International Electronics Manufacturing Technology Symposium,* pp. 177–184, September 1994.
125. Moresco, L., D. Love, V. Holalkeri, P. Boucher, W. Chou, C. Grilletto, and C. Wong, "Wire Interconnect Technology, a New Flip-Chip Technique," *Proceedings of NEPCON West,* pp. 947–955, February 1996.
126. Lee, C. H., and K. Loh, "Fine Pitch COG Interconnections Using Anisotropically Conductive Adhesives," *Proceedings of NEPCON West,* pp. 956–967, February 1996.
127. Degani, Y., T. D. Dudderar, and K. Tai, "High Density MCM Assembly Using Printed Solder Paste Process," *Proceedings of NEPCON West,* pp. 968–972, February 1996.
128. Vardaman, E. J., and T. Goodman, "Worldwide Trends in Flip Chip Developments," *Proceedings of NEPCON West,* pp. 973–975, February 1996.
129. Shock, G., "Flip Chip Assembly in High Volume, Hige Mix Products," *Proceedings of NEPCON West,* pp. 1433–1438, February 1996.
130. Slesinger, K., "Flip Chips on Laminates: High-Volume Assembly," *Proceedings of NEPCON West,* pp. 1439–1447, February 1996.
131. Han, B., Y. Guo, T. Chung, and D. Liu, "Reliability Assessment of Flip Chip Package with Encapsulation," *Proceedings of NEPCON West,* pp. 600–602, February 1995.
132. Gutentag, C., "Methods for Handling Bare Die for C4 Assembly/Placement," *Proceedings of NEPCON West,* pp. 603–609, February 1995.
133. Zakel, E., R. Aschenbrenner, J. Gwiasda, G. Azdasht, A. Ostmann, J. Eldring, and H. Reichl, "Fluxless Flip Chip Bonding on Flexible Substrates," *Proceedings of NEPCON West,* pp. 909–918, February 1995.
134. Degani, Y., T. D. Dudderar, R. C. Frye, and K. L. Tai, "A Cost Effective MCM Manufacturing Platform Achieved by Combining Surface Mount and Silicon Technologies," *Proceedings of NEPCON West,* pp. 932–937, February 1995.
135. Chanchani, R., K. Treece, and P. Dressendorfer, "Mini Ball Grid Array (mBGA) Technology," *Proceedings of NEPCON West,* pp. 938–945, February 1995.
136. Vardaman, E. J., "International Activities in Flip Chip on Board Technologies," *Proceedings of NEPCON West,* pp. 946–950, February 1995.
137. Jimarez, M., A. Smith, and J. Zdimal, "Development of a Rework Process for Unencapsulated Flip Chips on Organic Substrates," *Proceedings of NEPCON West,* pp. 1646–1669, February 1995.
138. Simon, J., J. Gwiasda, I. Kuhls, K. Werner, and H. Reichl, "Development of Display Module Using Flip Chip on Flex," *Proceedings of IEEE Japan IEMT Symposium,* pp. 48–51, December 1995.

139. Tsunetsugu, H., T. Hayashi, and K. Katsura, "Micro-alignment Technique using 26-um Diameter Microsolder Bumps and Its Shear Strength," *Proceedings of IEEE Japan IEMT Symposium*, pp. 52–55, December 1995.

140. Kloeser, J., A. Ostmann, R. Aschenbrenner, E. Zakel, and H. Reichl, "Approaches to Flip Chip Technology Using Electroless Nickel-Gold Bumps," *Proceedings of IEEE Japan IEMT Symposium*, pp. 60–66, December 1995.

141. Jung, E., R. Aschenbrenner, A. Ostmann, E. Zakel, and H. Reichl, "Flip Chip Soldering and Adhesive bonding on Organic Substrates," *Proceedings of IEEE Japan IEMT Symposium*, pp. 67–71, December 1995.

142. Lau, J. H., E. Schneider, and T. Baker, "Shock and Vibration of Solder Bumped Flip Chip on Organic Coated Copper Boards," *ASME Trans. J. Electronic Packaging*, 101–104, June 1996.

143. Fujiuchi, S., and K. Toriyama, "Collective Screen Printing for Carrier Bump and SMT Pads," *Proceedings of IEEE Japan IEMT Symposium*, pp. 109–112, December 1995.

144. Honma, S., K. Tateyama, H. Yamada, and M. Saito, "Evaluation of Barrier Metal of Solder Bumps for Flip-Chip Interconnection," *Proceedings of IEEE Japan IEMT Symposium*, pp. 113–116, December 1995.

145. Kato, Y., Y. Ueoka, E. Kono, and E. Hagimoto, "Solder Bump Forming Using Micro Punching Technology," *Proceedings of IEEE Japan IEMT Symposium*, pp. 117–120, December 1995.

146. Yanada, H., and M. Saito, "A Fine Pitch and High Aspect Ratio Bump Fabrication Process for Flip-Chip Interconnection," *Proceedings of IEEE Japan IEMT Symposium*, pp. 121–124, December 1995.

147. Ohunki, Y., H. Shibuya, J. Utsunomiya, and S. Iida, "Development of Low Cost Solder Bump on LSI," *Proceedings of IEEE Japan IEMT Symposium*, pp. 125–128, December 1995.

148. Homma, Y., T. Fujiki, K. Kobayashi, Y. Shirai, and K. Akazawa, "Fast-Cure Liquid Encapsulant for ICs," *Proceedings of IEEE Japan IEMT Symposium*, pp. 449–452, December 1995.

149. DeHaven, K., and J. Dietz, "Controlled Collapse Chip Connection (C4)—An Enabling Technology," *Proceedings of IEEE Electronic Components and Technology Conference*, pp. 1–6, May 1994.

150. Puttlitz, K., and W. Shutler, "C4/BGA Comparison with Other MLC Single Chip Package Alternatives," *Proceedings of IEEE Electronic Components and Technology Conference*, pp. 16–21, May 1994.

151. Kromann, G., D. Gerke, and W. Huang, "A Hi-Density C4/CBGA Interconnect Technology for a CMOS Microprocessor," *Proceedings of IEEE Electronic Components and Technology Conference*, pp. 22–28, May 1994.

152. Switky, A., V. Sajja, J. Darnauer, and W. Dai, "A 1024-Pin Plastic Ball Grid Array for Flip Chips Die," *Proceedings of IEEE Electronic Components and Technology Conference*, pp. 32–38, May 1994.

153. O'Malley, G., J. Giesler, and S. Machuga, "The Importance of Material Selection for Flip Chip on Board Assembly," *Proceedings of IEEE Electronic Components and Technology Conference*, pp. 387–394, May 1994.

154. Machuga, S., S. Lindsey, K. Moore, and A. Skipor, "Encapsulation of Flip Chip Structures," *Proceedings of IEEE International Electronics Manufacturing Technology Symposium*, pp. 53–58, September 1992.

155. Chu, D., C. Reber, and D. Palmer, "Screening ICs on the Bare Chip Level: Temporary Packaging," *Proceedings of IEEE International Electronics Manufacturing Technology Symposium*, pp. 223–226, September 1992.

156. Yamada, H., Y. Kondoh, and M. Saito, "A Fine Pitch and High Aspect Ratio Bump Array for Flip-Chip Interconnection," *Proceedings of IEEE International Electronics Manufacturing Technology Symposium*, pp. 288–292, September 1992.

157. Kromann, G., "Thermal Modeling and Experimental Characterization of the C4/Surface-Mount-Array Interconnect Technologies," *Proceedings of IEEE Electronic Components and Technology Conference*, pp. 395–402, May 1994.

158. Gasparini, N., and B. Bhattacharyya, "A Method of Designing a Group of Bumps for C4 packages to Maximize the Number of Bumps and Minimize the Number of Pack-

age Layers," *Proceedings of IEEE Electronic Components and Technology Conference,* pp. 695–699, May 1994.

159. Thompson, P., M. Begay, S. Lindsey, D. Vanoverloop, B. Vasquez, S. Walker, and B. Williams, "Mechanical and Electrical Evaluation of a Bumped-Substrate Die-Level Burn-In Carrier," *Proceedings of IEEE Electronic Components and Technology Conference,* pp. 700–703, May 1994.

160. Metzfer, D., U. Beutler, J. Eldring, and H. Reichl, "Laser Bumping for Flip Chip and TAB Applications," *Proceedings of IEEE Electronic Components and Technology Conference,* pp. 910–916, May 1994.

161. Gupta, D., "A Novel Active Area Bumped Flip Chip Technology for Convergent Heat Transfer from Gallium Arsenide Power Devices," *Proceedings of IEEE Electronic Components and Technology Conference,* pp. 917–921, May 1994.

162. Ito, S., M. Kuwamura, S. Sudo, M. Mizutani, T. Fukushima, H. Noro, S. Akizuki, and A. Prabhu, "Study of Encapsulating System for Diversified Area Bump Packages," *Proceedings of IEEE Electronic Components and Technology Conference,* pp. 46–53, San Jose, CA, May 1997.

163. Pascarella, N., and D. Baldwin, "Advanced Encapsulation Processing for Low Cost Electronics Assembly—A Cost Analysis," *The 3rd International Symposium and Exhibition on Advanced Packaging Materials, Processes, Properties, and Interfaces,* pp. 50–53, Braselton, GA, March 1997.

164. Nguyen, L., L. Hoang, P. Fine, Q. Tong, B. Ma, R. Humphreys, A. Savoca, C. P. Wong, S. Shi, M. Vincent, and L. Wang, "High Performance Underfills Development—Materials, Processes, and Reliability," *IEEE 1st International Symposium on Polymeric Electronics Packaging,* pp. 300–306, Norrkoping, Sweden, October 1997.

165. Erickson, M., and K. Kirsten, "Simplifying the Assembly Process with a Reflow Encapsulant," *Electronic Packaging and Production,* 81–86, February 1997.

166. Wong, C. P., M. B. Vincent, and S. Shi, "Fast-Flow Underfill Encapsulant: Flow Rate and Coefficient of Thermal Expansion," *Proceedings of the ASME—Advances in Electronic Packaging,* vol. 19-1, pp. 301–306, 1997.

167. Wong, C. P., S. H. Shi, and G. Jefferson, "High Performance No Flow Underfills for Low-Cost Flip-Chip Applications," *Proceedings of IEEE Electronic Components and Technology Conference,* pp. 850–858, San Jose, CA, May 1997.

168. Tummala, R., E. Rymaszewski, and A. Klopfenstein, *Microelectronics Packaging Handbook,* Chapman & Hall, New York, 1997.

169. Lau, J. H., C. Chang, and R. Chen, "Effects of Underfill Encapsulant on the Mechanical and Electrical Performance of a Functional Flip Chip Device," *J. Electronics Manufacturing,* **7** (4): 269–277, December 1997.

170. Lau, J. H., and C. Chang, "How to Select Underfill Materials for Solder Bumped Flip Chip on Low Cost Substrates?" *Int. J. Microcircuits and Electronic Packaging,* **22,** (1): 20–28, 1st Quarter 1999.

171. Nguyen, L., C. Quentin, P. Fine, B. Cobb, S. Bayyuk, H. Yang, and S. A. Bidstrup-Allen, "Underfill of Flip Chip on Laminates: Simulation and Validation," *Proceedings of the International Symposium on Adhesives in Electronics,* pp. 27–30, Binghamton, NY, September 1998.

172. Pascarella, N., and Baldwin, D., "Compression Flow Modeling of Underfill Encapsulants for Low Cost Flip Chip Assembly," *Proceedings of IEEE Electronic Components and Technology Conference,* pp. 463–470, Seattle, WA, May 1998.

173. Nguyen, L., P. Fine, B. Cobb, Q. Tong, B. Ma, and A. Savoca, "Reworkable Flip Chip Underfill—Materials and Processes," *Proceedings of the International Symposium on Microelectronics,* pp. 707–713, San Diego, CA, November 1998.

174. Capote, M. A., and S. Zhu, "No-Underfill Flip-Chip Encapsulation," *Proceedings of Surface Mount International Conference,* pp. 291–293, San Jose, CA, August 1998.

175. Capote, M. A., W. Johnson, S. Zhu, L. Zhou, and B. Gao, "Reflow-Curable Polymer Fluxes for Flip Chip Encapsulation," *Proceedings of the International Conference on Multichip Modules and High Density Packaging,* pp. 41–46, Denver, CO, April 1998.

176. Vincent, M. B., and C. P. Wong, "Enhancement of Underfill Encapsulants for Flip-Chip Technology," *Proceedings of Surface Mount International Conference,* pp. 303–312, San Jose, CA, August 1998.

177. Vincent, M. B., L. Meyers, and C. P. Wong, "Enhancement of Underfill Performance for Flip-Chip Applications by Use of Silane Additives," *Proceedings of IEEE Electronic Components and Technology Conference,* pp. 125–131, Seattle, WA, May 1998.
178. Wang, L., and C. P. Wong, "Novel Thermally Reworkable Underfill Encapsulants for Flip-Chip Applications," *Proceedings of IEEE Electronic Components and Technology Conference,* pp. 92–100, Seattle, WA, May 1998.
179. Shi, S. H., and C. P. Wong, "Study of the Fluxing Agent Effects on the Properties of No-Flow Underfill Materials for Flip-Chip Applications," *Proceedings of IEEE Electronic Components and Technology Conference,* pp. 117–124, Seattle, WA, May 1998.
180. Wong, C. P., D. Baldwin, M. B. Vincent, B. Fennell, L. J. Wang, and S. H. Shi, "Characterization of a No-Flow Underfill Encapsulant During the Solder Reflow Process," *Proceedings of IEEE Electronic Components and Technology Conference,* pp. 1253–1259, Seattle, WA, May 1998.
181. Ito, S., M. Mizutani, H. Noro, M. Kuwamura, and A. Prabhu, "A Novel Flip Chip Technology Using Non-Conductive Resin Sheet," *Proceedings of IEEE Electronic Components and Technology Conference,* pp. 1047–1051, Seattle, WA, May 1998.
182. Gilleo, K., and D. Blumel, "The Great Underfill Race," *Proceedings of the International Symposium on Microelectronics,* pp. 701–706, San Diego, CA, November 1998.
183. Lau, J. H., C. Chang, T. Chen, D. Cheng, and E. Lao, "A Low-Cost Solder-Bumped Chip Scale Package—NuCSP," *Circuit World,* **24** (3): 11–25, April 1998.
184. Elshabini-Riad, A., and F. Barlow, III, *Thin Film Technology Handbook,* McGraw-Hill, New York, 1998.
185. Garrou, P. E., and I. Turlik, *Multichip Module Technology Handbook,* McGraw-Hill, New York, 1998.
186. Lau, J. H., C. Chang, and O. Chien, "SMT Compatible No-Flow Underfill for Solder Bumped Flip Chip on Low-Cost Substrates," *J. Electronics Manufacturing,* **8** (3, 4): 151–164, December 1998.
187. Lau, J. H., and C. Chang, "Characterization of Underfill Materials for Functional Solder Bumped Flip Chips on Board Applications," *IEEE Trans. Components and Packaging Technology,* part A, **22** (1): 111–119, March 1999.
188. Thorpe, R., and D. F. Baldwin, "High Throughput Flip Chip Processing and Reliability Analysis Using No-Flow Underfills," *Proceedings of IEEE Electronic Components and Technology Conference,* pp. 419–425, San Diego, CA, June 1999.
189. Qian, Z., M. Lu, W. Ren, and S. Liu, "Fatigue Life Prediction of Flip-Chips in Terms of Nonlinear Behaviors of Solder and Underfill," *Proceedings of IEEE Electronic Components and Technology Conference,* pp. 141–148, San Diego, CA, June 1999.
190. Wang, L., and C. P. Wong, "Epoxy-Additive Interaction Studies of Thermally Reworkable Underfills for Flip-Chip Applications," *Proceedings of IEEE Electronic Components and Technology Conference,* pp. 34–42, San Diego, CA, June 1999.
191. Lau, J. H., S.-W. Lee, C. Chang, and O. Chien, "Effects of Underfill Material Properties on the Reliability of Solder Bumped Flip Chip on Board with Imperfect Underfill Encapsulants," *Proceedings of IEEE Electronic Components and Technology Conference,* pp. 571–582, San Diego, CA, June 1999.
192. Lau, J. H., C. Chang, and O. Chien, "No-Flow Underfill for Solder Bumped Flip Chip on Low-Cost Substrates," *Proceedings of NEPCON West,* pp. 158–181, February 1999.
193. Tong, Q., A. Savoca, L. Nguyen, P. Fine, and B. Cobb, "Novel Fast Cure and Reworkable Underfill Materials," *Proceedings of IEEE Electronic Components and Technology Conference,* pp. 43–48, San Diego, CA, June 1999.
194. Benjamin, T. A., A. Chang, D. A. Dubois, M. Fan, D. L. Gelles, S. R. Iyer, S. Mohindra, P. N. Tutunjian, P. K. Wang, and W. J. Wright, "CARIVERSE Resin: A Thermally Reversible Network Polymer for Electronic Applications," *Proceedings of IEEE Electronic Components and Technology Conference,* pp. 49–55, San Diego, CA, June 1999.
195. Wada, M., "Development of Underfill Material with High Valued Performance," *Proceedings of IEEE Electronic Components and Technology Conference,* pp. 56–60, San Diego, CA, June 1999.

196. Houston, P. N., D. F. Baldwin, M. Deladisma, L. N. Crane, and M. Konarski, "Low Cost Flip Chip Processing and Reliability of Fast-Flow, Snap-Cure Underfills," *Proceedings of IEEE Electronic Components and Technology Conference,* pp. 61–70, San Diego, CA, June 1999.

197. Kulojarvi, K., S. Pienimaa, and J. K. Kivilahti, "High Volume Capable Direct Chip Attachment Methods," *Proceedings of IEEE Electronic Components and Technology Conference,* pp. 441–445, San Diego, CA, June 1999.

198. Shi, S. H., and C. P. Wong, "Recent Advances in the Development of No-Flow Underfill Encapsulants—A Practical Approach towards the Actual Manufacturing Application," *Proceedings of IEEE Electronic Components and Technology Conference,* pp. 770–776, San Diego, CA, June 1999.

199. Rao, Y., S. H. Shi, and C. P. Wong, "A Simple Evaluation Methodology of Young's Modulus—Temperature Relationship for the Underfill Encapsulants," *Proceedings of IEEE Electronic Components and Technology Conference,* pp. 784–789, San Diego, CA, June 1999.

200. Fine, P., and L. Nguyen, "Flip Chip Underfill Flow Characteristics and Prediction," *Proceedings of IEEE Electronic Components and Technology Conference,* pp. 790–796, San Diego, CA, June 1999.

201. Johnson, C. H., and D. F. Baldwin, "Wafer Scale Packaging Based on Underfill Applied at the Wafer Level for Low-Cost Flip Chip Processing," *Proceedings of IEEE Electronic Components and Technology Conference,* pp. 950–954, San Diego, CA, June 1999.

202. DeBarros, T., P. Neathway, and Q. Chu, "The No-Flow Fluxing Underfill Adhesive for Low Cost, High Reliability Flip Chip Assembly," *Proceedings of IEEE Electronic Components and Technology Conference,* pp. 955–960, San Diego, CA, June 1999.

203. Shi, S. H., T. Yamashita, and C. P. Wong, "Development of the Wafer Level Compressive-Flow Underfill Process and Its Required Materials," *Proceedings of IEEE Electronic Components and Technology Conference,* pp. 961–966, San Diego, CA, June 1999.

204. Chau, M. M., B. Ho, T. Herrington, and J. Bowen, "Novel Flip Chip Underfills," *Proceedings of IEEE Electronic Components and Technology Conference,* pp. 967–974, San Diego, CA, June 1999.

205. Feustel, F., and A. Eckebracht, "Influence of Flux Selection and Underfill Selection on the Reliability of Flip Chips on FR-4," *Proceedings of IEEE Electronic Components and Technology Conference,* pp. 583–588, San Diego, CA, June 1999.

206. Okura, J. H., K. Drabha, S. Shetty, and A. Dasgupta, "Guidelines to Select Underfills for Flip Chip on Board Assemblies," *Proceedings of IEEE Electronic Components and Technology Conference,* pp. 589–594, San Diego, CA, June 1999.

207. Anderson, B., "Development Methodology for a High-Performance, Snap-Cure Flip-Chip Underfill," *Proceedings of NEPCON West,* pp. 135–143, February 1999.

208. Wyllie, G., and B. Miquel, "Technical Advancements in Underfill Dispensing," *Proceedings of NEPCON West,* pp. 152–157, February 1999.

209. Crane, L., A. Torres-Filho, E. Yager, M. Heuel, C. Ober, S. Yang, J. Chen, and R. Johnson, "Development of Reworkable Underfills, Materials, Reliability and Proceeding," *Proceedings of NEPCON West,* pp. 144–151, February 1999.

210. Gilleo, K., "The Ultimate Flip Chip-Integrated Flux/Underfill," *Proceedings of NEPCON West,* pp. 1477–1488, February 1999.

211. Miller, M., I. Mohammed, X. Dai, N. Jiang, and P. Ho, "Analysis of Flip-Chip Packages using High Resolution Moire Interferometry," *Proceedings of IEEE Electronic Components and Technology Conference,* pp. 979–986, San Diego, CA, June 1999.

212. Hanna, C., S. Michaelides, P. Palaniappan, D. Baldwin, and S. Sitaraman, "Numerical and Experimental Study of the Evolution of Stresses in Flip Chip Assemblies During Assembly and Thermal Cycling," *Proceedings of IEEE Electronic Components and Technology Conference,* pp. 1001–1009, San Diego, CA, June 1999.

213. Emerson, J., and L. Adkins, "Techniques for Determining the Flow Properties of Underfill Materials," *Proceedings of IEEE Electronic Components and Technology Conference,* pp. 777–781, San Diego, CA, June 1999.

214. Yao, Q., J. Qu, J. Wu, and C. P. Wong, "Quantitative Characterization of Underfill/Substrate Interfacial Toughness Enhancement by Silane Additives," *Pro-*

ceedings of IEEE Electronic Components and Technology Conference, pp. 1079–1082, San Diego, CA, June 1999.

215. Guo, Y., G. Lehmann, T. Driscoll, and E. Cotts, "A Model of the Underfill Flow Process: Particle Distribution Effects," *Proceedings of IEEE Electronic Components and Technology Conference,* pp. 71–76, San Diego, CA, June 1999.

216. Mercado, L., V. Sarihan, Y. Guo, and A. Mawer, "Impact of Solder Pad Size on Solder Joint Reliability in Flip Chip PBGA Packages," *Proceedings of IEEE Electronic Components and Technology Conference,* pp. 255–259, San Diego, CA, June 1999.

217. Gektin, V., A. Bar-Cohen, and S. Witzman, "Thermo-Structural Behavior of Underfilled Flip-Chips," *Proceedings of IEEE Electronic Components and Technology Conference,* pp. 440–447, Orlando, FL, May 1996.

218. Wu, T., Y. Tsukada, and W. Chen, "Materials and Mechanics Issues in Flip-Chip Organic Packaging," *Proceedings of IEEE Electronic Components and Technology Conference,* pp. 524–534, Orlando, FL, May 1996.

219. Peterson, D. W., J. S. Sweet, S. N. Burchett, and A. Hsia, "Stresses from Flip-Chip Assembly and Underfill: Measurements with the ATC4.1 Assembly Test Chip and Analysis by Finite Element Method," *Proceedings of IEEE Electronic Components and Technology Conference,* pp. 134–143, San Jose, CA, May 1997.

220. Zhou, T., M. Hundt, C. Villa, R. Bond, and T. Lao, "Thermal Study for Flip Chip on FR-4 Boards," *Proceedings of IEEE Electronic Components and Technology Conference,* pp. 879–884, San Jose, CA, May 1997.

221. Ni, G., M. H. Gordon, W. F. Schmidt, and R. P. Selvam, "Flow Properties of Liquid Underfill Encapsulations and Underfill Process Considerations," *Proceedings of IEEE Electronic Components and Technology Conference,* pp. 101–108, Seattle, WA, May 1998.

222. Hoang, L., A. Murphy, and K. Desai, "Methodology for Screening High Performance Underfill Materials," *Proceedings of IEEE Electronic Components and Technology Conference,* pp. 111–116, Seattle, WA, May 1998.

223. Qian, Z., J. Yang, and S. Liu, "Visco-Elastic-Plastic Properties and Constitutive Modeling of Underfills," *Proceedings of IEEE Electronic Components and Technology Conference,* pp. 969–974, Seattle, WA, May 1998.

224. Dai, X., M. V. Brillhart, and P. S. Ho, "Polymer Interfacial Adhesion in Microelectronic Assemblies," *Proceedings of IEEE Electronic Components and Technology Conference,* pp. 132–137, Seattle, WA, May 1998.

225. Zhao, J., X. Dai, and P. Ho, "Analysis and Modeling Verification for Thermal-mechanical Deformation in Flip-chip Packages," *Proceedings of IEEE Electronic Components and Technology Conference,* pp. 336–344, Seattle, WA, May 1998.

226. Matsushima, H., S. Baba, and Y. Tomita, "Thermally Enhanced Flip-chip BGA with Organic Substrate," *Proceedings of IEEE Electronic Components and Technology Conference,* pp. 685–691, Seattle, WA, May 1998.

227. Gurumurthy, C., L. G. Norris, C. Hui, and E. Kramer, "Characterization of Underfill/Passivation Interfacial Adhesion for Direct Chip Attach Assemblies using Fracture Toughness and Hydro-Thermal Fatigue Measurements," *Proceedings of IEEE Electronic Components and Technology Conference,* pp. 721–728, Seattle, WA, May 1998.

228. Palaniappan, P., P. Selman, D. Baldwin, J. Wu, and C. P. Wong, "Correlation of Flip Chip Underfill Process Parameters and Material Properties with In-Process Stress Generation," *Proceedings of IEEE Electronic Components and Technology Conference,* pp. 838–847, Seattle, WA, May 1998.

229. Qu, J., and C. P. Wong, "Effective Elastic Modulus of Underfill Material for Flip-Chip Applications," *Proceedings of IEEE Electronic Components and Technology Conference,* pp. 848–850, Seattle, WA, May 1998.

230. Sylvester, M., D. Banks, R. Kern, and R. Pofahl, "Thermomechanical Reliability Assessment of Large Organic Flip-Chip Ball Grid Array Packages," *Proceedings of IEEE Electronic Components and Technology Conference,* pp. 851–860, Seattle, WA, May 1998.

231. Wiegele, S., P. Thompson, R. Lee, and E. Ramsland, "Reliability and Process Characterization of Electroless Nickel-Gold/Solder Flip Chip Interconnect," *Proceedings*

of *IEEE Electronic Components and Technology Conference,* pp. 861–866, Seattle, WA, May 1998.

232. Caers, J., R. Oesterholt, R. Bressers, T. Mouthaan, and J. Verweij, "Reliability of Flip Chip on Board, First Order Model for the Effect on Contact Integrity of Moisture Penetration in the Underfill," *Proceedings of IEEE Electronic Components and Technology Conference,* pp. 867–871, Seattle, WA, May 1998.

233. Roesner, B., X. Baraton, K. Guttmann, and C. Samin, "Thermal Fatigue of Solder Flip-Chip Assemblies," *Proceedings of IEEE Electronic Components and Technology Conference,* pp. 872–877, Seattle, WA, May 1998.

234. Pang, J., T. Tan, and S. Sitaraman, "Thermo-Mechanical Analysis of Solder Joint Fatigue and Creep in a Flip Chip On Board Package Subjected to Temperature Cycling Loading," *Proceedings of IEEE Electronic Components and Technology Conference,* pp. 878–883, Seattle, WA, May 1998.

235. Gopalakrishnan, L., M. Ranjan, Y. Sha, K. Srihari, and C. Woychik, "Encapsulant Materials for Flip-Chip Attach," *Proceedings of IEEE Electronic Components and Technology Conference,* pp. 1291–1297, Seattle, WA, May 1998.

236. Yang, H., S. Bayyuk, A. Krishnan, and A. Przekwas, L. Nguyen, and P. Fine, "Computional Simulation of Underfill Encapsulation of Flip-Chip ICs, Part I: Flow Modeling and Surface-Tension Effects," *Proceedings of IEEE Electronic Components and Technology Conference,* pp. 1311–1317, Seattle, WA, May 1998.

237. Liu, S., J. Wang, D. Zou, X. He, Z. Qian, and Y. Guo, "Resolving Displacement Field of Solder Ball in Flip-Chip Package by Both Phase Shifting Moire Interferometry and FEM Modeling," *Proceedings of IEEE Electronic Components and Technology Conference,* pp. 1345–1353, Seattle, WA, May 1998.

238. Hong, B., and T. Yuan, "Integrated Flow—Thermomechanical and Reliability Analysis of a Low Air Cooled Flip Chip-PBGA Package," *Proceedings of IEEE Electronic Components and Technology Conference,* pp. 1354–1360, Seattle, WA, May 1998.

239. Wang, J., Z. Qian, D. Zou, and S. Liu, "Creep Behavior of a Flip-Chip Package by Both FEM Modeling and Real Time Moire Interferometry," *Proceedings of IEEE Electronic Components and Technology Conference,* pp. 1439–1445, Seattle, WA, May 1998.

7

Flip Chip on Board with No-Flow Underfills

7.1 Introduction

Solder-bumped flip chips on expensive substrates have been used since the 1960s. However, the past few years have witnessed explosive growth in research efforts devoted to solder-bumped flip chips on low-cost substrate technology.[1–109] There are at least two major reasons why the technology works. One is the existence of high-density substrates with fine lines and spaces, such as the printed circuit board (PCB) with sequential built-up circuits with microvias such as the DYCOstrate, plasma-etched redistribution layers (PERL), surface laminar circuits (SLCs), film redistribution layer (FRL), interpenetrating polymer buildup structure system (IBSS), high-density interconnect (HDI), conductive adhesive bonded flex, sequential bonded films, sequential bonded sheets, and filled microvia technologies, some of which have been discussed in Chap. 4.

The other, probably more important, reason is the underfill epoxy encapsulant used to reduce the effect of the global thermal expansion mismatch between the silicon chip and the low-cost organic substrate. (Since the chip, underfill, and substrate are deformed together as a unit—that is, the relative deformation between the chip and the substrate is very small—the shear deformation of the solder joint is very small.) The other advantages of underfill encapsulant are that it protects the chip from moisture, ionic contaminants, radiation, and hostile operating environments such as thermal and mechanical conditions, shock, and vibration. With proper application of the underfill encapsulant, solder-bumped flip chip on board is reliable to use, as is discussed in Chap. 6 of this book and in Lau and Pao.[34]

The important disadvantages of underfill encapsulant are that it makes rework very difficult and reduces manufacturing throughput. Even though research efforts on reworkable underfill are very active,[6, 17, 27, 34, 37, 40, 49, 51, 52, 55] in most of these studies solvent chemicals are used and most of the chips (such as passivation) and substrates (such as solder mask, via, and copper pads) are degraded or even damaged after rework. These issues indicate that more research needs to be done in this area.

As to the manufacturing throughput issue, fast-flow and fast-cure underfill encapsulants are on their way.[21, 27–29, 31, 34–37, 44, 49, 51, 52, 55, 61, 64, 75] However, the material properties of these underfills can be degraded (due to excessive/large voids, too high a thermal coefficient of expansion, and too low a Young's modulus) and thus affect the mechanical and physical properties of the solder-bumped flip chip on board assembly. Meantime, a class of no-flow underfill encapsulant materials is emerging.[25, 26, 28, 30, 32, 37, 41, 42, 46–49, 53, 56, 60, 66, 70] The advantages of no-flow underfills are that they reduce manufacturing processing steps and increase production throughput.

Basically, no-flow underfills come in two forms: epoxy-based no-clean flux liquids and epoxy-based nonconductive film sheets. The major assembly process for solder-bumped flip chips on low-cost substrate with conventional and no-flow underfill materials is shown in Fig. 7.1. It can be seen that the process flow with the no-flow (liquidlike) underfill material (Figs. 7.1b and e) is simpler than that with the conventional underfill material (Figs. 7.1a and d), and that the former reflows the solder bumps and partially (or fully) cures the underfill material at the same time. On the other hand, the process flow with the no-flow (filmlike) underfill material (Fig. 7.1c) is not SMT compatible and is more complicated than that with the conventional underfill material. In this chapter, the focus is on the epoxy-based no-clean flux liquidlike no-flow underfills. The filmlike no-flow underfill material will only be discussed briefly at the end of this chapter.

7.2   No-Flow Liquidlike Underfill Materials

Three different SMT-compatible no-flow underfills (underfills A through C) are considered. Each consists of a no-clean flux, an epoxy, and an anhydride hardener. The function of the no-clean flux is to prepare and protect the metal surfaces (solder bumps and copper pads) to be soldered by removing surface oxides. This provides a clean metallic surface and prevents further oxidation during the soldering process. The function of the cured epoxy and hardener is to act as a structural adhesive. This provides the necessary strength to hold the chip and the substrate together such that the solder joints are subjected to the min-

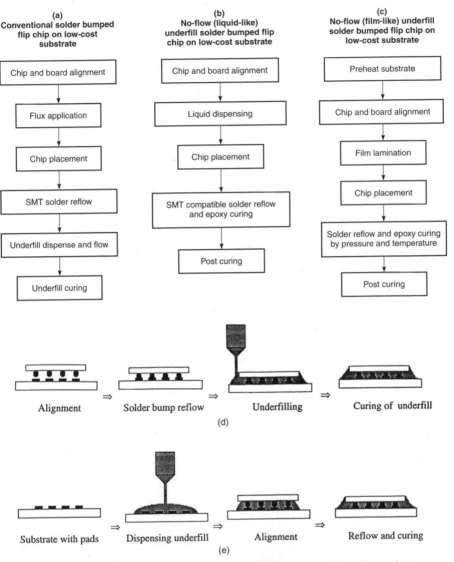

Figure 7.1 (*a–c*) Three different process steps for solder-bumped flip chip on low-cost substrates. (*d*) Schematic of process in *a*. (*e*) Schematic of process in *b*.

imum relative displacements. Since there is no filler in the material, the Young's modulus is very small and the thermal coefficient of expansion is very large.

Underfills A and B are reflowed under normal SMT reflow temperature profiles (dotted lines in Figs. 7.2*a* and *b*), while underfill C is

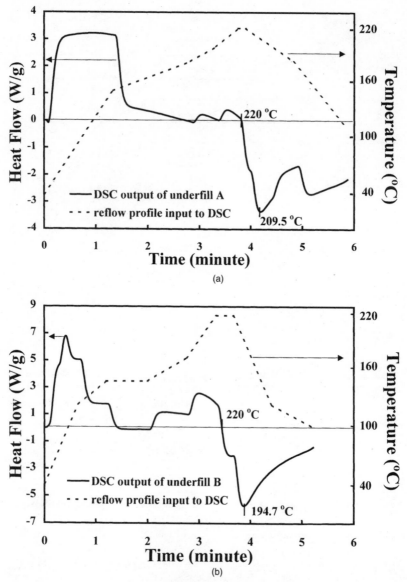

Figure 7.2 DSC thermal scan curves for (*a*) underfill A, (*b*) underfill B. (Dotted lines are solder reflow temperature profiles.)

reflowed under a special reflow temperature profile (dotted line in Fig. 7.2*c*). These reflow temperature profiles are recommended by vendors and are executed in DSC scanning from 40° to 204°C for underfill C and from 40 to 220°C for underfills A and B. Underfills A and B are meant for either solder-bumped flip chips on board (DCA) or on a substrate (flip

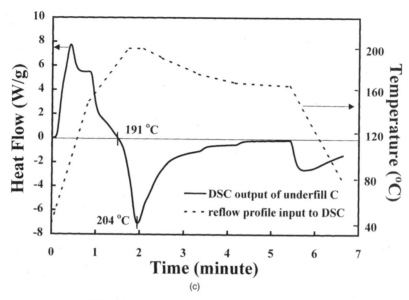

Figure 7.2   (*Continued*) DSC thermal scan curves for (*c*) underfill C. (Dotted lines are solder reflow temperature profiles.)

chip in a package), while underfill C is useful only for solder-bumped flip chip in a package. According to the vendors, after reflow, the postcuring condition for underfill A is 150°C for 30 min and that for underfill B is 160°C for 60 min. Postcure is not required for underfill C.

7.3   Curing Conditions of Liquidlike Underfills

To determine the curing conditions for underfills A through C, the underfills are placed into an aluminum pan (which will form a disc sample 6.4 ± 0.2 mm in diameter and 1.6 ± 0.1 mm thick), weighed, and then placed into DSC equipment. In order to aid in understanding the characteristics of no-flow underfills, the thermal scan is carried out according to the solder reflow temperature profile (dotted lines in Figs. 7.2 *a* to *c* for underfills A through C, respectively). These are the actual (heating rate) inputs to the DSC equipment, which is why they are not smooth curves. It can be seen that the temperature profiles in Figs. 7.2*a* to *b* are very similar to the customary SMT-compatible solder reflow temperature profile (preheat, preflow, reflow, and cooldown). On the other hand, the temperature profile in Fig. 7.2*c* rises very quickly to the peak reflow temperature and then cools down slowly.

The solid lines in Figs. 7.2*a* to *c* show the DSC thermal scan curve for underfills A through C, respectively. First of all, it can be seen that for all the underfills, the heat flow changes with heating rate changes. Second, all the values are not the same.

For underfill A (Fig. 7.2a), during the early stage of heating (within the first 1.5 min of the reflow temperatures) the no-clean flux agent is evaporating—an endothermic process. Then the polymer chains begin to rearrange and react with the anhydride hardener. Finally, the underfill begins to cure near the temperatures at which the heat flows are changing from endothermic to exothermic, i.e., heat is released during the curing process. (The heat releases from high to low temperatures have already been taken care of by substrating a baseline based on the same temperature profile.) It should be noted that underfill A starts curing not only after the solder begins to reflow (at 183°C) but also near the peak reflow temperature (220°C). The peak curing temperature of the raw (unreflowed) underfill A under a typical SMT reflow temperature profile is 209.5°C.

The DSC thermal scan curve for underfill B (Fig. 7.2b) is similar to that for underfill A, except (1) there is a peak of heat flow during no-clean flux evaporation (at about 96°C) and (2) underfill B absorbs more heat prior to curing near the peak reflow temperature (220°C). The peak curing temperature of the raw underfill B is 194.7°C.

The DSC thermal scan curve for underfill C (Fig. 7.2c) is quite different from those for underfills A and B. During no-clean flux evaporation, there is a heat-flow peak similar to that for underfill B but with a higher value. Also, underfill C starts to cure (near 191°C) much earlier than underfills A or B (almost at the same time as the solder starts to reflow). The peak curing temperature of the raw underfill C is 204°C.

In order to determine the percentage of curing of the raw underfills A through C after reflow (i.e., under postcuring conditions), another two sets of DSC measurements are performed. For one set of measurements, the samples used to obtain the solid lines in Figs. 7.2a to c are placed into the DSC equipment, and a thermal scan is carried out at a heating rate of 10°C/min, ranging from 40 to 300°C. For another set of measurements, the raw underfills A through C are placed into the DSC equipment and subjected to the same heating rate. The DSC thermal scan curves for underfills A through C are shown in Fig. 7.3, a–c, respectively. The solid lines are for the underfills after reflow, and the dotted lines are for the raw underfills.

From Fig. 7.3a, it can be seen that underfill A is not fully cured (solid line) after SMT solder reflow temperature. By comparing this curve with the DSC thermal scan curve of the raw underfill A (dotted line), it can be determined that $(334 - 86.8)/334 = 74$ percent of the underfill A is cured due to reflow temperatures. The peak curing temperature of the reflowed underfill A is 219°C.

Similarly, it can be seen from Figs. 7.3b and c that 61 percent of underfill B and 96 percent of underfill C are cured due to the reflow

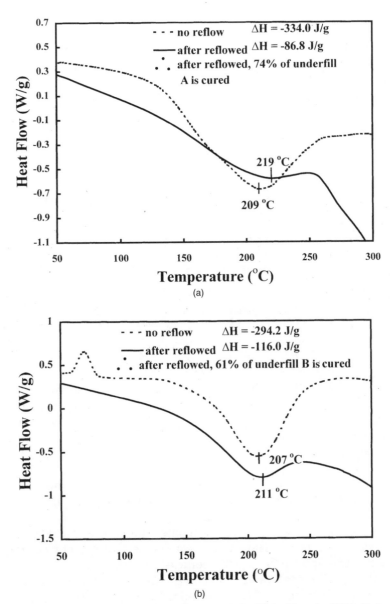

Figure 7.3   DSC thermal scan curve for (*a*) underfill A, (*b*) underfill B. (Dotted lines are for raw underfills. Solid lines are for reflowed underfills.)

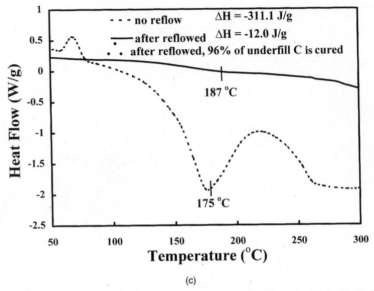

(c)

Figure 7.3 *(Continued)* DSC thermal scan curve for *(c)* underfill C. (Dotted lines are for raw underfills. Solid lines are for reflowed underfills.)

temperatures (dotted lines) shown in Figs. 7.2*b* and *c*, respectively. It should be noted that since 96 percent of underfill C is cured during solder reflow, postcuring of this underfill may not be necessary. The peak temperatures for curing the reflowed underfills B and C are 211 and 187°C, respectively.

The lines in Figs. 7.4*a* to *c* show the typical degree of conversion (reaction)–versus–time curves of the reflowed underfills A through C, respectively. It can be seen that the reflowed underfill A can be fully cured in less than 10 min if the applied curing temperature is 219°C. Also, this underfill can be fully cured in about 30 min if the applied curing temperature is 180°C. However, it cannot be 100 percent cured even in 1 h if the applied curing temperature is less than 150°C. For example, after solder is reflowed, underfill A is only 80 percent cured at a postcuring condition of 150°C for 30 min. The degrees of curing of underfill B (Fig. 7.4*b*) are very similar to those of underfill A. The degrees of curing of underfill C (Fig. 7.4*c*) are very different from those of underfills A and B: underfill C cures much faster. For example, underfill C can be fully cured at a postcuring condition of 204°C in less than 4 min.

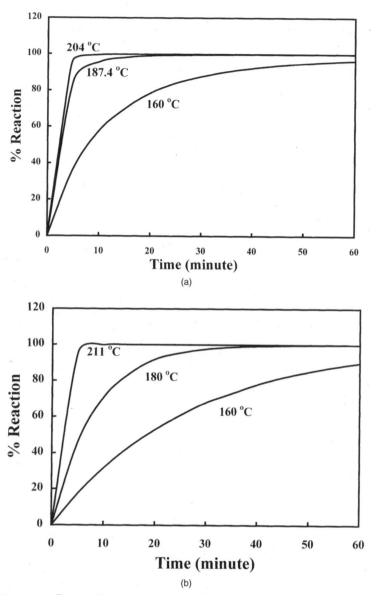

Figure 7.4   Degree of conversion–versus–time curves for (*a*) underfill A, (*b*) underfill B.

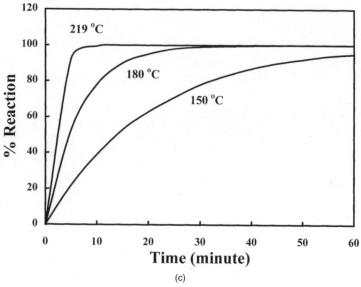

Figure 7.4 (*Continued*) Degree of conversion–versus–time curves for (*c*) underfill C.

7.4   Material Properties of Liquidlike Underfills

The material properties of underfills A through C, such as the thermal coefficient of expansion (TCE), storage modulus, loss modulus, tan δ, glass transition temperature (T_g), and moisture uptake, are determined in the following text.

7.4.1   TCE

The TCEs of underfills A through C (with sample dimensions of 6.4 ± 0.2 mm in diameter and 1.6 ± 0.1 mm in height) are determined by the TMA in an expansion quartz system (50 to 200°C) at a heating rate of 5°C/min. The TCE is obtained by the first slope of the dimensional change–versus–temperature curve.

Figure 7.5 shows the typical expansion curves of underfills A through C. It can be seen that the TCEs of underfills A through C are $70 \times 10^{-6}/°C$, $79 \times 10^{-6}/°C$, and $71 \times 10^{-6}/°C$, respectively, and are very large compared to those of the conventional underfills (20 to $40 \times 10^{-6}/°C$), as shown in Chap. 6. This is because there is no filler in the no-flow underfills. From better solder joint thermal fatigue reliability points of view it is preferable to have lower-TCE ($<27 \times 10^{-6}/°C$) underfill materials. For the materials under consideration, the TCE of underfill B is larger than those of underfills A and C.

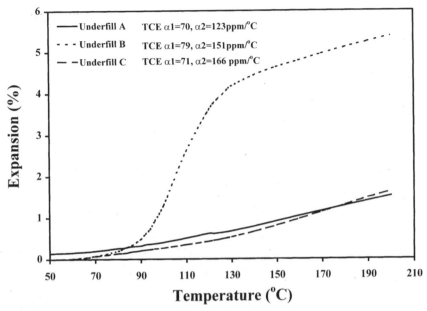

Figure 7.5 Expansion curves for underfills A, B, and C.

7.4.2 Storage modulus and loss modulus

The storage modulus and loss modulus of underfills A through C can be determined with a three-point bending specimen (3.0 ± 0.3 mm $\times 2.9 \pm 0.3$ mm $\times 19 \pm 3$ mm) in a DMA unit (50 to 200°C) at a heating rate of 5°C/min.

Figure 7.6 shows a set of typical flexural storage modules for underfills A through C. It can be seen that the storage modulus of all the no-flow underfill materials is temperature dependent—the higher the temperature, the lower the modulus. Also, the storage modulus of all the no-flow underfill materials is smaller than that of the conventional underfill materials (4 to 7 GPa), as shown in Chap. 6, because there is no filler in the no-flow underfills. It should be pointed out that not only is the TCE of underfill B larger than those of underfills A and C, but also the storage modulus of underfill B is smaller than those of underfills A and C. High TCE and low modulus are not favorable for flip chip solder joint thermal fatigue life.

7.4.3 T_g

Tangent delta (tan δ), a measure of the material-related damping properties of underfills A through C, can be obtained by dividing the loss modu-

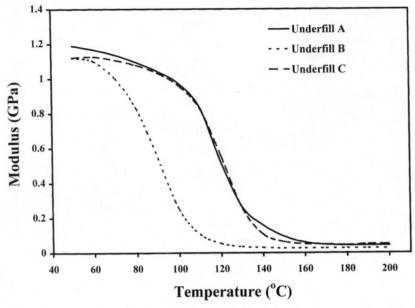

Figure 7.6   Flexural storage modulus of underfills A, B, and C.

lus by the storage modulus. The temperature at the peak of a tan δ curve is often reported in the literature as glass transition temperature (T_g).

Figure 7.7 shows the typical tangent delta curves for underfills A through C. It can be seen that since the T_g curve for underfill B is wider than those for underfills A and C, underfill B has the poorest stiffness. This is because the underfills become softer during a wider transition temperature range. The largest T_g is for underfill C and the smallest T_g is for underfill B. Finally, it should be noted that the T_g values for underfills A through C are lower than those for the conventional underfills (140 to 180°C), as shown in Chap. 6. A high T_g is favorable for endurance in higher-temperature environments.

7.4.4   Moisture content

Two sets of tests are carried out to determine the moisture content of underfills A through C. One is for dry specimens and the other is for steam-aged specimens. The steam-aged specimens are prepared under steam evaporation for 20 h in a closed hot-water bath. (It has been shown in Chap. 6 that this condition is equivalent to 85°C/85% RH for 168 h.) All the specimens are 6.4 ± 0.2 mm in diameter and 1.6 ± 0.1 mm in height. Weight loss for underfills A through C is measured with the TGA equipment at 104°C for 4 h.

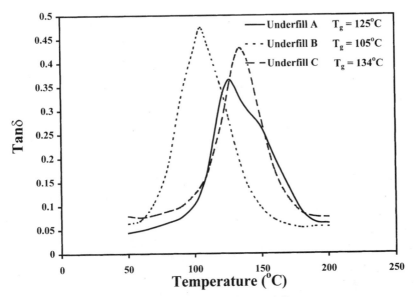

Figure 7.7 Tangent delta curves for underfills A, B, and C.

The change in mass during thermal scan can be expressed as $(W_f - W_i)/W_i$, where W_f is the final weight after thermal scan and W_i is the initial weight before thermal scan. Figures 7.8a to c show the typical percent weight loss (moisture content) for underfills A through C before and after 20 h of steam aging. It can be seen that the moisture

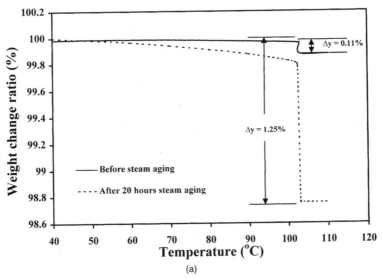

(a)

Figure 7.8 Moisture absorption for underfill A.

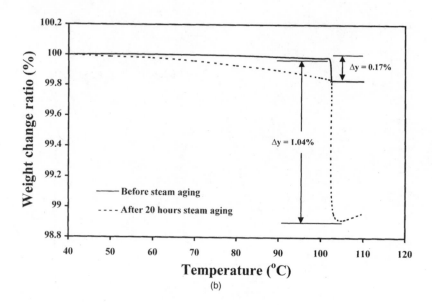

(b)

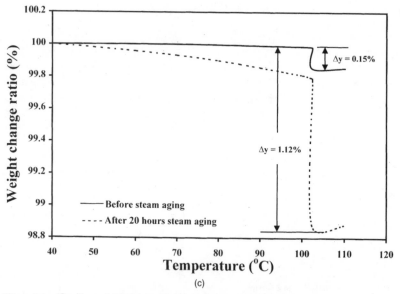

(c)

Figure 7.8 *(Continued)* Moisture absorption for underfills B (*b*) and C (*c*).

content of underfills A through C after 20 h of steam aging is higher than before aging (the dry condition). Also, these underfills absorb more moisture than the conventional underfills (less than 0.1 percent for dry conditions and 0.3 to 0.5 percent for steam aging conditions), as shown in Chap. 6. Low moisture absorption in underfills can extend shelf life.

7.5   FCOB Assembly with Liquidlike No-Flow Underfills

Figure 7.1*b* shows the major process steps with all the liquidlike no-flow underfills. It can be seen that after the 63wt%Sn-37wt%Pb solder-bumped chip (5 × 5 mm with 40 bumps on an 0.18-mm pitch) is aligned with the substrate with look-up and look-down cameras, the no-flow

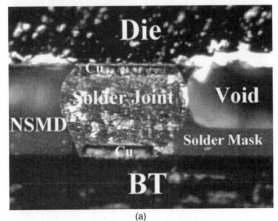

(a)

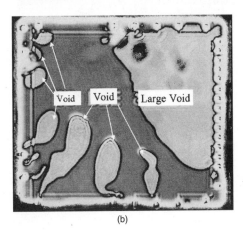

(b)

Figure 7.9   (*a*) Cross-sectional views of flip chip assembly with underfill A. (*b*) TAMI 2-D and (*c*) TAMI 3-D views of the FCOB with underfill A. There are many voids.

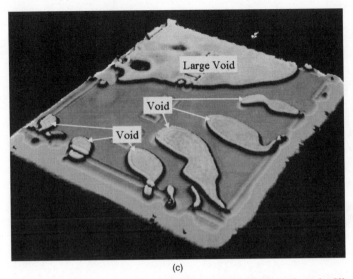

(c)

Figure 7.9 *(Continued)* (c) TAMI 3-D view of the FCOB with underfill A. There are many voids.

underfill is applied on the substrate via a syringe dispenser. (It should be noted that higher manufacturing throughput could be achieved with the screen printing or metal bump transfer methods.) Then the chip is placed face down on the substrate with a very minimal force. During this step, some of the no-flow underfill is squeezed outside the chip.

After placement, the chip is put on the conveyor belt of a reflow oven with a nitrogen gas environment. (It should be pointed out that nitrogen gas is not necessary.) The dotted lines in Figs. 7.2a to c show the temperature profiles for underfills A through C, which are obtained by mounting thermal couples on the substrate during solder reflow. It has been shown in Sec. 7.3 that these no-flow underfills are only partially cured (except underfill C, which is almost fully cured) during solder reflow; thus postcuring (according to the conditions shown in Figs. 7.4a to c) is necessary in order to achieve fully cured physical and mechanical material properties. However, in this study, vendors' recommendations are used (30 min at 150°C for underfill A, 60 min at 160°C for underfill B, and no postcure for underfill C).

Figures 7.9a and b show cross sections of a solder-bumped flip chip on low-cost substrate with underfill A. It can be seen that the solder bumps did solder on the copper pads. However, there are many voids present.

Figures 7.10a to d show cross sections with underfill B. First of all, the solder joint (sharp angle) is not as smooth as that with underfill

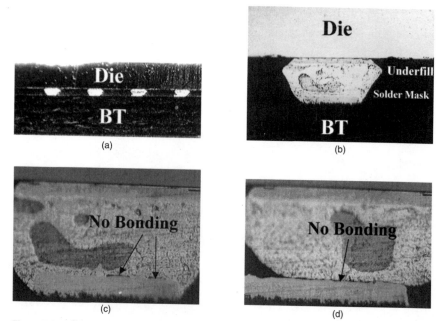

(a)   (b)   (c)   (d)

Figure 7.10   Cross-sectional views of flip chip assembly with underfill B.

A. Figure 7.10*b* shows good bonding between the solder and the copper pad. However, Fig. 7.10*c* shows that there are no bondings between the solder and the copper pad at two locations. Also, Fig. 7.10*d* shows that there is almost no bonding between the solder and the copper pad.

Figure 7.11 shows cross sections with underfill C. Just like those with underfill B, the solder joints are not in the customary truncated

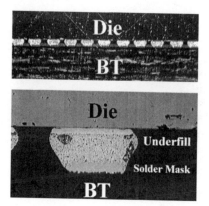

Figure 7.11   Cross-sectional views of flip chip assembly with underfill C.

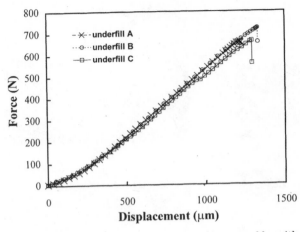

Figure 7.12 Shear test results of flip chip assembly with underfills A, B, and C.

smooth spherical shape. However, most of the solder joints did make good bondings with the copper pads.

7.6   Reliability Testing of FCOB with Liquidlike No-Flow Underfills

7.6.1   Shear test

The no-flow underfill flip chip assemblies are subjected to shear test. A set of typical force displacement curves is shown in Fig. 7.12 for underfills A through C. It can be seen that the curves are very much the same, and that the maximum shear force is about 700 N. The fracture surfaces on the substrate are shown in Figs. 7.13a to c, respectively. It can be seen that some of the solder masks are sheared off from the substrate, which shows that these underfill materials provide very good adhesion. However, there are many voids.

7.6.2   Thermal cycling test

The flip chip assemblies are also subjected to the thermal cycling test with the on-substrate temperature profile shown in Fig. 7.14. All have passed 1600 cycles at the time of writing of this book. The typical cross sections of the tested samples with underfills A through C are shown in Figs. 7.15a to c, respectively. These tested solder joints look very similar to those of the flip chip assembly with conventional underfills after 1500 thermal cycles.[34]

Figure 7.13 Fracture surface of flip chip assembly with underfills A (top), B (middle), and C (bottom).

7.7 Nonlinear Finite Element Analysis of Liquidlike Underfills

Nonlinear finite element analysis of the solder-bumped flip chip assemblies with the no-flow liquidlike underfill material is presented. The calculated stresses and strains are compared with those for the conventional underfill material. The material properties of the analysis are shown in Table 7.1. It can be seen that the Young's modulus of the no-flow liquidlike underfill (3.9 GPa) is much smaller than that of the conventional underfill (~6 GPa). Also, the thermal coefficient of expansion of the no-flow liquidlike underfill (75 ppm/°C) is much larger than that of the conventional underfill (~30 ppm/°C). This is expected, since there is no filler in the no-flow liquidlike underfill material. The Young's modulus of the 63wt%Sn-37wt%Pb solder bumps (Fig. 7.16) and the stress-strain relations (Fig. 7.17) are temperature dependent.

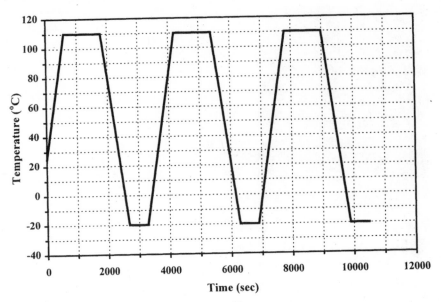

Figure 7.14   Thermal cycling temperature profile.

Two different BT substrate thicknesses are considered: 0.5 and 1.57 mm. The loading condition is that the entire assembly is subjected to a temperature change from 25 to 110°C. The whole-field deformation is shown in Fig. 7.18a for the thinner substrate (0.5 mm). It can be seen that the chip and substrate deform as a unit and that their relative displacement is very small. (Thus, the shear strain in the solder joints is very small.) This is because of the underfill material, which cements the chip and the substrate together. On the other hand, if there is no underfill material (Fig. 7.18b), then the relative displacement is very large and the solder joints are subjected to a very large shear deformation, especially the corner solder joint. This is because of the very large global thermal expansion mismatch between the silicon chip (2.8 ppm/°C) and the BT substrate (15 ppm/°C).

The von Mises stress distributions at the corner solder joint in the thinner-substrate flip chip assembly with the conventional and no-flow liquidlike underfill materials are shown, respectively, in Figs. 7.19a and b. It can be seen that, due to the large local thermal expansion mismatch between the no-flow underfill and the chip, solder joint (21 ppm/°C), and BT substrate, the stresses are higher in the solder joint with the no-flow underfill material (Fig. 7.19b). Similarly, the accumulated effective plastic strains in the solder bump with the no-

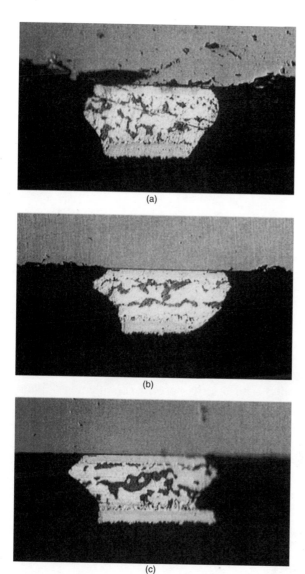

(a)

(b)

(c)

Figure 7.15 Cross-sectional views of solder joints after 1600 thermal cycles (*a*) with underfill A, (*b*) with underfill B, and (*c*) with underfill C.

flow underfill material are higher (Fig. 7.19*d*) than those in the bumps with the conventional underfill material (Fig. 7.19*c*). (The maximum stress and strain are summarized in Table 7.2, which also includes the results for the thicker substrate.) Thus, the thermal reliability of solder joints with the no-flow liquidlike underfill material is not as good as that of solder joints with the conventional underfill material.

TABLE 7.1   Material Properties of the Solder-Bumped Flip Chip Assembly

Material properties	Young's modulus (GPa)	Poisson's ratio (v)	Thermal expansion coefficient (α) ppm/°C
BT substrate	26	0.39	15
Eutectic solder	Temperature dependent	0.4	21
FR4 substrate	22	0.28	18
Si	131	0.30	2.8
Al	69	0.33	22.8
Conventional underfill	6	0.35	30
No-flow liquidlike underfill	3.9	0.35	75
Copper	76	0.35	17
Silicon nitride	314	0.3	3.0
Solder mask	6.9	0.35	19

It is not the intention of this book to predict the thermal fatigue life of the solder joints on the BT substrate. However, it is possible to compare the solder-joint thermal fatigue life with fatigue life values different underfill materials. For 60Sn-40Pb eutectic solders, Solomon[34] (page 323) fit his data at −50, +35, and +125°C into the Coffin-Manson equa-

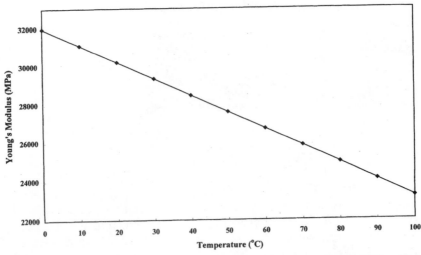

Figure 7.16   Temperature-dependent Young's modulus of the 63wt%Sn-37wt%Pb solder.

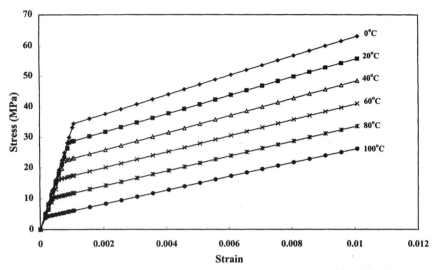

Figure 7.17 Temperature-dependent stress-strain curves for the 63wt%Sn-37wt%Pb solder.

tion and calculated the average thermal fatigue life (N_p) as 1.29 times the accumulated equivalent plastic strain range ($\Delta\gamma_p$) to the power of −1.96, i.e., $N_p = 1.29(\Delta\gamma_p)^{-1.96}$. Based on this equation, it can be shown that the average thermal fatigue life of the solder-bumped flip chip with the conventional underfill material is 7.23 times that of the chip with the liquidlike underfill material on the 0.5-mm (thinner) BT substrate. For the 1.57-mm (thicker) substrate, it is 5.4 times larger. The effect of underfill material is more profound for the thinner substrate.

Figures 7.20a to d shows the von Mises stresses and accumulated effective plastic strains in the corner solder joint on the thicker substrate (1.57 mm) with the no-flow and conventional underfill materials. The trends are the same as those shown in Figs. 7.19a to d, except that the stresses and strains in the solder joints with both underfill materials are larger for the thicker substrate (Table 7.2). This is due to the ability of the stiff thicker substrate to resist the bending of the whole assembly. Thus, it can be shown that the average thermal fatigue life of the solder-bumped flip chip with the conventional underfill material on the 0.5-mm (thinner) BT substrate is 1.87 times that for the chip with the 1.57-mm (thicker) substrate. Also, the average thermal fatigue life of the solder-bumped flip chip with the liquidlike underfill material on the 0.5-mm BT substrate is 1.39 times that of the chip with the 1.57-mm substrate. The effect of substrate thickness is more profound for the conventional underfill material.

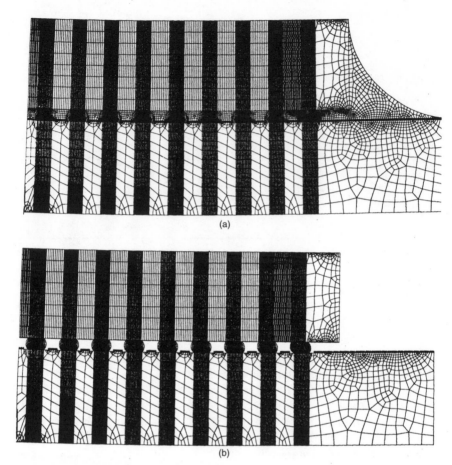

Figure 7.18 (*a*) Deformed shape of the solder-bumped flip chip with underfill encapsulant. The chip and substrate deform as a unit and their relative displacement is very small. (*b*) Deformed shape of the solder-bumped flip chip without underfill encapsulant. The solder joints are subjected to a very large shear deformation, especially the corner solder joint.

7.8 Summary and Recommendations for Liquidlike Underfills

The characteristics of three different epoxy-based no-clean flux liquid-like no-flow underfills used for solder-bumped flip chips on low-cost substrates have been studied. Emphasis is placed on curing temperature and time, thermal coefficient of expansion, storage modulus, loss modulus, tan δ, glass transition temperature, moisture absorption, solder reflow, and postcuring condition. The solder-bumped flip chip

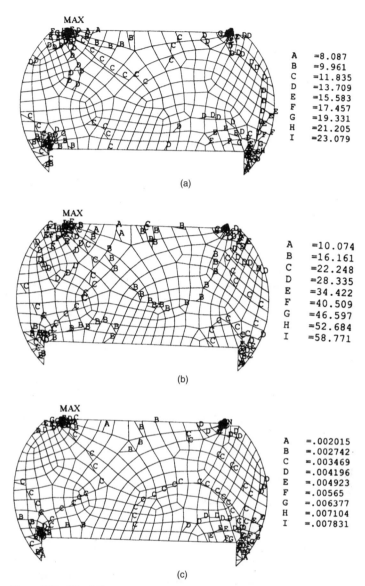

(a)

A =8.087
B =9.961
C =11.835
D =13.709
E =15.583
F =17.457
G =19.331
H =21.205
I =23.079

(b)

A =10.074
B =16.161
C =22.248
D =28.335
E =34.422
F =40.509
G =46.597
H =52.684
I =58.771

(c)

A =.002015
B =.002742
C =.003469
D =.004196
E =.004923
F =.00565
G =.006377
H =.007104
I =.007831

Figure 7.19 Von Mises stress distribution in the corner solder joint on the thinner substrate (a) with the conventional underfill material and (b) with the no-flow (liquidlike) underfill material. Effective plastic strain in the corner solder joint on the thinner substrate (c) with the conventional underfill material.

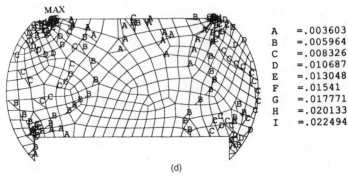

A	=.003603
B	=.005964
C	=.008326
D	=.010687
E	=.013048
F	=.01541
G	=.017771
H	=.020133
I	=.022494

(d)

Figure 7.19 (*Continued*) Effective plastic strain in the corner solder joint on the thinner substrate (*d*) with the no-flow (liquidlike) underfill material.

assemblies with these three no-flow underfills are subjected to shear (destructive) tests and thermal cycling (nondestructive) tests. Cross sections of these no-flow assemblies (before and after tests) have been examined for a better understanding of the effect of such tests on solder joints. Some important results are summarized as follows.

- For all the no-flow underfills considered, the heat flow changes with heating rate changes.

- All the no-flow underfills considered cure faster at higher temperatures.

- For all the no-flow underfills considered, the TCE is much higher than that of the conventional underfills.

- For all the no-flow underfills considered, the storage modulus is temperature dependent and is much lower than that of the conventional underfills.

TABLE 7.2   Maximum von Mises Stress and Effective Plastic Strain for Flip Chip Assemblies with the Conventional and No-Flow Liquidlike Underfill Materials

	Maximum values at the corner solder joint			
	Von Mises stress (MPa)		Effective plastic strain (%)	
	Conventional underfill	Liquidlike underfill	Conventional underfill	Liquidlike underfill
Thin substrate (0.5mm)	23.08	58.77	0.008	0.022
Thick substrate (1.57mm)	31.24	66.45	0.011	0.026

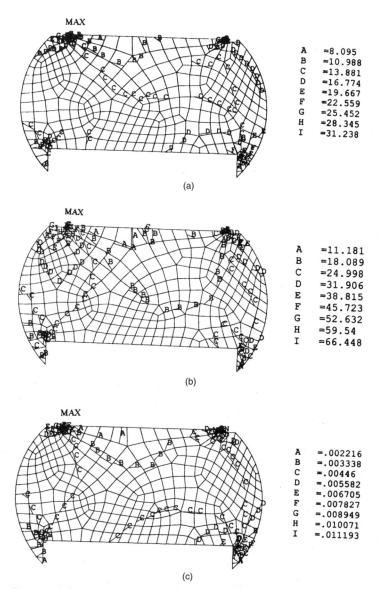

<table>
<tr><td>A</td><td>=8.095</td></tr>
<tr><td>B</td><td>=10.988</td></tr>
<tr><td>C</td><td>=13.881</td></tr>
<tr><td>D</td><td>=16.774</td></tr>
<tr><td>E</td><td>=19.667</td></tr>
<tr><td>F</td><td>=22.559</td></tr>
<tr><td>G</td><td>=25.452</td></tr>
<tr><td>H</td><td>=28.345</td></tr>
<tr><td>I</td><td>=31.238</td></tr>
</table>

(a)

A	=11.181
B	=18.089
C	=24.998
D	=31.906
E	=38.815
F	=45.723
G	=52.632
H	=59.54
I	=66.448

(b)

A	=.002216
B	=.003338
C	=.00446
D	=.005582
E	=.006705
F	=.007827
G	=.008949
H	=.010071
I	=.011193

(c)

Figure 7.20   Von Mises stress distribution in the corner solder joint on the thicker substrate (*a*) with the conventional underfill material and (*b*) with the no-flow (liquidlike) underfill material. Effective plastic strain in the corner solder joint on the thicker substrate (*c*) with the conventional underfill material.

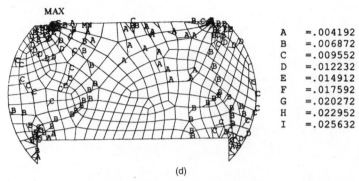

MAX

A	=.004192
B	=.006872
C	=.009552
D	=.012232
E	=.014912
F	=.017592
G	=.020272
H	=.022952
I	=.025632

(d)

Figure 7.20 *(Continued)* Effective plastic strain in the corner solder joint on the thicker substrate (*d*) with the no-flow (liquidlike) underfill material.

- For all the no-flow underfills considered, the tangent delta curve is wider than that of the conventional underfills. This means the no-flow underfills are less stiff than the conventional underfills.

- For all the no-flow underfills considered, the glass transition temperature is lower than that of the conventional underfills.

- For all the no-flow underfills considered, the moisture absorption is higher than that of the conventional underfills.

- Since low TCE and high modulus are favorable for better solder joint thermal fatigue reliability, it is recommended that vendors lower the TCE and raise the modulus of their no-flow underfills.

- Since high T_g is favorable for endurance in higher-temperature environments, it is recommended that vendors raise the T_g of their materials.

- Vendors should lower the moisture absorption of their no-flow underfills to extend shelf life.

- The solder joint thermal life of the flip chip assemblies with the epoxy-based no-clean flux liquidlike underfill material is not as good as that of assemblies with the conventional underfill material.

- The effect of conventional and liquidlike underfill materials on solder joint reliability is more profound for the thinner (0.5-mm) substrate.

- The effect of 0.5- and 1.57-mm substrate thicknesses on solder joint reliability is more profound for the conventional underfill material.

7.9   FCOB with Filmlike No-Flow Underfills

Figure 7.1c shows the process flow of the solder-bumped flip chip on low-cost substrates with epoxy-based nonconductive filmlike underfill encapsulant. Again, it should be noted that this is not an SMT-compatible process. In comparison with the conventional solder-bumped flip chip on low-cost substrate process (Fig. 7.1a), it appears that, with the filmlike underfill materials, additional equipment (such as the lamination machine and hot-press bonder) is necessary and higher manufacturing costs are incurred. (The conventional process is SMT compatible and only requires the additional dispensers for flux and underfill.)

7.9.1   Material

The no-flow filmlike underfill material considered in this study consists of an epoxy and a nonconductive filler such as silica. There is no flux! The film is sandwiched between two layers of release paper. The intended function of the fully cured epoxy is to provide structural integrity to the assembly such that the solder bumps and copper pads can remain in contact. The intended function of the nonconductive fillers is to enhance the mechanical and physical material properties (e.g., higher Young's modulus and lower thermal coefficient of expansion) such that the solder joints are more reliable.

7.9.2   Process

Figure 7.1c shows the major process steps with the no-flow filmlike underfill. It can be seen that the very first step is to preheat the substrate at about 125°C for 24 h. (This could be very inconvenient for a fully populated substrate.) The next step is to align the solder-bumped chip (with a look-up camera) with the substrate (with a look-down camera).

The film is cut slightly larger than the size of the chip and one of the release papers is removed. The film is placed on the substrate with the other release paper facing upward at room temperature, and the substrate is heated on the bonder to about 90°C. At this stage, the film becomes soft and can be laminated onto the substrate with a pressure of 4 kgf/cm^2 or 3.92×10^5 Pa for 5 s (film lamination).

After lamination, both chunks (one holding the chip and the other the substrate) of the pick and place machine are heated up to 215°C, and at the same time the bonding begins with a pressure rising from 0 to 4.9 kgf/cm^2 (4.8×10^5 Pa), as shown in Fig. 7.21. The 63wt%Sn-37wt%Pb solder bumps are melted at 183°C and are subjected to pressure in order to make connection to the copper pads on the substrate

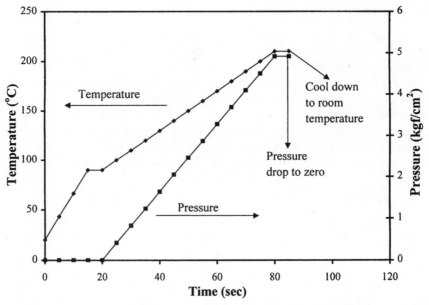

Figure 7.21   A temperature and pressure profile for assembling the solder-bumped flip chip on low-cost substrate with no-flow (filmlike) underfill material.

without flux. Just as with the liquidlike underfill material, postcuring (30 min at 150°C) is required.

Figures 7.22*a* and *b* show top and side views of a flip chip on board with the no-flow filmlike underfill material. It can be seen that some of the solder bumps are pushed outside of the chip. This could be due to either oxidation of the solder bump and copper pad surfaces, excessive pressure, incorrect temperature, material, or all of the above. Figures 7.23*a* to *c* show a cross section of the flip chip assembly with the no-flow filmlike underfill material. It can be seen that the solder bump interconnects are poor. There are many opens (i.e., no connection between the solder bumps and the copper pads) and voids, and some of the solder has been pushed around. Figure 7.24 shows a 3-D tomographic acoustic microimaging (TAMI) view of an assembly with the filmlike underfill. It can be seen that there is a crack on the chip due to the applied pressure.

7.9.3   Shear test

The no-flow filmlike underfill flip chip assemblies are subjected to shear test. A typical force displacement curve is shown in Fig. 7.25. The

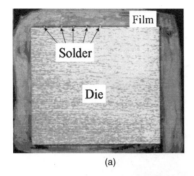

(a)

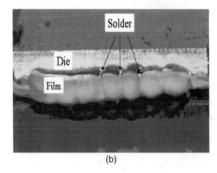

(b)

Figure 7.22 (*a*) Top view and (*b*) side view of the solder-bumped flip chip on low-cost substrate with the no-flow (filmlike) underfill material. The solder bumps have been pushed outside the chip.

maximum shear force is about 62 kgf (607.6 *N*), which shows very good adhesion of the material. The fracture surface is shown in Fig. 7.26. It can be seen that there are many solder bumps pushed outside of the chip that do not make any connection at all.

7.9.4   Summary and recommendations

The 63wt%Sn-37wt%Pb solder-bumped flip chips have been assembled on low-cost organic substrates with the epoxy-based nonconductive (filmlike) material. Cross sections and 3-D scanning images of the chip, solder bumps, substrate, and underfill encapsulants have been examined for a better understanding of the effects of the material on the interconnects of the flip chip assemblies. Some important results are summarized as follows:

- Good solder joints are very difficult to make with the filmlike underfill material.

- The flip chip assembly process with the filmlike underfill material is not SMT compatible.

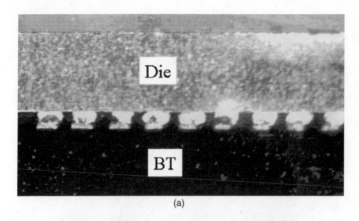

(a)

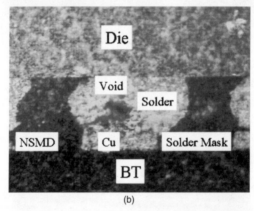

(b)

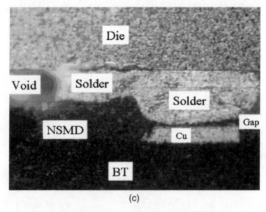

(c)

Figure 7.23 Cross section of a solder-bumped flip chip on low-cost substrate with the no-flow (filmlike) underfill. (*a*) Overall view. (*b*) Close-up view of one solder joint with a large void. (*c*) Close-up view of one solder joint with solder flows into the nearby void. Also, there is a gap between the badly shaped solder joint and copper pad.

Figure 7.24   3-D TAMI view of the solder-bumped flip chip on low-cost substrate with the no-flow (filmlike) underfill material. There are some voids in the underfill and a crack on the chip.

- The flip chip assembly process with the liquidlike material is much simpler than that with the filmlike material.

- The additional equipment and manufacturing floor space required by the flip chip assembly with the filmlike material are greater than those needed with the liquidlike material.

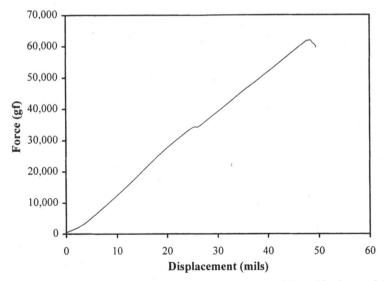

Figure 7.25   A typical shear force–displacement curve of the solder-bumped flip chip on low-cost substrate with the no-flow (filmlike) underfill material.

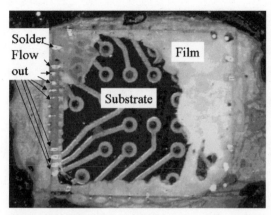

Figure 7.26 Substrate fracture surface of a solder-bumped flip chip on low-cost substrate with the no-flow (filmlike) underfill material after shear test. A good deal of solder is flowed off the copper pads.

- The production cost of flip chip assemblies with the filmlike material is higher than that of assemblies with the liquidlike material.
- Many and large voids exist in the flip chip assemblies with filmlike underfill material.
- The adhesion of filmlike material is excellent.

Acknowledgments

The author thanks C. P. Wong of GIT for promoting the no-flow underfill technology, and Joel Sigmund, Damon Rachell, and Jim William of SONIX and C. Chang and C. Chen of EPS for their useful help and constructive comments.

References

1. Tsukada, Y., Y. Mashimoto, T. Nishio, and N. Mii, "Reliability and Stress Analysis of Encapsulated Flip Chip Joint on Epoxy Base Printed Circuit Board," *Proceedings of the 1st ASME/JSME Advances in Electronic Packaging Conference,* pp. 827–835, Milpitas, CA, April 1992.
2. Guo, Y., W. T. Chen, and K. C. Lim, "Experimental Determinations of Thermal Strains in Semiconductor Packaging Using Moire Interferometry," *Proceedings of the 1st ASME/JSME Advances in Electronic Packaging Conference,* pp. 779–784, Milpitas, CA, April 1992.
3. Tsukada, Y., S. Tsuchida, and Y. Mashimoto, "Surface Laminar Circuit Packaging," *Proceedings of IEEE Electronic Components and Technology Conference,* pp. 22–27, San Diego, CA, May 1992.

4. Lau, J. H., "Thermal Fatigue Life Prediction of Encapsulated Flip Chip Solder Joints for Surface Laminar Circuit Packaging," ASME Paper No. 92W/EEP-34, ASME Winter Annual Meeting, Anaheim, CA, November 1992.
5. Lau, J. H., T. Krulevitch, W. Schar, M. Heydinger, S. Erasmus, and J. Gleason, "Experimental and Analytical Studies of Encapsulated Flip Chip Solder Bumps on Surface Laminar Circuit Boards," Circuit World, 19 (3): 18–24, March 1993.
6. Tsukada, Y., S. Tsuchida, and Y. Mashimoto, "A Novel Chip Replacement Method for Encapsulated Flip Chip Bonding," Proceedings of IEEE Electronic Components and Technology Conference, pp. 199–204, Orlando, FL, June 1993.
7. Powell, D. O., and A. K. Trivedi, "Flip-Chip on FR-4 Integrated Circuit Packaging," Proceedings of IEEE Electronic Components and Technology Conference, pp. 182–186, Orlando, FL, June 1993.
8. Wong, C. P., Polymers for Electronic and Photonic Applications, Academic Press, New York, 1993.
9. Wang, D. W., and K. I. Papathomas, "Encapsulant for Fatigue Life Enhancement of Controlled Collapse Chip Connection (C4)," IEEE Trans. Components, Hybrids, and Manufacturing Technology, 16: 863–867, 1993.
10. Tsukada, Y., "Solder Bumped Flip Chip Attach on SLC Board and Multichip Module," in Chip On Board Technologies for Multichip Modules, Lau, J. H., ed., Van Nostrand Reinhold, New York, pp. 410–443, 1994.
11. Wong, C. P., J. M. Segelken, and C. N. Robinson, "Chip on Board Encapsulation," in Chip On Board Technologies for Multichip Modules, Lau, J. H., ed., Van Nostrand Reinhold, New York, pp. 470–503, 1994.
12. Suryanarayana, D., and D. S. Farquhar, "Underfill Encapsulation for Flip Chip Applications," in Chip On Board Technologies for Multichip Modules, Lau, J. H., ed., Van Nostrand Reinhold, New York, pp. 504–531, 1994.
13. Lau, J. H., M. Heydinger, J. Glazer, and D. Uno, "Design and Procurement of Eutectic Sn/Pb Solder-Bumped Flip Chip Test Die and Organic Substrates," Proceedings of the IEEE International Manufacturing Technology Symposium, pp. 132–138, San Diego, CA, September 1994.
14. Wun, K. B., and J. H. Lau, "Characterization and Evaluation of the Underfill Encapsulants for Flip Chip Assembly," Proceedings of the IEEE International Manufacturing Technology Symposium, pp. 139–146, San Diego, CA, September 1994.
15. Kelly, M., and J. H. Lau, "Low Cost Solder Bumped Flip Chip MCM-L Demonstration," Proceedings of the IEEE International Manufacturing Technology Symposium, pp. 147–153, San Diego, CA, September 1994.
16. Pompeo, F. L., A. J. Call, J. T. Coffin, and S. Buchwalter, "Reworkable Encapsulation for Flip Chip Packaging," Proceedings of the International Intersociety Electronic Packaging Conference, pp. 781–787, Maui, HI, March 1995.
17. Suryanarayana, D., J. A. Varcoe, and J. V. Ellerson, "Reparability of Underfill Encapsulated Flip-Chip Packages," Proceedings of IEEE Electronic Components and Technology Conference, pp. 524–528, Las Vegas, NV, May 1995.
18. Schwiebert, M. K., and W. H. Leong, "Underfill Flow as Viscous Flow Between Parallel Plates Driven by Capillary Action," Proceedings of the IEEE International Manufacturing Technology Symposium, pp. 8–13, Austin, TX, October 1995.
19. Lau, J. H., Ball Grid Array Technology, McGraw-Hill, New York, 1995.
20. Han, S., K. Wang, and S. Cho, "Experimental and Analytical Study on the Flow of Encapsulant During Underfill Encapsulation of Flip-Chips," Proceedings of IEEE Electronic Components and Technology Conference, pp. 327–334, Orlando, FL, May 1996.
21. Wun, K. B., and G. Margaritis, "The Evaluation of Fast-Flow, Fast-Cure Underfills for Flip Chip on Organic Substrates," Proceedings of IEEE Electronic Components and Technology Conference, pp. 540–545, Orlando, FL, May 1996.
22. Hwang, J. S., Modern Solder Technology for Competitive Electronics Manufacturing, McGraw-Hill, New York, 1996.
23. Lau, J. H., "Solder Joint Reliability of Flip Chip and Plastic Ball Grid Array Assemblies Under Thermal, Mechanical, and Vibration Conditions," IEEE Trans. Component, Packaging, and Manufacturing Technology, part B, 19 (4): 728–735, November 1996.

24. Lau, J. H., E. Schneider, and T. Baker, "Shock and Vibration of Solder Bumped Flip Chip on Organic Coated Copper Boards," *ASME Trans. J. Electronic Packaging,* **118:** 101–104, June 1996.
25. Gamota, D., and C. Melton, "Reflowable Material Systems to Integrate the Reflow and Encapsulant Dispensing Process for Flip Chip on Board Assemblies," IPC-TP-1098, 1996.
26. Ito, S., M. Kuwamura, S. Sudo, M. Mizutani, T. Fukushima, H. Noro, S. Akizuki, and A. Prabhu, "Study of Encapsulating System for Diversified Area Bump Packages," *Proceedings of IEEE Electronic Components and Technology Conference,* pp. 46–53, San Jose, CA, May 1997.
27. Lau, J. H., *Flip Chip Technologies,* McGraw-Hill, New York, 1996.
28. Pascarella, N., and D. Baldwin, "Advanced Encapsulation Processing for Low Cost Electronics Assembly—A Cost Analysis," *The 3rd International Symposium and Exhibition on Advanced Packaging Materials, Processes, Properties, and Interfaces,* pp. 50–53, Braselton, GA, March 1997.
29. Nguyen, L., L. Hoang, P. Fine, Q. Tong, B. Ma, R. Humphreys, A. Savoca, C. P. Wong, S. Shi, M. Vincent, and L. Wang, "High Performance Underfills Development— Materials, Processes, and Reliability," *IEEE 1st International Symposium on Polymeric Electronics Packaging,* pp. 300–306, Norrkoping, Sweden, October 1997.
30. Erickson, M., and K. Kirsten, "Simplifying the Assembly Process with a Reflow Encapsulant," *Electronic Packaging and Production,* 81–86, February 1997.
31. Wong, C. P., M. B. Vincent, and S. Shi, "Fast-Flow Underfill Encapsulant: Flow Rate and Coefficient of Thermal Expansion," *Proceedings of the ASME—Advances in Electronic Packaging,* Vol. 19-1, pp. 301–306, 1997.
32. Wong, C. P., S. H. Shi, and G. Jefferson, "High Performance No Flow Underfills for Low-Cost Flip-Chip Applications," *Proceedings of IEEE Electronic Components and Technology Conference,* pp. 850–858, San Jose, CA, May 1997.
33. Tummala, R., E. Rymaszewski, and A. Klopfenstein, *Microelectronics Packaging Handbook,* Chapman & Hall, New York, 1997.
34. Lau, J. H., and Y. H. Pao, *Solder Joint Reliability of BGA, CSP, Flip Chip, and Fine Pitch SMT Assemblies,* McGraw-Hill, New York, 1997.
35. Lau, J. H., C. Chang, and R. Chen, "Effects of Underfill Encapsulant on the Mechanical and Electrical Performance of a Functional Flip Chip Device," *J. Electronics Manufacturing,* **7** (4): 269–277, December 1997.
36. Lau, J. H., and C. Chang, "How to Select Underfill Materials for Solder Bumped Flip Chip on Low Cost Substrates?" *Proceedings of the International Symposium on Microelectronics,* pp. 693–700, San Diego, CA, November 1998.
37. Lau, J. H., C. P. Wong, J. L. Prince, and W. Nakayama, *Electronic Packaging: Design, Materials, Process, and Reliability,* McGraw-Hill, New York, 1998.
38. Nguyen, L., C. Quentin, P. Fine, B. Cobb, S. Bayyuk, H. Yang, and S. A. Bidstrup-Allen, "Underfill of Flip Chip on Laminates: Simulation and Validation," *Proceedings of the International Symposium on Adhesives in Electronics,* pp. 27–30, Binghamton, NY, September 1998.
39. Pascarella, N., and Baldwin, D., "Compression Flow Modeling of Underfill Encapsulants for Low Cost Flip Chip Assembly," *Proceedings of IEEE Electronic Components and Technology Conference,* pp. 463–470, Seattle, WA, May 1998.
40. Nguyen, L., P. Fine, B. Cobb, Q. Tong, B. Ma, and A. Savoca, "Reworkable Flip Chip Underfill—Materials and Processes," *Proceedings of the International Symposium on Microelectronics,* pp. 707–713, San Diego, CA, November 1998.
41. Capote, M. A., and S. Zhu, "No-Underfill Flip-Chip Encapsulation," *Proceedings of Surface Mount International Conference,* pp. 291–293, San Jose, CA, August 1998.
42. Capote, M. A., W. Johnson, S. Zhu, L. Zhou, and B. Gao, "Reflow-Curable Polymer Fluxes for Flip Chip Encapsulation," *Proceedings of the International Conference on Multichip Modules and High Density Packaging,* pp. 41–46, Denver, CO, April 1998.
43. Vincent, M. B., and C. P. Wong, "Enhancement of Underfill Encapsulants for Flip-Chip Technology," *Proceedings of Surface Mount International Conference,* pp. 303–312, San Jose, CA, August 1998.

44. Vincent, M. B., L. Meyers, and C. P. Wong, "Enhancement of Underfill Performance for Flip-Chip Applications by Use of Silane Additives," *Proceedings of IEEE Electronic Components and Technology Conference*, pp. 125–131, Seattle, WA, May 1998.

45. Wang, L., and C. P. Wong, "Novel Thermally Reworkable Underfill Encapsulants for Flip-Chip Applications," *Proceedings of IEEE Electronic Components and Technology Conference*, pp. 92–100, Seattle, WA, May 1998.

46. Shi, S. H., and C. P. Wong, "Study of the Fluxing Agent Effects on the Properties of No-Flow Underfill Materials for Flip-Chip Applications," *Proceedings of IEEE Electronic Components and Technology Conference*, pp. 117–124, Seattle, WA, May 1998.

47. Wong, C. P., D. Baldwin, M. B. Vincent, B. Fennell, L. J. Wang, and S. H. Shi, "Characterization of a No-Flow Underfill Encapsulant During the Solder Reflow Process," *Proceedings of IEEE Electronic Components and Technology Conference*, pp. 1253–1259, Seattle, WA, May 1998.

48. Ito, S., M. Mizutani, H. Noro, M. Kuwamura, and A. Prabhu, "A Novel Flip Chip Technology Using Non-Conductive Resin Sheet," *Proceedings of IEEE Electronic Components and Technology Conference*, pp. 1047–1051, Seattle, WA, May 1998.

49. Gilleo, K., and D. Blumel, "The Great Underfill Race," *Proceedings of the International Symposium on Microelectronics*, pp. 701–706, San Diego, CA, November 1998.

50. Lau, J. H., C. Chang, T. Chen, D. Cheng, and E. Lao, "A Low-Cost Solder-Bumped Chip Scale Package—NuCSP," *Circuit World*, 24 (3): 11–25, April 1998.

51. Elshabini-Riad, A., and F. Barlow, III, *Thin Film Technology Handbook*, McGraw-Hill, New York, 1998.

52. Garrou, P. E., and I. Turlik, *Multichip Module Technology Handbook*, McGraw-Hill, New York, 1998.

53. Lau, J. H., C. Chang, and O. Chien, "SMT Compatible No-Flow Underfill for Solder Bumped Flip Chip on Low-Cost Substrates," *J. Electronics Manufacturing*, 8 (3, 4): 151–164, December 1998.

54. Lau, J. H., and C. Chang, "Characterization of Underfill Materials for Functional Solder Bumped Flip Chips on Board Applications," *IEEE Transactions on Components and Packaging Technology*, part A, 22 (1): pp. 111–119, March 1999.

55. Lau, J. H., and S.-W. Ricky Lee, *Chip Scale Package, Design, Materials, Process, Reliability, and Applications*, McGraw-Hill, New York, 1999.

56. Thorpe, R., and D. F. Baldwin, "High Throughput Flip Chip Processing and Reliability Analysis Using No-Flow Underfills," *Proceedings of IEEE Electronic Components and Technology Conference*, pp. 419–425, San Diego, CA, June 1999.

57. Qian, Z., M. Lu, W. Ren, and S. Liu, "Fatigue Life Prediction of Flip-Chips in Terms of Nonlinear Behaviors of Solder and Underfill," *Proceedings of IEEE Electronic Components and Technology Conference*, pp. 141–148, San Diego, CA, June 1999.

58. Wang, L., and C. P. Wong, "Epoxy-Additive Interaction Studies of Thermally Reworkable Underfills for Flip-Chip Applications," *Proceedings of IEEE Electronic Components and Technology Conference*, pp. 34–42, San Diego, CA, June 1999.

59. Lau, J. H., S.-W. Lee, C. Chang, and O. Chien, "Effects of Underfill Material Properties on the Reliability of Solder Bumped Flip Chip on Board with Imperfect Underfill Encapsulants," *Proceedings of IEEE Electronic Components and Technology Conference*, pp. 571–582, San Diego, CA, June 1999.

60. Lau, J. H., C. Chang, and O. Chien, "No-Flow Underfill for Solder Bumped Flip Chip on Low-Cost Substrates," *Proceedings of NEPCON West*, pp. 158–181, February 1999.

61. Tong, Q., A. Savoca, L. Nguyen, P. Fine, and B. Cobb, "Novel Fast Cure and Reworkable Underfill Materials," *Proceedings of IEEE Electronic Components and Technology Conference*, pp. 43–48, San Diego, CA, June 1999.

62. Benjamin, T. A., A. Chang, D. A. Dubois, M. Fan, D. L. Gelles, S. R. Iyer, S. Mohindra, P. N. Tutunjian, P. K. Wang, and W. J. Wright, "CARIVERSE Resin: A Thermally Reversible Network Polymer for Electronic Applications," *Proceedings of IEEE Electronic Components and Technology Conference*, pp. 49–55, San Diego, CA, June 1999.

63. Wada, M., "Development of Underfill Material with High Valued Performance," *Proceedings of IEEE Electronic Components and Technology Conference*, pp. 56–60, San Diego, CA, June 1999.

64. Houston, P. N., D. F. Baldwin, M. Deladisma, L. N. Crane, and M. Konarski, "Low Cost Flip Chip Processing and Reliability of Fast-Flow, Snap-Cure Underfills," *Proceedings of IEEE Electronic Components and Technology Conference*, pp. 61–70, San Diego, CA, June 1999.

65. Kulojarvi, K., S. Pienimaa, and J. K. Kivilahti, "High Volume Capable Direct Chip Attachment Methods," *Proceedings of IEEE Electronic Components and Technology Conference*, pp. 441–445, San Diego, CA, June 1999.

66. Shi, S. H., and C. P. Wong, "Recent Advances in the Development of No-Flow Underfill Encapsulants—A Practical Approach towards the Actual Manufacturing Application," *Proceedings of IEEE Electronic Components and Technology Conference*, pp. 770–776, San Diego, CA, June 1999.

67. Rao, Y., S. H. Shi, and C. P. Wong, "A Simple Evaluation Methodology of Young's Modulus—Temperature Relationship for the Underfill Encapsulants," *Proceedings of IEEE Electronic Components and Technology Conference*, pp. 784–789, San Diego, CA, June 1999.

68. Fine, P., and L. Nguyen, "Flip Chip Underfill Flow Characteristics and Prediction," *Proceedings of IEEE Electronic Components and Technology Conference*, pp. 790–796, San Diego, CA, June 1999.

69. Johnson, C. H., and D. F. Baldwin, "Wafer Scale Packaging Based on Underfill Applied at the Wafer Level for Low-Cost Flip Chip Processing," *Proceedings of IEEE Electronic Components and Technology Conference*, pp. 950–954, San Diego, CA, June 1999.

70. DeBarros, T., P. Neathway, and Q. Chu, "The No-Flow Fluxing Underfill Adhesive for Low Cost, High Reliability Flip Chip Assembly," *Proceedings of IEEE Electronic Components and Technology Conference*, pp. 955–960, San Diego, CA, June 1999.

71. Shi, S. H., T. Yamashita, and C. P. Wong, "Development of the Wafer Level Compressive-Flow Underfill Process and Its Required Materials," *Proceedings of IEEE Electronic Components and Technology Conference*, pp. 961–966, San Diego, CA, June 1999.

72. Chau, M. M., B. Ho, T. Herrington, and J. Bowen, "Novel Flip Chip Underfills," *Proceedings of IEEE Electronic Components and Technology Conference*, pp. 967–974, San Diego, CA, June 1999.

73. Feustel, F., and A. Eckebracht, "Influence of Flux Selection and Underfill Selection on the Reliability of Flip Chips on FR-4," *Proceedings of IEEE Electronic Components and Technology Conference*, pp. 583–588, San Diego, CA, June 1999.

74. Okura, J. H., K. Drabha, S. Shetty, and A. Dasgupta, "Guidelines to Select Underfills for Flip Chip on Board Assemblies," *Proceedings of IEEE Electronic Components and Technology Conference*, pp. 589–594, San Diego, CA, June 1999.

75. Anderson, B., "Development Methodology for a High-Performance, Snap-Cure Flip-Chip Underfill," *Proceedings of NEPCON WEST* pp. 135–143, February 1999.

76. Wyllie, G., and B. Miquel, "Technical Advancements in Underfill Dispensing," *Proceedings of NEPCON WEST* pp. 152–157, February 1999.

77. Crane, L., A. Torres-Filho, E. Yager, M. Heuel, C. Ober, S. Yang, J. Chen, and R. Johnson, "Development of Reworkable Underfills, Materials, Reliability and Proceeding," *Proceedings of NEPCON WEST* pp. 144–151, February 1999.

78. Gilleo, K., "The Ultimate Flip Chip-Integrated Flux/Underfill," *Proceedings of NEPCON WEST* pp. 1477–1488, February 1999.

79. Miller, M., I. Mohammed, X. Dai, N. Jiang, and P. Ho, "Analysis of Flip-Chip Packages using High Resolution Moire Interferometry," *Proceedings of IEEE Electronic Components and Technology Conference*, pp. 979–986, San Diego, CA, June 1999.

80. Hanna, C., S. Michaelides, P. Palaniappan, D. Baldwin, and S. Sitaraman, "Numerical and Experimental Study of the Evolution of Stresses in Flip Chip Assemblies During Assembly and Thermal Cycling," *Proceedings of IEEE Electronic Components and Technology Conference*, pp. 1001–1009, San Diego, CA, June 1999.

81. Emerson, J., and L. Adkins, "Techniques for Determining the Flow Properties of Underfill Materials," *Proceedings of IEEE Electronic Components and Technology Conference*, pp. 777–781, San Diego, CA, June 1999.

82. Guo, Y., G. Lehmann, T. Driscoll, and E. Cotts, "A Model of the Underfill Flow Process: Particle Distribution Effects," *Proceedings of IEEE Electronic Components and Technology Conference*, pp. 71–76, San Diego, CA, June 1999.

83. Mercado, L., V. Sarihan, Y. Guo, and A. Mawer, *Proceedings of IEEE Electronic Components and Technology Conference*, pp. 255–259, San Diego, CA, June 1999.

84. Qian, Z., M. Lu, W. Ren, and S. Liu, "Fatigue Life Prediction of Flip-Chips in Terms of Nonlinear Behaviors of Solder and Underfill," *Proceedings of IEEE Electronic Components and Technology Conference*, pp. 141–148, San Diego, CA, June 1999.

85. Gektin, V., A. Bar-Cohen, and S. Witzman, "Thermo-Structural Behavior of Underfilled Flip-Chips," *Proceedings of IEEE Electronic Components and Technology Conference*, pp. 440–447, Orlando, FL, May 1996.

86. Wu, T. Y., Y. Tsukada, W. T. Chen, "Materials and Mechanics Issues in Flip-Chip Organic Packaging," *Proceedings of IEEE Electronic Components and Technology Conference*, pp. 524–534, Orlando, FL, May 1996.

87. Doot, R. K., "Motorola's First DCA Product: The Gold Line Pen Pager," *Proceedings of IEEE Electronic Components and Technology Conference*, pp. 535–539, Orlando, FL, May 1996.

88. Greer, S. T., "An Extended Eutectic Solder Bump for FCOB," *Proceedings of IEEE Electronic Components and Technology Conference*, pp. 546–551, Orlando, FL, May 1996.

89. Peterson, D. W., J. S. Sweet, S. N. Burchett, and A. Hsia, "Stresses From Flip-Chip Assembly and Underfill: Measurements with the ATC4.1 Assembly Test Chip and Analysis by Finite Element Method," *Proceedings of IEEE Electronic Components and Technology Conference*, pp. 134–143, San Jose, CA, May 1997.

90. Zhou, T., M. Hundt, C. Villa, R. Bond, and T. Lao, "Thermal Study for Flip Chip on FR-4 Boards," *Proceedings of IEEE Electronic Components and Technology Conference*, pp. 879–884, San Jose, CA, May 1997.

91. Ni, G., M. H. Gordon, W. F. Schmidt, and R. P. Selvam, "Flow Properties of Liquid Underfill Encapsulations and Underfill Process Considerations," *Proceedings of IEEE Electronic Components and Technology Conference*, pp. 101–108, Seattle, WA, May 1998.

92. Hoang, L., A. Murphy, and K. Desai, "Methodology for Screening High Performance Underfill Materials," *Proceedings of IEEE Electronic Components and Technology Conference*, pp. 111–116, Seattle, WA, May 1998.

93. Dai, X., M. V. Brillhart, and P. S. Ho, "Polymer Interfacial Adhesion in Microelectronic Assemblies," *Proceedings of IEEE Electronic Components and Technology Conference*, pp. 132–137, Seattle, WA, May 1998.

94. Zhao, J., X. Dai, and P. Ho, "Analysis and Modeling Verification for Thermal-mechanical Deformation in Flip-chip Packages," *Proceedings of IEEE Electronic Components and Technology Conference*, pp. 336–344, Seattle, WA, May 1998.

95. Matsushima, H., S. Baba, and Y. Tomita, "Thermally Enhanced Flip-chip BGA with Organic Substrate," *Proceedings of IEEE Electronic Components and Technology Conference*, pp. 685–691, Seattle, WA, May 1998.

96. Gurumurthy, C., L. G. Norris, C. Hui, and E. Kramer, "Characterization of Underfill/Passivation Interfacial Adhesion for Direct Chip Attach Assemblies using Fracture Toughness and Hydro-Thermal Fatigue Measurements," *Proceedings of IEEE Electronic Components and Technology Conference*, pp. 721–728, Seattle, WA, May 1998.

97. Palaniappan, P., P. Selman, D. Baldwin, J. Wu, and C. P. Wong, "Correlation of Flip Chip Underfill Process Parameters and Material Properties with In-Process Stress Generation," *Proceedings of IEEE Electronic Components and Technology Conference*, pp. 838–847, Seattle, WA, May 1998.

98. Qu, J., and C. P. Wong, "Effective Elastic Modulus of Underfill Material for Flip-Chip Applications," *Proceedings of IEEE Electronic Components and Technology Conference*, pp. 848–850, Seattle, WA, May 1998.

99. Sylvester, M., D. Banks, R. Kern, and R. Pofahl, "Thermomechanical Reliability Assessment of Large Organic Flip-Chip Ball Grid Array Packages," *Proceedings of IEEE Electronic Components and Technology Conference,* pp. 851–860, Seattle, WA, May 1998.
100. Wiegele, S., P. Thompson, R. Lee, and E. Ramsland, "Reliability and Process Characterization of Electroless Nickel-Gold/Solder Flip Chip Interconnect," *Proceedings of IEEE Electronic Components and Technology Conference,* pp. 861–866, Seattle, WA, May 1998.
101. Caers, J., R. Oesterholt, R. Bressers, T. Mouthaan, J. Verweij, "Reliability of Flip Chip on Board, First Order Model for the Effect on Contact Integrity of Moisture Penetration in the Underfill," *Proceedings of IEEE Electronic Components and Technology Conference,* pp. 867–871, Seattle, WA, May 1998.
102. Roesner, B., X. Baraton, K. Guttmann, and C. Samin, "Thermal Fatigue of Solder Flip-Chip Assemblies," *Proceedings of IEEE Electronic Components and Technology Conference,* pp. 872–877, Seattle, WA, May 1998.
103. Pang, J., T. Tan, and S. Sitaraman, "Thermo-Mechanical Analysis of Solder Joint Fatigue and Creep in a Flip Chip On Board Package Subjected to Temperature Cycling Loading," *Proceedings of IEEE Electronic Components and Technology Conference,* pp. 878–883, Seattle, WA, May 1998.
104. Gopalakrishnan, L., M. Ranjan, Y. Sha, K. Srihari, and C. Woychik, "Encapsulant Materials for Flip-Chip Attach," *Proceedings of IEEE Electronic Components and Technology Conference,* pp. 1291–1297, Seattle, WA, May 1998.
105. Yang, H., S. Bayyuk, A. Krishnan, and A. Przekwas, L. Nguyen, and P. Fine, "Compuptional Simulation of Underfill Encapsulation of Flip-Chip ICs, Part I: Flow Modeling and Surface-Tension Effects," *Proceedings of IEEE Electronic Components and Technology Conference,* pp. 1311–1317, Seattle, WA, May 1998.
106. Liu, S., J. Wang, D. Zou, X. He, Z. Qian, and Y. Guo, "Resolving Displacement Field of Solder Ball in Flip-Chip Package by Both Phase Shifting Moire Interferometry and FEM Modeling," *Proceedings of IEEE Electronic Components and Technology Conference,* pp. 1345–1353, Seattle, WA, May 1998.
107. Hong, B., and T. Yuan, "Integrated Flow—Thermomechanical and Reliability Analysis of a Low Air Cooled Flip Chip-PBGA Package," *Proceedings of IEEE Electronic Components and Technology Conference,* pp. 1354–1360, Seattle, WA, May 1998.
108. Wang, J., Z. Qian, D. Zou, and S. Liu, "Creep Behavior of a Flip-Chip Package by Both FEM Modeling and Real Time Moire Interferometry," *Proceedings of IEEE Electronic Components and Technology Conference,* pp. 1439–1445, Seattle, WA, May 1998.
109. Lau, J. H., C. Chang, and C. Chen, "Characteristics and Reliability of No-Flow Underfills for Solder Bumped Flip Chips on Low Cost Substrates," *Microcircuits and Electronic Packaging,* 4 (22/4), 1999.

Flip Chip on Board
with Imperfect Underfills

8.1 Introduction

The past decade has witnessed explosive growth in the research and development efforts devoted to solder-bumped flip chips on low-cost organic substrates[1-30] as a direct result of the higher requirements for package density and performance and of cost advantages over the ceramic substrates. Just like many other new technologies, low-cost solder-bumped flip chips still face many critical issues. In the development of flip chip on board or on substrate in a plastic package, as mentioned earlier, the following (compared with the wire bonding and conventional surface mount technologies) must be noted and understood:

The infrastructure of flip chip is not well established.

Flip chip expertise is not commonly available.

Wafer bumping is still too costly.

Bare die/wafer is not commonly available.

Bare die/wafer handling is not easy.

Pick and place is more difficult.

Fluxing is more critical.

Underfill is too troublesome for system makers.

Rework is more difficult.

Solder joint reliability is more critical.

Inspection is more difficult.

Flip chip assembly testability is not well established.

Die shrinkage and expansion force changes in PCBs.

There are known-good-die issues.

Chip cracking during solder reflow has been reported.

In this chapter, the solder joint reliability of solder-bumped flip chip on board with different imperfect underfills (in terms of delaminations, voids, and cracks) is presented.

8.2   Possible Failure Modes of FCOB with Imperfect Underfills[17, 26, 27, 30]

Solder joint thermal fatigue reliability of a solder-bumped flip chip on low-cost substrate with a perfect underfill encapsulant has been demonstrated by many researchers through thermal cycling testing and mathematical modeling. Also, it has been shown by three different thermal cycling test results in Chap. 6 that, if the underfill is perfectly applied, there is no solder joint reliability problem. However, because of manufacturing defects in areas such as fluxing, dispensing, and curing, and the outgasing of PCB, imperfections such as delaminations, voids and cracks, and missing fillets are not uncommon. In those cases, solder joint reliability becomes an issue.

Figure 8.1 schematically shows the possible failure modes of solder-bumped flip chip on board or on substrate in a package with imperfect underfills. The first kind of failure mode is a crack that begins at the

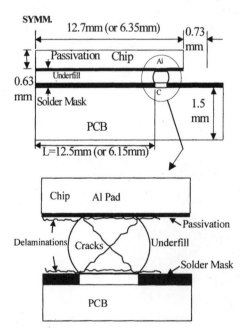

Figure 8.1 Failure modes for solder-bumped flip chip on PCB with imperfect underfill.

upper right or left corner of the solder joint. These initial cracks may propagate either in the horizontal direction near the under-bump metallurgy or down the diagonals of the solder joint. The cause of this failure could be delamination between the underfill and the passivation of the silicon chip. (Based on the author's experience, very often the device fails before the solder joints.) The second kind of failure mode is a crack that begins in the lower right or left corner of the solder joint. These initial cracks may propagate either in the horizontal direction near the copper pad or up the diagonals of the solder joint. This could be due to delamination between the underfill and the solder mask on the organic substrate. Finally, the solder joint could have cracks in all directions due to both delaminations. In addition to these failure modes, there are possible failures due to missing fillet or the presence of large voids in the underfill.

8.3   Fracture Mechanics in Finite Element Analysis

In this chapter, the eight-node plane strain elements are used to model the solder-bumped flip chip assembly with imperfect underfill. The crack (delamination or void) along the interface between the solder mask and the underfill encapsulant (or between the passivation and the underfill encapsulant) is simulated by an array of double nodes (see Fig. 8.2).

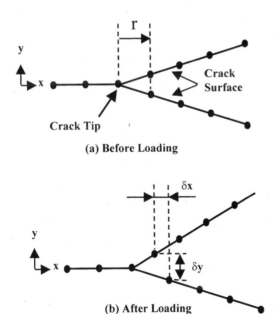

(a) Before Loading

(b) After Loading

Figure 8.2   Schematic diagram for the crack tip opening displacement (CTOD).

The calculation of fracture parameters is based on the crack tip opening displacement (CTOD). According to Hutchinson and Suo,[31] the relationship between stress intensity factors (K_I and K_{II}) and CTOD (δ_x and δ_y, see Fig. 8.2) can be expressed in a complex variable form as

$$\delta_y + i\delta_x = \frac{8}{(1 + 2i\varepsilon)\cosh(\pi\varepsilon)} \frac{(K_I + iK_{II})}{E^*} \left(\frac{r}{2\pi}\right)^{1/2} r^{i\varepsilon} \tag{8.1}$$

where i is the imaginary unity, r is the distance from the crack tip, and x and y are the directions parallel to and perpendicular to the crack, respectively. E^* and ε denote the effective properties of materials adjacent to the crack. For a bimaterial interface crack (in this chapter, assuming underfill as material 1 and solder mask or passivation as material 2), these effective material properties are defined by Rice[32] as

$$E^* = \frac{2E_1E_2}{(E_1 + E_2)} \tag{8.2}$$

$$\varepsilon = \frac{1}{2\pi} \ln\left(\frac{1 - \beta}{1 + \beta}\right) \tag{8.3}$$

And, for the plane strain case,

$$\beta = \frac{1}{2} \frac{\mu_1(1 - 2v_2) - \mu_2(1 - 2v_1)}{\mu_1(1 - v_2) + \mu_2(1 - v_1)} \tag{8.4}$$

where μ and v denote shear modulus and Poisson's ratio, respectively, and the subscript 1 and 2 indicate the corresponding materials. After certain mathematical operations, the stress intensity factors of the opening mode (K_I) and shearing mode (K_{II}) can be obtained from Eq. 8.1 as

$$K_I = [A \cos(\varepsilon \ln r) + B \sin(\varepsilon \ln r)]/D \tag{8.5}$$

$$K_{II} = [B \cos(\varepsilon \ln r) - A \sin(\varepsilon \ln r)]/D \tag{8.6}$$

where

$$A = \delta_y - 2\varepsilon\delta_x \tag{8.7}$$

$$B = \delta_x + 2\varepsilon\delta_y \tag{8.8}$$

$$D = \frac{8}{\cosh(\pi\varepsilon)} \frac{1}{E^*} \left(\frac{r}{2\pi}\right)^{1/2} \tag{8.9}$$

Subsequently, by the formula of Malyshev and Salganik,[33] the strain energy release rate can be obtained as

$$G = \frac{(1 - \beta^2)}{E^*} (K_I^2 + K_{II}^2) = \frac{1}{E^* \cosh^2 (\pi\varepsilon)} K_{eff}^2 \qquad (8.10)$$

It should be noted that

$$(1 - \beta^2) = \frac{1}{\cosh^2 (\pi\varepsilon)} \qquad (8.11)$$

and

$$K_{eff} = \sqrt{K_I^2 + K_{II}^2} \qquad (8.12)$$

is the effective stress intensity factor at the crack tip. Furthermore, a phase angle is defined by Hutchinson and Suo[31] as

$$\psi = \tan^{-1} \left(\frac{K_{II}}{K_I} \right) \qquad (8.13)$$

This quanity is considered an index for measuring the dominance of mode I (opening, $\psi < 45°$) or mode II (shearing, $\psi > 45°$) fracture.

8.4 FCOB with Imperfect Underfills near the Fillet Areas[30]

8.4.1 Problem definition

To simulate the imperfections of underfill fillet, three interfacial delamination cases, namely (1) separation between the underfill encapsulant and the solder mask on the PCB (crack initiated at the tip of underfill fillet), (2) separation between the chip and the underfill encapsulant (crack initiated at the chip corner), and (3) separation between the chip and the underfill encapsulant (crack initiated at the chip corner), but without the underfill fillet, are modeled. The problem under investigation is a flip chip on board assembly. However, the present methodology could be applied to problems with flip chip in package (FCIP) configuration. The detailed dimensions of the FCOB assembly are given in Fig. 8.3.

To study the present problem, ANSYS, a commercial finite element code, is employed. A two-dimensional model is established as shown in Fig. 8.3c, using eight-node plane strain elements. It should be noted that all detailed assembly structures such as chip, Al pads, passivation, UBM, solder joints, Cu pads, and solder mask are modeled in the finite element analysis. Due to the symmetry in the assembly structure, only half of the cross section is considered.

The material properties used in the computational modeling are given in Table 8.1. Except that the eutectic solder (63Sn-37Pb) is a

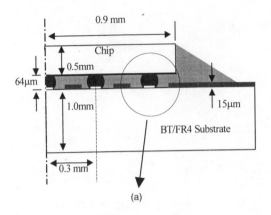

(a)

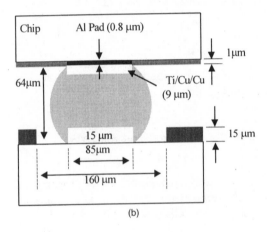

(b)

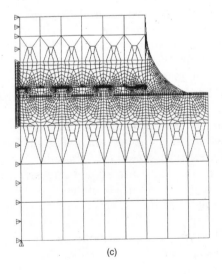

(c)

Figure 8.3 (*a*) Cross section of FCOB assembly showing dimensions of various constituents. (*b*) Close-up diagram around a solder joint. (*c*) Meshes of finite element model.

TABLE 8.1 Material Properties of the Solder-Bumped FCOB Assemblies with Imperfect Underfills

Properties	E (GPa)	ν	α (ppm/°C)
FR4	22	0.28	18
63Sn-37Pb	32–0.088*T(°C)	0.4	21
Si	131	0.3	2.8
Cu	82.7	0.35	17
Al	69	0.33	22.8
Underfill	See Table 8.2	0.35	See Table 8.2
Passivation	4.2	0.35	47
Solder mask	6.9	0.35	19

temperature-dependent elastoplastic material, as shown in Figs. 7.16 and 7.17, all other constituents are considered as linear elastic materials. Five different combinations of TCE and Young's modulus are studied, as shown in Table 8.2.

The main objective of this study is to investigate the behaviors of imperfect underfill encapsulants. The imperfections under consideration are the three types of interfacial delaminations illustrated in Fig. 8.4. These defects might be introduced by surface contamination and poor underfill encapsulation processes. The initial defect could be just a localized flaw, but it might extend to large area of delamination after the application of thermal and mechanical loadings.

Crack propagation is simulated in the computational model, and the strain energy release rate at the crack tip (G) of each crack length is calculated. In addition, the maximum accumulated equivalent plastic strain (ε_p) in the solder joints at each crack length is also evaluated as an index of reliability. In total, there are 15 cases (5 combinations of underfill material properties × 3 types of imperfections) in the present study. For each case under investigation, a temperature drop of 110 → 25°C is applied. G and ε_p are evaluated with respect to progressive crack growth for the comparison of reliability.

TABLE 8.2 Combination of Underfill Material Properties

Case	E (GPa)	α (ppm/°C)
1	6	30
2	3	30
3	9	30
4	6	20
5	6	40

SYMM.

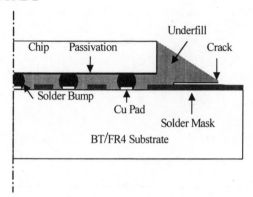

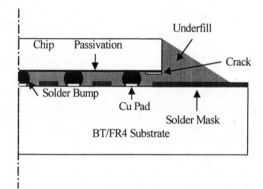

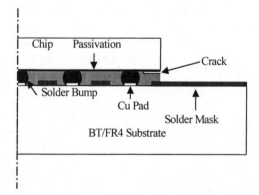

Figure 8.4 Three types of underfill imperfections under investigation. (*a*) Delamination between the underfill and the solder mask. (*b*) Delamination between the underfill and the chip passivation (with fillet). (*c*) Delamination between the underfill and the chip passivation (without fillet).

8.4.2   Effects of imperfect fillet underfills on solder joint reliability

The results of computational analysis are presented in Figs. 8.5 through 8.10 for three types of underfill imperfections. Figure 8.5 shows progressive crack growth along the interface between underfill encapsulant and solder mask. The scale factor for the presentation of deformation is 50. The material properties of the underfill are $E = 6$

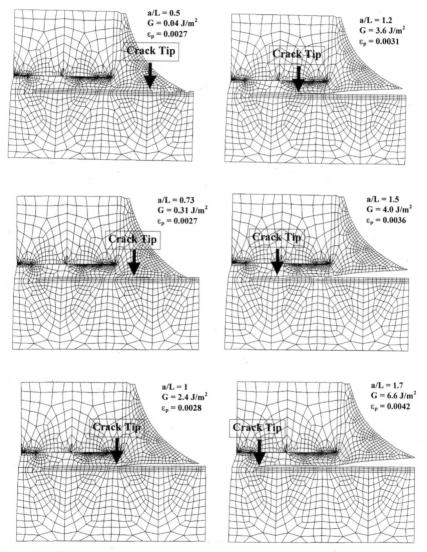

Figure 8.5   Progressive crack growth along the interface between the underfill encapsulant and the solder mask (a = crack length, L = solder joint pitch, G = strain energy release rate, ε_p = maximum accumulated equivalent plastic strain).

GPa and TCE = 30 ppm/°C. It should be noted that the shown mesh is a cutout from the global model for better demonstration. The growth of delamination can be clearly observed.

Figures 8.6 and 8.7 illustrate progressive crack growth along the interface between underfill and chip for the cases with and without encapsulant fillet, respectively. The attributes of demonstration of these two figures are the same as those in Fig. 8.5. It is observed that the crack opening in the case with encapsulant fillet is not significant,

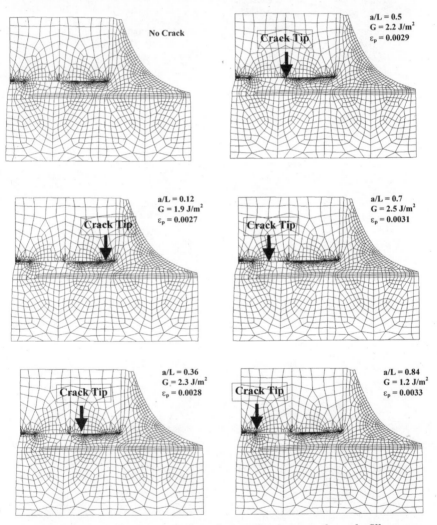

Figure 8.6 Progressive crack growth along the interface between the underfill encapsulant and the chip passivation (a = crack length, L = solder joint pitch, G = strain energy release rate, ε_p = maximum accumulated equivalent plastic strain).

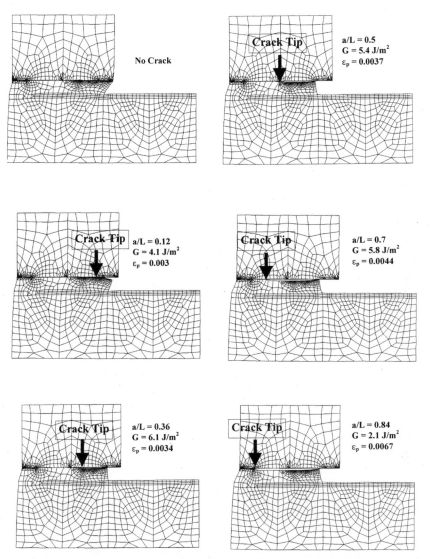

Figure 8.7   Progressive crack growth along the interface between the underfill encapsulant and the chip passivation (no fillet) (a = crack length, L = solder joint pitch, G = strain energy release rate, ε_p = maximum accumulated equivalent plastic strain).

while the case without fillet has a larger crack opening. Judging from the shown deformation, it is obvious that the FCOB assembly without encapsulant fillet is much more critical than is the case with fillet.

Detailed comparisons of the effects of underfill material properties are given in Figs. 8.8 through 8.10. From Fig. 8.8a, for the cases with delamination between underfill encapsulant and solder mask, it can be

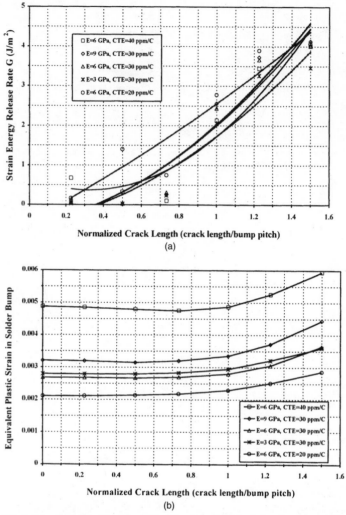

Figure 8.8 Results of computational analysis with delamination along the interface between the underfill and the solder mask. (*a*) Strain energy release rate at the crack tip. (*b*) Maximum accumulated equivalent plastic strain in the solder joint.

seen that the strain energy release rate always increases with respect to the crack growth. This is an indication of an unstable crack and will lead to catastrophic failure at the end. Therefore, such damaging mechanisms should be avoided.

Figure 8.8*b* shows that the maximum plastic strain in the solder increases with respect to the crack growth. This is due to the fact that a longer crack reduces the load transfer to the underfill and thus leads to more loading on the solder joints. This figure also reveals that underfills with higher Young's modulus (9 GPa) and higher TCE (40 ppm/°C) values should result in poorer solder joint reliability.

Figure 8.9a shows a convex shape for the strain energy release rate. This phenomenon indicates that, under a fixed loading, the crack will be arrested after propagating to a certain length. In other words, the delamination will not grow further unless the loading is increased.

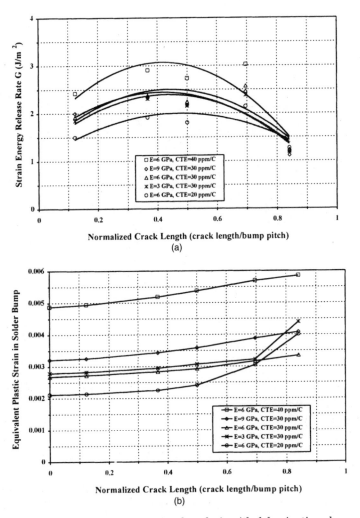

Figure 8.9 Results of computational analysis with delamination along the interface between the underfill and the passivation. (a) Strain energy release rate at the crack tip. (b) Maximum accumulated equivalent plastic strain in the solder joint.

Besides, the values of G are much lower than those shown in Fig. 8.8a. Therefore, this type of imperfection is not as damaging as the previous case. On the other hand, the trends and magnitudes of maximum accumulated equivalent plastic strain in Fig. 8.9b seem similar to those in Fig. 8.8b. It should be noted, however, that the higher values of ε_p are based on the assumption that the delamination would grow to the corresponding length. If the crack were arrested beforehand, the larger ε_p values would have never been reached.

Figure 8.10 presents the results for the case without encapsulant fillet. The general trends are similar to those for the case with fillet. How-

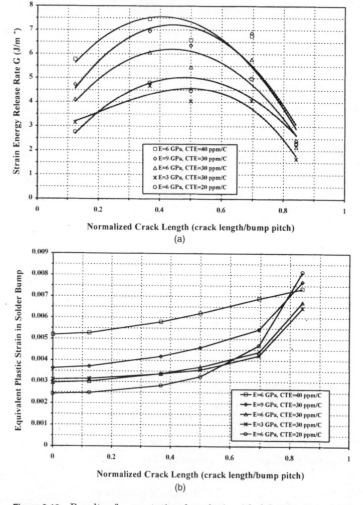

Figure 8.10 Results of computational analysis with delamination along the interface between the underfill and the passivation (no fillet). (a) Strain energy release rate at the crack tip. (b) Maximum accumulated equivalent plastic strain in the solder joint.

ever, although the strain energy release rate shows possible crack arrest behavior, the values of G are much higher than in the other two cases. Furthermore, the values of ε_p are the highest among the three cases under investigation. Therefore, this type of imperfection is definitely unfavorable and should be avoided.

Among all combinations of material properties investigated in this series of parametric studies, the best case (based on the smallest ε_p) is the FCOB assembly with underfill of $E = 6$ GPa and TCE = 20 ppm/°C. When the TCE of the underfill becomes larger, the ε_p increases. However, due to the filler contents, most underfill materials available nowadays have TCEs larger than 20 ppm/°C. Therefore, the next best case with underfill of $E = 6$ GPa and TCE = 30 ppm/°C should be more practical.

On the other hand, for fixed TCE (30 ppm/°C), the ε_p will be raised if the Young's modulus of the underfill is larger than 6 GPa in the present study. This is because the encapsulant becomes too stiff. As a result, due to the local thermal expansion/contraction of underfill material, excessive loading will be exerted on the solder joints (considering that the thermal stress is the product of thermal strain and Young's modulus).

It should be pointed out that for all cases considered in the present study, the maximum accumulated equivalent plastic strain is less than 1 percent and the strain energy release rate is relatively small. It seems that, even with the presence of the three types of imperfections under investigation, the solder joints should be reliable for use at most operating conditions. This is because the chip size in the present analysis is rather small (see Fig. 8.3). Once the chip size becomes larger, because of the increase in the distance from the neutral point (DNP), the solder joint reliability of FCOB assemblies with the aforementioned imperfections may become an issue.

8.5 FCOB with Imperfect Underfills near the Corner Solder Joints (Chip Size Effect)[29]

8.5.1 Problem definition

In this section, before we investigate the problem of delamination (cracking) between the underfill and the solder mask on the PCB near the corner solder joint, we will study the effects of chip size on the solder joint reliability of flip chip on board with a perfect underfill and without underfill. It is well known that without underfill, the larger the chip the higher the stresses and strains at the corner solder joints due to the thermal expansion mismatch between the silicon chip and epoxy glass substrate. However, one of the objectives of this section is to show that, with a perfect underfill, the chip size does not significantly affect the solder joint reliability.

For flip chip assemblies with imperfect underfill, the objectives are to show that (1) for the same delamination (crack length), the chip size does not significantly affect the solder joint reliability, (2) the delamination will not become unstable under the present thermal loading, and (3) the stresses and strains at the corner solder joint increase as the crack length increases. It should be noted that the aim of this study is not to predict the thermal fatigue life of the solder joint, but to investigate the effects of chip size on solder joint reliability with perfect and imperfect underfills.

The loading condition in this study is a temperature drop from 110 to 25°C. The material properties used in the analyses are given in Table 8.3. It can be seen that the eutectic solder (63Sn-37Pb) is considered a temperature-dependent elastoplastic material as shown in Figs. 7.16 and 7.17, and all other constituents are considered linear elastic materials. The chip sizes under consideration are 2.12×1.4 mm, 6.36×4.2 mm, 10.6×7 mm, and 21.2×14 mm. All the chips have solder bumps on the two short edges only. Detailed dimensions of the Al pads and passivation on the chip, the copper pad and solder mask on the PCB, the solder joint, and the PCB are shown in Fig. 8.11a.

8.5.2   Effects of chip size on solder joint reliability without underfill

Rows 2 through 4 of Table 8.4 show the von Mises stress, accumulated equivalent plastic strain, and shear plastic strain in a corner solder joint for different chip sizes on PCB without underfill. It can be seen that (1) all the stresses and plastic strains are very large because of the thermal expansion mismatch between the silicon chip and the FR-4 PCB, (2) the shear plastic strain is larger than the accumulated equivalent plastic strain, since it is dominated by shear deformation, and (3) the stresses and strains increase as the chip size increases.

TABLE 8.3   Material Properties of the FCOB Assemblies

Material properties	Young's modulus (GPa)	Poisson's ratio (v)	Thermal expansion coefficient (α) ppm/°C
FR4	22	0.28	18.5
Copper	76	0.35	17
63Sn-37Pb	See note	0.4	21
Silicon chip	131	0.3	2.8
Solder mask	6.9	0.35	19
AL	69	0.33	22.8
Si$_3$N$_4$	314	0.33	3
Underfill	6	0.35	30

Both the Young's modulus and the stress-strain relationship of solder are temperature dependent.

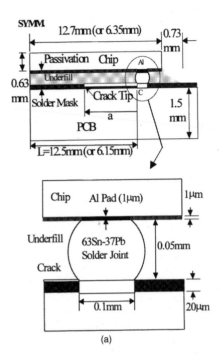

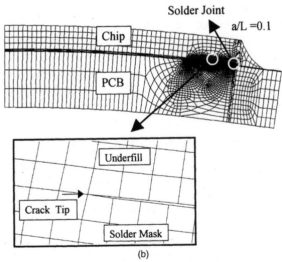

Figure 8.11   (*a*) Solder-bumped flip chip on board with cracks between underfill and solder mask near the corner solder joints. (*b*) Deformed shape of FCOB with crack initiated between underfill and solder mask near the corner solder joint.

TABLE 8.4   Stress and Strains in the Corner Solder Joint with Different Chip Sizes
(with and Without Perfect Underfill)

Chip size (mm)	2.12 × 1.4	6.36 × 4.2	10.6 × 7.0	21.2 × 14
Without underfill				
von Mises stress (MPa)	138	210	248	292
Equivalent plastic strain (%)	3.34	5.53	6.68	8.00
Shear plastic strain (%)	5.56	9.12	11.0	13.1
With underfill				
von Mises stress (MPa)	44.6	44.8	44.8	44.8
Equivalent plastic strain (%)	0.537	0.544	0.544	0.544
Shear plastic strain (%)	0.851	0.862	0.862	0.862

It is interesting to note that the maximum shear stress at the interface between the underfill and the passivation is 23 MPa and that this
value at the interface between the underfill and the solder mask is 11.5
MPa (not shown herein). These values are lower than those measured
in Lau et al.,[30] where the former is over 50 MPa and the latter is over
30 MPa. Thus, cracks at these locations will not be initiated under the
present thermal loading.

8.5.3 Effects of chip size on solder joint reliability with perfect underfill

The last three rows of Table 8.4 show the von Mises stress, accumulated equivalent plastic strain, and shear plastic strain in the corner
solder joint for different chip sizes on PCB with a perfect underfill. It
can be seen that (1) the stresses and plastic strains are almost the
same for all the chip sizes, since the chip and the PCB adhere tightly to
the perfect underfill, and thus the relative deformation of the solder
joint is very small, and (2) the stresses and plastic strains are much
smaller than for chips without underfill.

8.5.4 Effects of chip size on solder joint reliability with imperfect underfill

Figure 8.11b shows the deformed shape of the 10.6 × 7-mm flip chip
assembly with an *imperfect* underfill (crack length = 0.615 mm). The

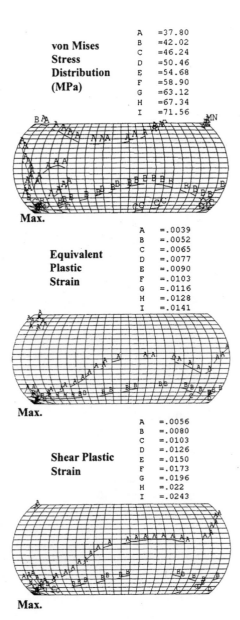

	A	=37.80
von Mises	B	=42.02
Stress	C	=46.24
Distribution	D	=50.46
(MPa)	E	=54.68
	F	=58.90
	G	=63.12
	H	=67.34
	I	=71.56

Max.

	A	=.0039
	B	=.0052
Equivalent	C	=.0065
Plastic	D	=.0077
Strain	E	=.0090
	F	=.0103
	G	=.0116
	H	=.0128
	I	=.0141

Max.

	A	=.0056
	B	=.0080
	C	=.0103
	D	=.0126
Shear Plastic	E	=.0150
Strain	F	=.0173
	G	=.0196
	H	=.022
	I	=.0243

Figure 8.12 Typical stress and strain contours in the corner solder joint.

Max.

von Mises stress, accumulated equivalent plastic strain, and shear plastic strain in the corner solder joint are shown in Fig. 8.12. It can be seen that the maximum values of these parameters are at the lower left corner of the solder joint, where the crack between the underfill and the solder mask began.

TABLE 8.5 Stresses and Strains in the Corner Solder Joint
with Different Chip Sizes and Crack Lengths

Quantities	Crack length (mm)	Chip size (mm)	
		10.6×7	21.2×14
von Mises	0	44.8	44.8
stress	1.27	89.7	89.7
(MPa)	2.54	107	107
	5.14	120	120
Equivalent	0	0.544	0.544
plastic	1.27	1.90	1.90
strain	2.54	2.42	2.42
(%)	5.14	2.82	2.82
Shear	0	0.862	0.862
plastic	1.27	3.27	3.27
strain	2.54	4.16	4.16
(%)	5.14	4.84	4.84

The simulation results for other crack lengths and chip sizes are shown in Table 8.5. It can be seen that (1) for a given crack length, the von Mises stress, accumulated equivalent plastic strain, and shear plastic strain in the corner solder joint are the same for different chip sizes (again, this is because the chip, underfill, and PCB are deformed as a unit and thus the relative displacement of the solder joint is very small), and (2) as the crack lengths increase, all the von Mises stress, accumulated equivalent plastic strain, and shear strain at the corner solder joint increase. Figures 8.13 through 8.15 show, respectively, the von Mises stress, accumulated equivalent plastic strain, and shear plastic strain in the corner solder joint with different crack lengths.

The strain energy release rate at the crack tip for different crack lengths is shown in Fig. 8.16. It can be seen that the strain energy release rate decreases as the crack length increases. This indicates that, for a given crack length and with the current temperature change, the crack will not become unstable. Figure 8.17 shows the phase angle with different crack lengths. It can be seen that the crack between the underfill and the solder mask is dominated by the shearing mode of fracture.

8.5.5 Summary

The effects of chip size on the solder joint thermal fatigue reliability of solder-bumped flip chip on PCB with *perfect* and *imperfect* underfills and without underfill have been investigated. The von Mises stress, accumulated equivalent plastic strain, and shear plastic strain in the corner solder joint, and the strain energy release rate at the crack tip with different crack lengths, are also provided to aid in understanding

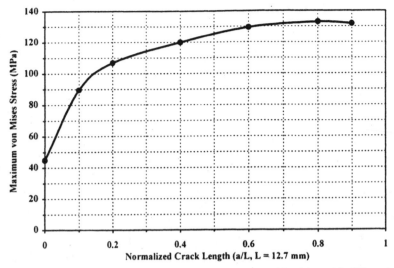

Figure 8.13 Maximum von Mises stress in the corner solder joint with different crack lengths.

the thermal-mechanical behavior of flip chip assemblies. Some important results are summarized as follows.

- For solder-bumped flip chip assembly without underfill, the stress and strain increase as the chip size increases.

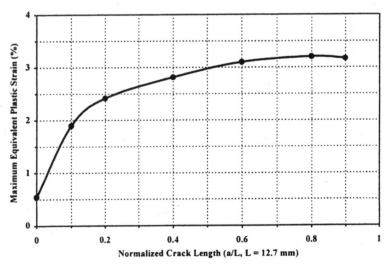

Figure 8.14 Maximum equivalent plastic strain in the corner solder joint with different crack lengths.

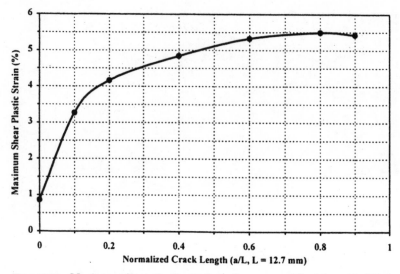

Figure 8.15 Maximum shear plastic strain in the corner solder joint with different crack lengths.

- For solder-bumped flip chip assembly with *perfect* underfill, the chip size does not affect the solder joint reliability. Also, the stresses and strains are much smaller than for chips without underfill.

- For solder-bumped flip chip assembly with *imperfect* underfill and for a given crack length, the chip size does not affect the solder joint reliability.

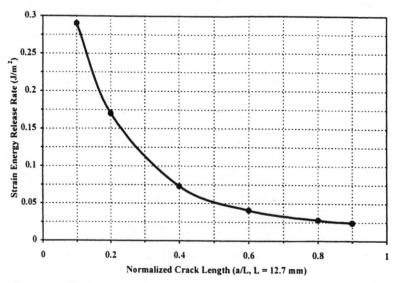

Figure 8.16 Strain energy release rate at the crack tip of the underfill with different crack length.

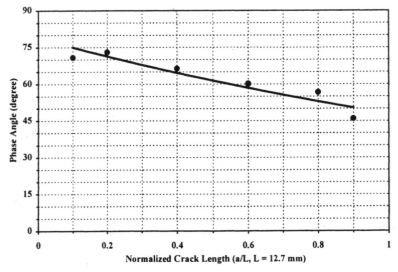

Figure 8.17   Phase angle at the crack tip of the underfill with different crack length.

- For solder-bumped flip chip assembly with *imperfect* underfill, the stresses and strains in the corner solder joint increase as the crack length increases.

- For solder-bumped flip chip assembly with *imperfect* underfill, the strain energy release rate at the crack tip between the underfill and the solder mask decreases as the crack length increases. Thus, the crack along the interface will be arrested under the present thermal loading.

- For solder-bumped flip chip assembly with *imperfect* underfill, the crack between the underfill and the solder mask is dominated by the shearing mode of fracture.

8.6   FCOB with Imperfect Underfill near the Corner Solder Joints (PCB Thickness Effect)

8.6.1   Problem definition

The FCOB assemblies under study are exactly the same as those in Sec. 8.5 except that the focus of this section is on the PCB thickness effects. As shown in Fig. 8.18, three different kinds of PCB thickness are considered, namely, 1.5, 1, and 0.5 mm.

8.6.2   Stresses and strains at the corner solder joint

Figure 8.19 shows the deformed shape of the solder-bumped flip chip assembly with an *imperfect* underfill (crack length = 7.5 mm, PCB

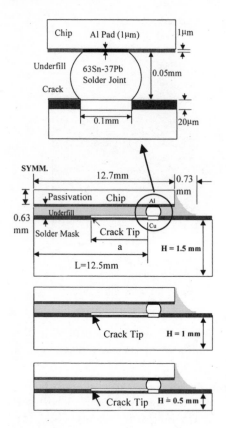

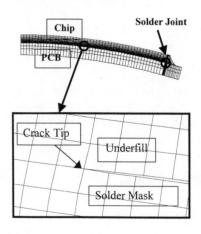

Figure 8.18 Solder-bumped flip chip assembly with cracks (delaminations) between underfill and solder mask. Three different kinds of PCB thickness are considered, $H = 1.5$, 1, and 0.5 mm, respectively.

thickness = 1 mm). The contours of von Mises stress, accumulated equivalent plastic strain, and shear plastic strain in the corner solder joint are shown in Fig. 8.20. It can be seen that the maximum value of these parameters is at the lower left corner of the solder joint, which is near the crack initiation site between the underfill and the solder mask.

Figure 8.19 Deformed shape of a solder-bumped flip chip with imperfect underfill (crack length = 7.5 mm).

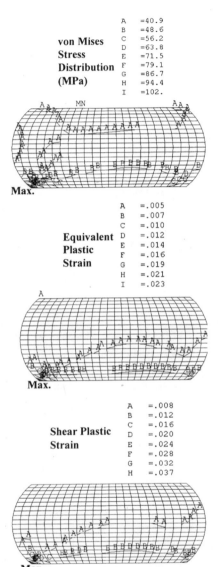

von Mises
Stress
Distribution
(MPa)

A	=40.9
B	=48.6
C	=56.2
D	=63.8
E	=71.5
F	=79.1
G	=86.7
H	=94.4
I	=102.

Max.

Equivalent
Plastic
Strain

A	=.005
B	=.007
C	=.010
D	=.012
E	=.014
F	=.016
G	=.019
H	=.021
I	=.023

Max.

Shear Plastic
Strain

A	=.008
B	=.012
C	=.016
D	=.020
E	=.024
F	=.028
G	=.032
H	=.037

Max.

Figure 8.20 Typical stress and strain contours in the corner solder joint.

For other PCB thicknesses and crack lengths, the simulation results are shown in Figs. 8.21, 8.22, and 8.23, respectively, for the maximum von Mises stress, accumulated equivalent plastic strain, and shear plastic strain in the corner solder joint. Two conclusions can be obtained from these results. (1) For all the crack lengths between the underfill and solder mask considered, the thinner the PCB the smaller the stress and strain in the solder joint. This is because the thinner PCB reduces the bending and stretching stiffness and thus leads to a more compliant

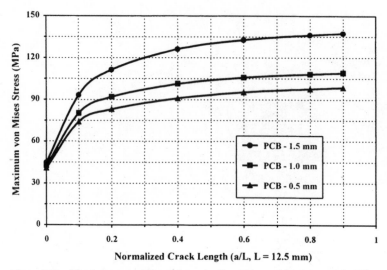

Figure 8.21   Maximum von Mises stress in the corner solder joint with different crack lengths and PCB thicknesses.

structural system. (2) As the crack lengths between the underfill and solder mask increase, for all the PCB thicknesses considered, the von Mises stress, accumulated equivalent plastic strain, and shear plastic strain in the corner solder joint increase. This is because there is less area to support the thermal expansion mismatch between the chip and the PCB.

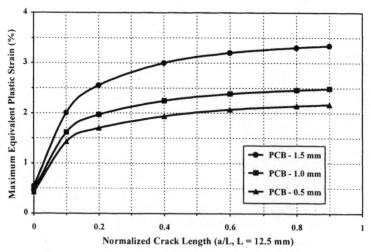

Figure 8.22   Maximum equivalent plastic strain in the corner solder joint with different crack lengths and PCB thicknesses.

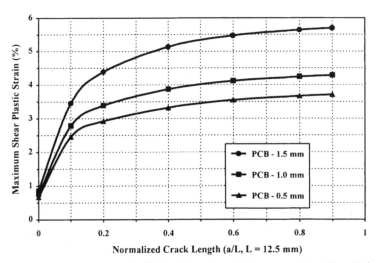

Figure 8.23   Maximum shear plastic strain in the corner solder joint with different crack lengths and PCB thicknesses.

8.6.3   Strain energy release rate and phase angle at the crack tip

The strain energy release rate at the crack tip between the underfill and solder mask for different crack lengths is shown in Fig. 8.24. It can be seen that the strain energy release rate decreases as the crack length increases. This indicates that, for a given crack length and with the cur-

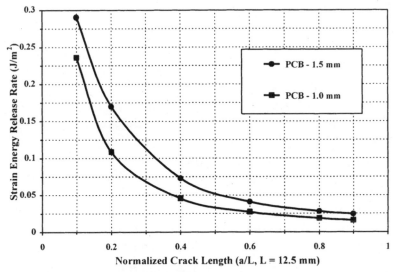

Figure 8.24   Strain energy release rate at the crack tip between the underfill and solder mask with different crack lengths and PCB thicknesses.

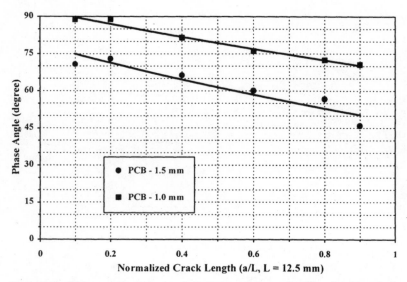

Figure 8.25 Phase angle at the crack tip between the underfill and solder mask with different crack lengths and PCB thicknesses.

rent temperature change, the crack will not become unstable. Figure 8.25 shows the phase angle with different crack lengths. It can be seen that the crack (delamination) between the underfill and the solder mask is dominated by the shearing mode of fracture.

8.6.4 Summary

Some important results are summarized as follows.

- For all the crack (delamination) lengths between the underfill and solder mask considered, the thinner PCB leads to smaller stress and plastic strain in the solder joint.

- For all the PCB thicknesses considered, as the crack length between the underfill and the solder mask increases, the stress and plastic strain in the corner solder joint increase.

- For all the PCB thicknesses under consideration, the strain energy release rate at the crack tip between the underfill and solder mask decreases as the crack length increases.

- For all the PCB thicknesses under consideration, for a given crack length and under the present temperature conditions, the crack will not become unstable, i.e., the crack will be arrested.

- The phase angle with different crack lengths shows that the crack between the underfill and the solder mask is dominated by the shearing mode of fracture.

8.7   Effects of Underfill Voids on Solder Joint Reliability[27]

The effects of underfill voids on flip chip solder joint reliability are studied in this section. The location of the void is at the center of the FCOB assembly.

8.7.1   Problem definition

Figure 8.26 shows the 2-D finite element model for the analysis of the present solder-bumped flip chip assembly along the diagonal direction.

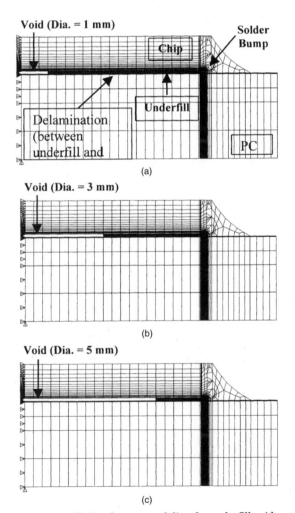

Figure 8.26   Finite element modeling for underfill voids of 1 mm (*a*), 3 mm (*b*), and 5 mm (*c*).

Due to symmetry, only one half of the assembly is considered. It can be seen that there is an underfill void at the center of the assembly. Three different underfill void sizes will be studied, namely 1, 3, and 5 mm in diameter. For each underfill void size, the crack is assumed to be initiated at the underfill void and then to propagate along the interface between the underfill and solder mask of the BT substrate.

It should be noted that the objective of this study is not to determine the thermal fatigue life, but to investigate the effects of underfill void sizes on solder joint reliability. The loading condition is a temperature drop from 110 to 25°C. The material properties used in the analyses are given in Table 8.3. It can be seen that the 63wt%Sn-37wt%Pb solder is considered a temperature-dependent elastoplastic material as shown in Figs. 7.16 and 7.17, and all other constituents are considered linear elastic materials.

8.7.2 Stresses and strains at the corner solder joint

Figure 8.27 shows the deformed shape of a flip chip assembly with an underfill void 1 mm in diameter with full crack length, i.e., the crack tip is at the interface between the underfill and the solder mask, and just approaches the solder joint. The contours of von Mises stress, accumulated equivalent plastic strain, and shear plastic strain in the corner solder joint are shown in Fig. 8.28. It can be seen that the maximum values of these parameters are at the lower left corner of the solder joint, where the final crack tip between the underfill and the solder mask is located.

For other underfill void sizes and crack lengths, the simulation results are shown in Tables 8.6, 8.7, and 8.8, respectively, for the maximum von Mises stress, maximum equivalent plastic strain, and maximum shear plastic strain in the corner solder joint. It can be seen that (1) for different void sizes under consideration, if there is no crack, the stress and strain at the corner solder joint are almost the same (because the remaining underfill and fillet are holding the chip and substrate together); (2) for a given underfill void size, the stress and

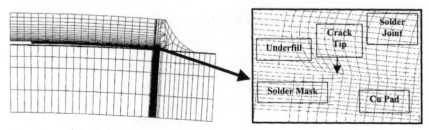

Figure 8.27 Deformed shape of a solder-bumped FCOB with a void of 1 mm and a crack between the underfill and solder mask.

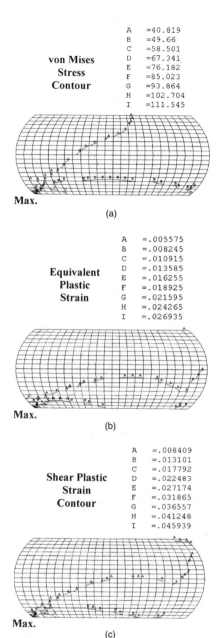

von Mises Stress Contour	A	=40.819
	B	=49.66
	C	=58.501
	D	=67.341
	E	=76.182
	F	=85.023
	G	=93.864
	H	=102.704
	I	=111.545

Max.

(a)

Equivalent Plastic Strain	A	=.005575
	B	=.008245
	C	=.010915
	D	=.013585
	E	=.016255
	F	=.018925
	G	=.021595
	H	=.024265
	I	=.026935

Max.

(b)

Shear Plastic Strain Contour	A	=.008409
	B	=.013101
	C	=.017792
	D	=.022483
	E	=.027174
	F	=.031865
	G	=.036557
	H	=.041248
	I	=.045939

Max.

(c)

Figure 8.28 Typical stress and strain contours in the corner solder joint.

strain in the corner solder joint are basically the same for different crack lengths until the crack reaches full length, and then the values of stress and strain jump to 4 to 5 times higher (because there is less cross-sectional area to support the thermal expansion mismatch); (3) for different void sizes under consideration, even with full crack length,

TABLE 8.6   Comparison of Maximum von Mises Stress in Solder Joint

Void size	von Mises stress (MPa)			
		Crack length		
	No crack	1 mm	2 mm	Full crack
No void	44.79	—	—	115.20
Diameter = 1 mm	44.79	44.81	45.33	115.97
Diameter = 3 mm	44.79	—	—	117.33
Diameter = 5 mm	45.26	—	—	118.66
No underfill	—	—	—	171.76

the stress and strain at the corner solder joint do not change much (because they have the same underfill fillet to support the solder joint); and (4) the stresses and strains considered in points 1, 2, and 3 are much smaller than those of the same flip chip assembly without underfill. Figures 8.29, 8.30, and 8.31 show, respectively, the maximum von Mises stress, accumulated equivalent plastic strain, and shear plastic strain in the corner solder joint with different crack lengths.

8.7.3   Strain energy release rate and phase angle at the crack tip

The strain energy release rate at the crack tip for different crack lengths is shown in Fig. 8.32. It can be seen that the strain energy release rate increases as the crack length increases. This indicates that, for a given crack length and with the current temperature change, the crack will become unstable. Figure 8.33 shows the phase angle with different crack lengths. It can be seen that the crack between the underfill and the solder mask is dominated by the shearing mode of fracture.

TABLE 8.7   Comparison of Maximum Equivalent Plastic Strain in Solder Joint

Void size	Equivalent plastic strain (%)			
		Crack length		
	No crack	1 mm	2 mm	Full crack
No void	0.54	—	—	2.67
Diameter = 1 mm	0.54	0.54	0.56	2.69
Diameter = 3 mm	0.54	—	—	2.73
Diameter = 5 mm	0.56	—	—	2.77
No underfill	—	—	—	4.38

TABLE 8.8 Comparison of Maximum Shear Plastic Strain in Solder Joint

| | Shear plastic strain (%) | | | |
| | Crack length | | | |
Void size	No crack	1 mm	2 mm	Full crack
No void	0.86	—	—	4.55
Diameter = 1 mm	0.86	0.86	0.89	4.59
Diameter = 3 mm	0.86	—	—	4.67
Diameter = 5 mm	0.89	—	—	4.74
No underfill	—	—	—	6.17

8.7.4 Summary

The effects of underfill void sizes on the flip chip solder joint reliability have been studied by nonlinear, temperature-dependent, finite element analyses and fracture mechanics. Some important results are summarized as follows.

- For different underfill void sizes under consideration, if there is no crack, the stress and strain in the corner solder joint are almost the same.

- For a given underfill void size, the stress and strain in the corner solder joint are basically the same for different crack lengths, until the

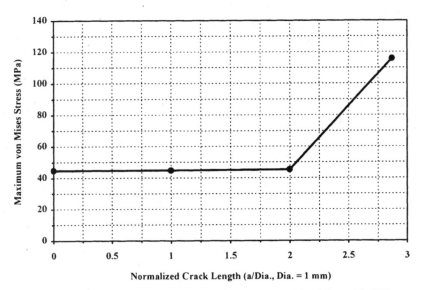

Figure 8.29 Maximum von Mises stress in the corner solder joint with different crack length (underfill void diameter = 1 mm).

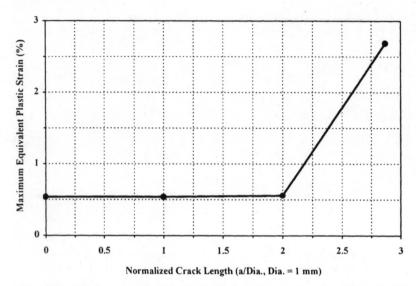

Figure 8.30 Maximum equivalent plastic strain in the corner solder joint with different crack length (underfill void diameter = 1 mm).

crack reaches full length. Then the stress and strain jump to values 4 to 5 times higher.

- For different underfill void sizes under consideration, even with full crack length, the stress and strain in the corner joint do not change much. Also, they are smaller than for chips without underfill.

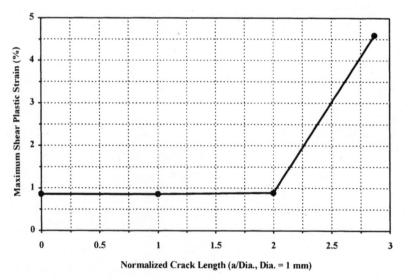

Figure 8.31 Maximum shear plastic strain in the corner solder joint with different crack length (underfill void diameter = 1 mm).

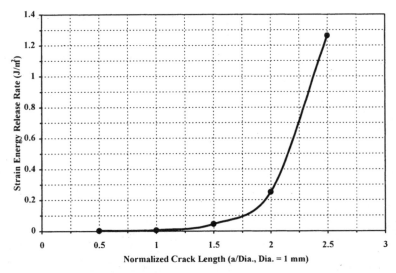

Figure 8.32 Strain energy release rate at the crack tip of the underfill with different crack length.

- Solder joints with perfect underfill most likely will not fail under most of the operating conditions. Thus, underfill materials that have strong adhesion with the passivation and with the solder mask are one of the key conditions necessary for solder-bumped flip chip on low-cost substrate to become popular.

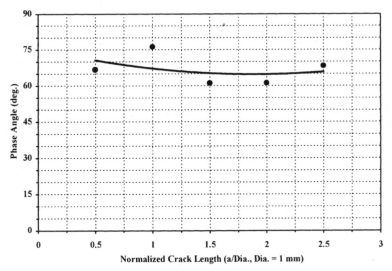

Figure 8.33 Phase angle at the crack tip of the underfill with different crack length.

Acknowledgments

The author would like to thank Professor S. W. Ricky of The Hong Kong University of Science and Technology, and Drs. Chris Chang, Arthur Chen, and Chien Ouyang of EPS, Inc., for their great contributions to this chapter.

References

1. Nguyen, L., L. Hoang, P. Fine, Q. Tong, B. Ma, R. Humphreys, A. Savoca, C. P. Wong, S. Shi, M. Vincent, and L. Wang, "High Performance Underfills Development—Materials, Processes, and Reliability," *IEEE 1st International Symposium on Polymeric Electronics Packaging*, pp. 300–306, Norrkoping, Sweden, October 1997.
2. Wong, C. P., M. B. Vincent, and S. Shi, "Fast-Flow Underfill Encapsulant: Flow Rate and Coefficient of Thermal Expansion," *Proceedings of the ASME—Advances in Electronic Packaging*, vol. 19-1, pp. 301–306, 1997.
3. Wong, C. P., S. H. Shi, and G. Jefferson, "High Performance No Flow Underfills for Low-Cost Flip-Chip Applications," *Proceedings of IEEE Electronic Components and Technology Conference*, pp. 850–858, San Jose, CA, May 1997.
4. Nguyen, L., C. Quentin, P. Fine, B. Cobb, S. Bayyuk, H. Yang, and S. A. Bidstrup-Allen, "Underfill of Flip Chip on Laminates: Simulation and Validation," *Proceedings of the International Symposium on Adhesives in Electronics*, pp. 27–30, Binghamton, NY, September 1998.
5. Pascarella, N., and Baldwin, D., "Compression Flow Modeling of Underfill Encapsulants for Low Cost Flip Chip Assembly," *Proceedings of IEEE Electronic Components and Technology Conference*, Seattle, WA, May 1998, pp. 463–470.
6. Nguyen, L., P. Fine, B. Cobb, Q. Tong, B. Ma, and A. Savoca, "Reworkable Flip Chip Underfill—Materials and Processes," *Proceedings of the International Symposium on Microelectronics*, pp. 707–713, San Diego, CA, November 1998.
7. Wang, L., and C. P. Wong, "Novel Thermally Reworkable Underfill Encapsulants for Flip-Chip Applications," *Proceedings of IEEE Electronic Components and Technology Conference*, pp. 92–100, Seattle, WA, May 1998.
8. Shi, S. H., and C. P. Wong, "Study of the Fluxing Agent Effects on the Properties of No-Flow Underfill Materials for Flip-Chip Applications," *Proceedings of IEEE Electronic Components and Technology Conference*, pp. 117–124, Seattle, WA, May 1998.
9. Wong, C. P., D. Baldwin, M. B. Vincent, B. Fennell, L. J. Wang, and S. H. Shi, "Characterization of a No-Flow Underfill Encapsulant During the Solder Reflow Process," *Proceedings of IEEE Electronic Components and Technology Conference*, pp. 1253–1259, Seattle, WA, May 1998.
10. Wang, J., Z. Qian, D. Zou, and S. Liu, "Creep Behavior of a Flip-Chip Package by Both FEM Modeling and Real Time Moire Interferometry," *Proceedings of IEEE Electronic Components and Technology Conference*, pp. 1439–1445, Seattle, WA, May 1998.
11. Gektin, V., A. Bar-Cohen, and S. Witzman, "Thermo-Structural Behavior of Underfilled Flip-Chips," *Proceedings of IEEE Electronic Components and Technology Conference*, pp. 440–447, Orlando, FL, May 1996.
12. Qian, Z., M. Lu, W. Ren, and S. Liu, "Fatigue Life Prediction of Flip-Chips in Terms of Nonlinear Behaviors of Solder and Underfill," *Proceedings of IEEE Electronic Components and Technology Conference*, pp. 141–148, San Diego, CA, June 1999.
13. Lau, J. H., C. Chang, and R. Chen, "Effects of Underfill Encapsulant on the Mechanical and Electrical Performance of a Functional Flip Chip Device," *J. Electronics Manufacturing*, 7 (4): 269–277, December 1997.
14. Lau, J. H., C. Chang, and C. Ouyang, "SMT Compatible No-Flow Underfill for Solder Bumped Flip Chip on Low-Cost Substrates," *J. Electronics Manufacturing*, 8 (3, 4): 151–164, September and December 1998.
15. Lau, J. H., and C. Chang, "CharaTCErization of Underfill Materials for Functional Solder Bumped Flip Chips on Board Applications," *IEEE Trans. Components and Packaging Technology*, part A, 22 (1): 111–119, March 1999.

16. Lau, J. H., and C. Chang, "How to Select Underfill Materials for Solder Bumped Flip Chip on Low Cost Substrates?" *Int. J. Microelectronics and Electronic Packaging,* **22** (1): 20–28, First quarter 1999.

17. Lau, J. H., S. W. R. Lee, and C. Ouyang, "Effects of Underfill Delamination and Chip Size on the Reliability of Solder Bumped Flip Chip on Board," *Proceedings of IMAPS International Symposium on Microelectronics,* October 1999.

18. Lau, J. H., and Y. H. Pao, *Solder Joint Reliability of BGA, CSP, Flip Chip, and Fine Pitch SMT Assemblies,* McGraw-Hill, New York, 1997.

19. Okura, J. H., K. Drabha, S. Shetty, and A. Dasgupta, "Guidelines to Select Underfills for Flip Chip on Board Assemblies," *Proceedings of IEEE Electronic Components and Technology Conference,* pp. 589–594, San Diego, CA, June 1999.

20. Wu, T. Y., Y. Tsukada, and W. T. Chen, "Materials and Mechanics Issues in Flip-Chip Organic Packaging," *Proceedings of IEEE Electronic Components and Technology Conference,* pp. 524–534, Orlando, FL, May 1996.

21. Liu, S., J. Wang, D. Zou, X. He, and Z. Qian, Y. Guo, "Resolving Displacement Field of Solder Ball in Flip-Chip Package by Both Phase Shifting Moire Interferometry and FEM Modeling," *Proceedings of IEEE Electronic Components and Technology Conference,* pp. 1345–1353, Seattle, WA, May 1998.

22. Lau, J., C. Chang, C. Chen, R. Lee, T. Chen, D. Cheng, T. Tseng, and D. Lin, "Via-In-Pad (VIP) Substrates for Solder Bumped Flip Chip Applications," *Proceedings of Surface Mount International Conference,* pp. 128–136, September 1999.

23. Lau, J. H., "Cost Analysis: Solder Bumped Flip Chip versus Wire Bonding," *IEEE Transactions on Electronics Packaging Manufacturing,* January 2000.

24. Lau, J. H., T. Chung, R. Lee, C. Chang, and C. Chen, "A Novel and Reliable Wafer-Level Chip Scale Package (WLCSP)," *Proceedings of the Chip Scale International Conference,* pp. H1–8, September 1999.

25. Lau, J. H., R. Lee, C. Chang, and C. Chen, "Solder Joint Reliability of Wafer Level Chip Scale Packages (WLCSP): A Time-Temperature-Dependent Creep Analysis," ASME Paper No. 99-IMECE/EEP-5, *International Mechanical Engineering Congress and Exposition,* November 1999.

26. Lau, J. H., and R. Lee, "Effects of Printed Circuit Board Thickness on Solder Joint Reliability of Flip Chip Assemblies with Imperfect Underfill," ASME Paper No. 99-IMECE/EEP-4, *International Mechanical Engineering Congress and Exposition,* November 1999.

27. Lau, J. H., C. Chang, and R. Lee, "Failure Analysis of Solder Bumped Flip Chip on Low-Cost Substrate," *Proceedings of the International Electronic Manufacturing Technology Symposium,* pp. 457–472, October 1999.

28. Lau, J. H., C. Chang, and C. Chen, "Characteristics and Reliability of No-Flow Underfills for Solder Bumped Flip Chips on Low Cost Substrates," *Proceedings of the International Symposium on Microelectronics,* pp. 439–449, October 1999.

29. Lau, J. H., and R. Lee, "Effects of Underfill Delamination and Chip Size on the Reliability of Solder Bumped Flip Chip on Board," *Proceedings of the International Symposium on Microelectronics,* pp. 592–598, October 1999.

30. Lau, J. H., R. Lee, C. Chang, and C. Ouyang, "Effects of Underfill Material Properties on Reliability of Solder Bumped Flip Chip on Board with Imperfect Underfill Encapsulants," *Proceedings of IEEE Electronic Components and Technology Conference,* pp. 571–582, San Diego, CA, June 1999.

31. Hutchinson, J. W., and Z. Suo, "Mixed Mode Cracking in Layered Materials," *Adv. Appl. Mechanics,* **29:** 64–187, 1992.

32. Rice, J. R., "Elastic Fracture Mechanics Concepts for Interfacial Cracks," *ASME Trans. J. Appl. Mechanics,* **55:** 98–103, 1988.

33. Malyshev, B. M. and R. L. Salganik, "The Strength of Adhesive Joints Using the Theory of Cracks," *Int. J. Fracture Mechanics,* **1:** 114–128, 1965.

Chapter

9

Thermal Management of Flip Chip on Board

9.1 Introduction

As mentioned in Chap. 1, one of the major functions of the electronic package is to remove the heat generated by the circuits on the IC chip. Thus, one of the most important characteristics of the electronic package is its ability to dissipate heat in order to ensure its proper operation and reliability.[1-12]

For solder-bumped flip chip on PCB, the power density is higher than that for the IC packages such as the PQFP and PBGA. Will the flip chip device become too hot? To answer this question, useful thermal characteristics of solder-bumped flip chip on FR-4 PCB have been determined by SGS-Thomson Microelectronics.[1-3] In this chapter the effects of PCB construction, air flow, chip size and power dissipation area, underfill, solder-joint population, and heat sink on the thermal resistance of the DCA assembly are determined by experimental measurements and numerical simulations. Also, the thermal paths are analyzed to aid in understanding of the flip chip heat dissipation mechanism.

9.2 The SGS-Thomson Test Chip[1-3]

The SGS-Thomson flip chip thermal test die is schematically shown in Fig. 9.1.[1] It can be seen that the basic cell is 4×4 mm with 48 peripheral pads. Larger chips can be obtained by combinations of the basic cell, e.g., 8×8 and 12×12 mm, which are the focus of this study. The thermal test structure on the die consists of power transistors for heating and diodes for temperature sensing. By powering up the selected transistors, uniform as well as nonuniform power dissipation can be achieved.

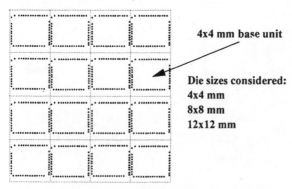

Figure 9.1 The footprint of SGS-Thomson's thermal die.
The die consists of 4 × 4-mm base units. (Not to scale.)

9.3   Effects of PCB Construction

Four different PCB constructions are considered as shown in Fig. 9.2.
They are referred to as A, B, C, and D. It can be seen that PCB A is a
two-layer board without power or ground planes. PCB B is a six-
layer board with one power plane and one ground plane. PCB C is a
six-layer board with one power plane and one ground plane. Also,
right under the chip there is a copper pad and some thermal vias
that are connected to the ground plane. PCB D has six layers, including two power planes and two ground planes. Also, on the top and bottom signal layers, a solid copper layer is added 10 mm away from the

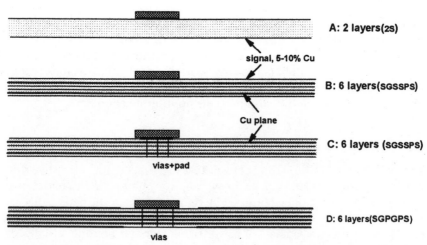

Figure 9.2   Cross sections of the four types of thermal test PCB (101.6 × 114.3 mm).

chip. Furthermore, the thermal vias under the chip are connected to the ground planes. For all the PCBs, the dimensions are 4 × 4.5 in. (101 × 114 mm). After the DCA assembly, the solder joint height is about 100 μm.

Figure 9.3 shows the junction to ambient thermal resistance for the 4 × 4-, 8 × 8-, and 12 × 12-mm solder-bumped flip chip assemblies as functions of PCB construction. As expected, PCB A (without power/ground planes) yields the highest thermal resistance and all boards with power/ground layers (B, C, and D) have much lower thermal resistance. Figure 9.4 shows the comparison between different PCBs for the 8 × 8-mm chip at still air. It can be seen that, with the power and ground planes (B, C, and D), the thermal resistance drops drastically. Two additional copper layers and some solid copper area on signal layers (D) further reduce the thermal resistance. However, PCB D may not be realistic.

9.4   Effects of Air Flow

The effect of air flow is to reduce the thermal resistance, as shown in Fig. 9.3. With 1-m/s air flow, the thermal resistance is dropped by 5 to 7°C/W for PCB B, C, and D and by 15°C/W for PCB A. Thus, the air flow helps more on the boards with low thermal conductivity than the board with high thermal conductivity.[1] The effect of air flow on thermal resistance reduction is slightly larger for larger chips (Fig. 9.5).

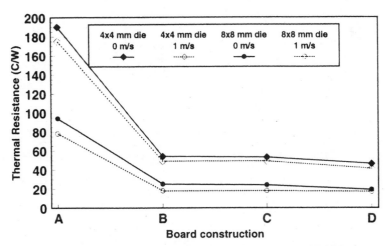

Figure 9.3   Still air junction to ambient thermal resistance of FCOB as a function of board construction. Uniform power dissipation is assumed.

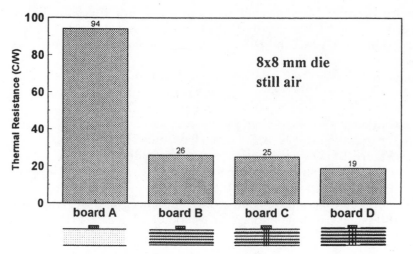

Figure 9.4   Effect of power and ground planes in PCB on thermal resistance.

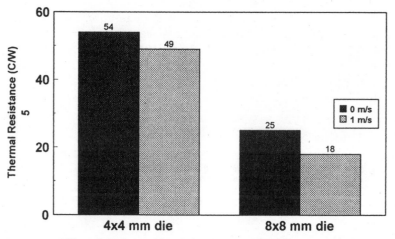

Figure 9.5   Effect of air flow on thermal resistance (PCB C).

9.5   Effects of Chip Size and Power Dissipation Area

The effects of chip size on the thermal resistance of the solder-bumped flip chip assemblies are shown in Fig. 9.6. As expected, the larger the chips, the smaller the thermal resistances (Fig. 9.7). Also, the reduction is more prominent for PCB A (Fig. 9.6). The three curves for PCB B, C, and D are approximately parallel, which indicates that the chip and board thermal resistances can be decoupled for multilayer boards.[1]

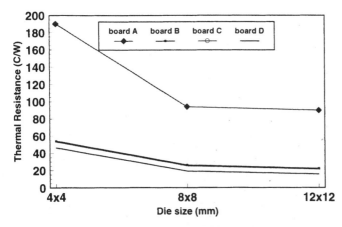

Figure 9.6 Still air junction to ambient thermal resistance of FCOB as a function of chip size and board construction. Uniform power dissipation is applied to the 4×4- and 8×8-mm chip. The power dissipation area for the 12×12-mm chip is 8×8 mm.

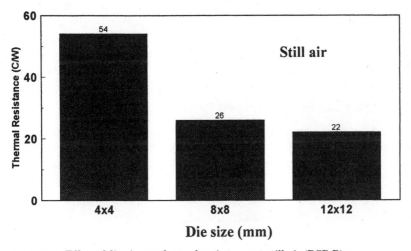

Figure 9.7 Effect of die size on thermal resistance at still air (PCB B).

Figure 9.8 shows the thermal resistance of the 8×8-mm solder-bumped flip chip on PCB B and PCB D as a function of power dissipation area. It can be seen that for both PCB constructions, the thermal resistance decreases significantly with the increase in power dissipation area. This is because smaller chip (power dissipation) areas lead to higher temperatures on the PCB, as shown in Figs. 9.9 and 9.10. This is further shown in Fig. 9.11 by measuring the thermal resis-

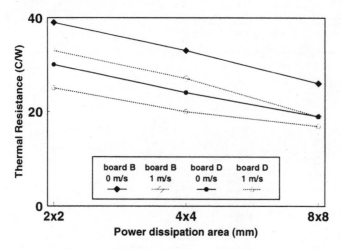

Figure 9.8 Measured junction to ambient thermal resistance of the
8 × 8-mm chip as a function of power dissipation.

tances of the 8 × 8-mm chip with various power dissipation areas on
PCB B in still air. Some closed-form solutions for the temperature and
stress distributions of a chip attach on a substrate have been given in
Refs. 4 and 5.

9.6   Heat Paths of Solder-Bumped Flip Chip on Board

The computational fluid dynamic (CFD) calculations have been performed to aid in the understanding of the effects of solder bump size and
population, underfill material, and heat sink on the thermal character-

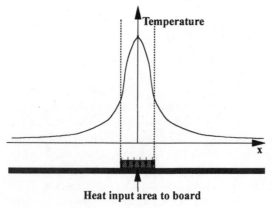

Figure 9.9   Temperature distribution on PCB.

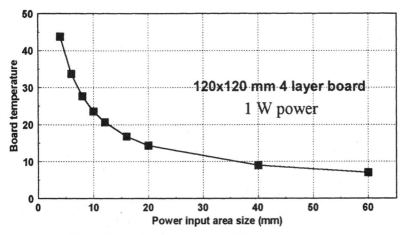

Figure 9.10   Effect of power input area size on PCB temperature.

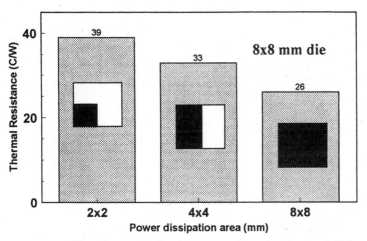

Figure 9.11   Effect of power dissipation area on thermal resistance (PCB B at still air).

istics of solder-bumped flip chip on board.[1] The material properties used in the thermal simulation are shown in Table 9.1. Table 9.2 shows the comparison between the simulation and measurement results for three different chips on PCB B. It can be seen that great confidence in the CDF simulation is achieved.

There are three heat paths in solder-bumped flip chip on board assembly: from the die to the solder joints to the PCB, from the die to the underfill to the PCB, and from the back of the die (die top) to the ambient. The CFD results are shown in Table 9.3. First of all, the amount of heat dis-

TABLE 9.1 Material Properties
Used in Thermal Modeling

Material	Thermal conductivity (W/m K)
Die	84
Underfill	0.6
Solder joints	40
Copper	396
FR-4	0.2

sipated through the underfill to the PCB is greater than that dissipated through the solder joints to the PCB. This is due to the fact that all the solder joints together only occupy 2.4 percent of the chip area. Second, without heat sink, 95 percent of the heat goes to the PCB (Fig. 9.12).

9.7 Effects of Solder Joint Population

With the 8×8-mm chip, eight different solder-bumped flip chip on board without heat sink assemblies are modeled. The results are summarized in Table 9.4 (the definitions of θ_{ja}, θ_{jb}, and θ_{ba} are shown in Fig. 9.13) and Figs. 9.14 through 9.16. It can be seen that the solder joint size has no significant effect on the thermal resistance (Fig. 9.14).

The effects of solder joint population are shown in models 1 and 2 in Table 9.4. The solder joint number of model 2 is 4 times that of the original model 1. It can be seen that the junction to ambient thermal resistance (θ_{ja}) is reduced by 3.4°C/W, with reductions of 1 and 2.4°C/W in θ_{jb} and θ_{ba}, respectively. The reduction is attributed to the better conductivity from the chip to PCB. The PCB resistance is also reduced. This is because when more bumps are presented, the power input to the board is more uniform, which reduces the local temperature gradient and thus the board spreading resistance.[1]

9.8 Effects of Signal Copper Content in PCB

One of the cheapest ways to reduce thermal resistance is to maximize the copper content in the signal layer in the PCB. This could be achieved

TABLE 9.2 Comparison Between Modeling and Measurements
for Three Different FCOBs

Die size	Calculated θ_{ja} (°C/W)	Measured θ_{ja} (°C/W)	Difference
8×8 mm	27	26	4%
12×12 mm	21.3	21.5	1%
4×4 mm	58.5	56	4.5%

TABLE 9.3   Heat Flow Through Different Thermal Paths

Die size	Conduction to PCB via joints	Conduction to PCB via underfill	Conduction to PCB via die top
8 × 8 mm	26.1%	72.4%	1.5%
12 × 12 mm	10%	85.7%	4.3%
4 × 4 mm	26.8%	69.8%	3.4%

by filling the spaces on the signal layers with copper and just leaving minimum spaces. Models 3 and model 4 in Table 9.4 show that if the signal copper content is increased to 60 percent, the thermal resistance is reduced by 11°C/W for the original solder bump pattern and by ~8°C/W for the fully populated solder bump pattern. This is better than adding two more copper planes and some solid copper area on the signal layers (Fig. 9.15).

The PCB thermal resistance is the combination of spreading and convection resistance. The convection resistance is calculated by assuming very high (1000 W/m K) board thermal conductivity (model 5 in Table 9.4). In this case, the board spreading resistance is negligible and the total board resistance (~6°C/W) equals the convection thermal resistance.

9.9   Effects of Underfill Materials

First of all, the underfill in a solder-bumped flip chip assembly helps to transfer the heat to the PCB by conduction. In most cases this is the

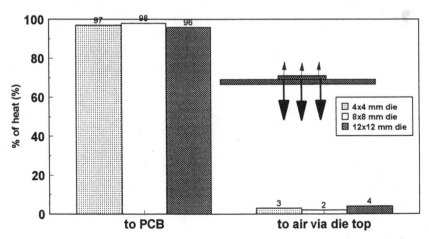

Figure 9.12   Heat paths of flip chip on board (without heat sink, more than 95 percent of heat goes to the PCB).

TABLE 9.4 Thermal Resistances of Eight Different FCOB Configurations
with the 8 × 8-mm Die

Model	Underfill K (W/m K)	Joint number	Standoff (mm)	PCB signal Cu content	θ_{ja} (°C/W)	θ_{jb} (°C/W)	θ_{ba} (°C/W)
1	0.5	1 row	0.1	5%	27	1.3	25.7
2	0.5	Full	0.1	5%	23.6	0.3	23.3
3	0.5	1 row	0.1	60%	16	0.6	15.4
4	0.5	Full	0.1	60%	15.7	0.3	15.4
5	0.5	Full	0.1	$K = 1000$	6.1	0.3	5.8
6	0.5	1 row	0.09	5%	26.5	1.2	25.3
7	0.026	1 row	0.1	5%	39	5.9	33.1
8	2.5	1 row	0.1	5%	25.5	0.8	23.7

major heat path. The effects of high-thermal-conductivity underfill are
shown in Table 9.4 and Fig. 9.16. It can be seen that even with an underfill having 5 times better thermal conductivity (model 8), the thermal
resistance is only reduced by 1.5°C/W (compared to model 1). This is
because for flip chip with underfill, the θ_{jb} is trivial.

9.10   Effects of Heat Sinks

The effects of heat sinks on the thermal resistance of solder-bumped
flip chip on board are shown in Table 9.5 and Figs. 9.17 through 9.20
for a 12 × 12-mm chip. It can be seen from Fig. 9.17 that the flip chip
offers the shortest thermal path to the heat sink from the back of the
chip. Also, as expected, all the heat sinks significantly reduce the θ_{ja}.
When a higher-performance heat sink is used, the heat dissipation
through the heat sink is more dominant. For example, with high-
performance fan heat sinks, more than 90 percent of the heat is dissipated through the heat sink (Figs. 9.19 and 9.20). Also, when a heat sink
is applied, the θ_{ja} is much less sensitive to the board thermal conductivity.

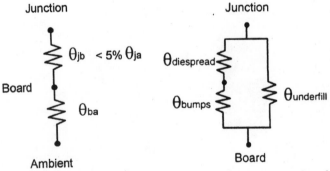

Figure 9.13   One-dimensional model for flip chip junction to board
thermal resistance. $\theta_{ja} \sim \theta_{ba}$ and $\theta_{ba} = \theta_{spreading} + \theta_{convection}$.

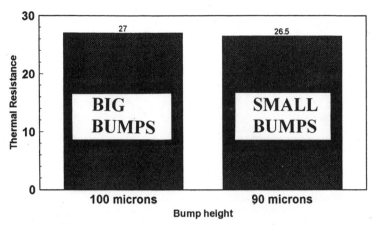

Figure 9.14 Effect of solder joint size on thermal resistance.

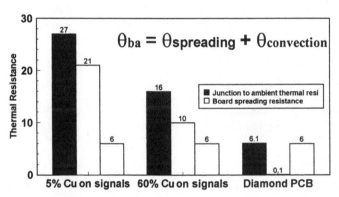

Figure 9.15 Effect of PCB thermal conductivity on thermal resistance (without heat sink).

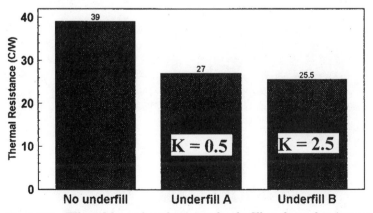

Figure 9.16 Effect of thermal conductivity of underfill on thermal resistance of FCOB assembly.

TABLE 9.5   Thermal Resistance of the 12 × 12-mm FCOB with Different Heat Sinks

Heat sink	Heat sink description	PCB signal Cu content	θ_{ja} (°C/W)	Heat to PCB	Heat to H/S
None	N/A	5%	21.3	95.7%	4.3%
Plate	50 × 50 × 3 mm	5%	11	32%	68%
Plate	50 × 50 × 3 mm	60%	9.8	51%	49%
Fin heat sink	50 × 50 × 25 mm, 11 × 8 pins	5%	6.9	20%	80%
Fin heat sink	50 × 50 × 25 mm, 11 × 8 pins	60%	6.5	33%	67%
Fan heat sink	51 × 53 × 16.5 mm, 21 × 21 pins	5%	1.6	8%	92%

9.11   Summary

Very useful thermal characteristic data for solder-bumped flip chip on PCB assemblies obtained by SGS-Thomson have been reported in this chapter. Some important results are summarized as follows.

- The thermal resistance cannot be significantly reduced by improving the solder joint geometry and thermal conductivity of underfill materials.

- Without heat sinks, the thermal resistance is dominated by the PCB resistance. PCBs with no power or ground planes have very high thermal resistance. Once two solid copper planes are added to the PCB, the thermal resistance is significantly reduced. Adding more copper will only slightly improve the thermal performance.

- Without heat sinks, the cheapest approach to reducing thermal resistance is to maximize the copper content on the signal layers.

- Without heat sinks, the thermal resistance strongly depends on the chip size and power dissipation area.

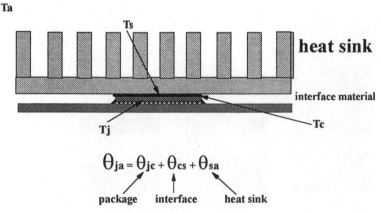

$$\theta_{ja} = \theta_{jc} + \theta_{cs} + \theta_{sa}$$

Figure 9.17   Solder-bumped FCOB with heat sink.

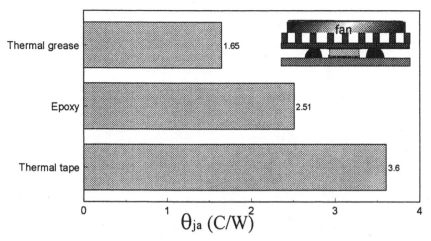

Figure 9.18   Effect of heat sink attach materials on thermal resistance of FCOB assembly.

- Without heat sinks, the air flow reduces the thermal resistance; this helps the low-thermal-conductivity PCB more than the high-conductivity board.

- Without heat sinks, the underfill helps to reduce the thermal resistance.

- Without heat sinks, the major heat path is from the chip to underfill and then to the PCB.

- With heat sinks, the major heat path is from the back of the chip to the heat sink and then to the ambient.

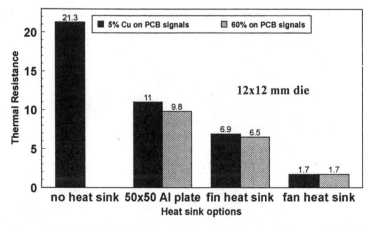

Figure 9.19   Effect of heat sink options on thermal resistance of FCOB assembly.

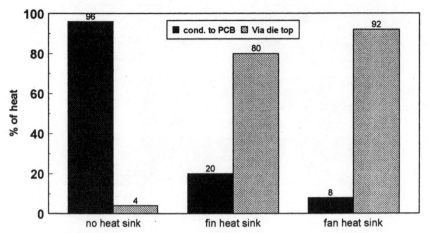

Figure 9.20 With high-efficiency fan heat sinks, more than 90 percent of heat will go through the heat sink to the air.

- With heat sinks, the thermal resistance is significantly reduced and the overall thermal performance is determined by the heat sink thermal performance. With a high-performance fan heat sink, more than 90 percent of the heat is dissipated from the back of the chip and heat sink.

Acknowledgments

The author would like to thank Tiao Zhou, Michael Hundt, Claudio Villa, Robert Bond, and Tom Lao of SGS-Thomson Microelectronics, Inc. for sharing their useful results with the electronic industry.

References

1. Zhou, T., M. Hundt, C. Villa, R. Bond, and T. Lao, "Thermal Study for Flip Chip on FR-4 Boards," *Proceedings of Electronic Components and Technology Conference,* pp. 879–884, May 1997.
2. Zhou, T., and M. Hundt, "Evaluation of Heat Sink Attach to Flip Chip," *Proceedings of NEPCON West,* pp. 909–914, February 1998.
3. Zhou, T., and M. Hundt, "Thermal Enhancement Guidelines for PQFP, BGA, and Flip Chip," *Proceedings of NEPCON West,* pp. 1139–1149, February 1998.
4. Lau, J. H., "Thermoelastic Solutions for a Semi-Infinite Substrate with a Powered Electronic Device," *J. Electronic Packaging, Trans. ASME,* **114:** 353–358, September 1992.
5. Lau, J. H., "Thermoelastic Solutions for a Finite Substrate with an Electronic Device," *J. Electronic Packaging, Trans. ASME,* **113:** 84–88, March 1991.
6. Lau, J. H., "Thermoelastic Problems for Electronic Packaging," *J. Int. Soc. Hybrid Microelectronics,* 25: 11–15, May 1991.
7. Lau, J. H., "Temperature and Stress Time History Responses in Electronic Packaging," *IEEE Trans. CPMT,* part B, **19** (1): 248–254, February 1996.

8. Lau, J. H., *Thermal Stress and Strain in Microelectronics Packaging,* Van Nostrand Reinhold, New York, 1993.
9. Lau, J. H., C. P. Wong, J. L. Prince, and W. Nakayama, *Electronic Packaging: Design, Materials, Process, and Reliability,* McGraw-Hill, New York, 1998.
10. Bar-Cohen, A., and A. D. Kraus, *Advances in Thermal Modeling of Electronic Components and Systems,* vol. 1, Hemisphere Publishing, New York, 1988.
11. Bar-Cohen, A., and A. D. Kraus, *Advances in Thermal Modeling of Electronic Components and Systems,* vol. 2, ASME Press, New York, 1990.
12. Pecht, M., *Handbook of Electronic Package Design,* Marcel Dekker, New York, 1991.

Wafer-Level Packaging

10.1 Introduction

As mentioned earlier, one of the most cost-effective packaging tech-nologies is DCA, such as the solder-bumped flip chip on PCB. However, because of the thermal expansion mismatch between the silicon chip and epoxy PCB, underfill encapsulant is usually needed for solder joint reliability, as shown in Chaps. 6 through 8. Due to the underfill opera-tion, the manufacturing cost is increased and the manufacturing throughput is reduced. In addition, the rework of underfilled flip chip on PCB is very difficult, if not impossible. This further complicates the KGD-related issues.

There is another reason why DCA is not very popular yet. Usually, the pitches and sizes of the pads on the peripheral-arrayed chip are very small and pose great demands on the supporting structure—the PCB. PCBs with high density and fine line width/spacing and sequen-tial built-up circuits with microvias are not commonly available at rea-sonable cost yet, as discussed in Chap. 4. Thus, alternative low-cost and high-performance electronic packaging techniques are needed.

Wafer-level chip scale package (WLCSP)[1–39] provides solutions to these problems. The unique feature of most WLCSPs is the use of a metal layer to redistribute the very fine-pitch peripheral-arrayed pads on the chip to much larger-pitch area-array pads (Fig. 10.1) with much larger solder joints on the PCB. Thus, with WLCSPs, the demands on PCB are relaxed, the underfill is not needed, and the KGD issues are much simpler. From the system makers' point of view, WLCSP is just another solder-bumped flip chip, except that (1) the solder bumps of WLCSP are much larger; (2) the PCB assembly of WLCSP is more robust; and (3) the system makers do not have to struggle with the underfill encapsulant.

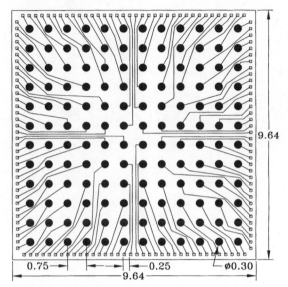

Figure 10.1   WLCSP of a peripheral-arrayed pad chip.

In Lau and Lee,[26] the author discusses the wafer-level packages from companies like ChipScale, EPIC, Flip Chip Technologies, Fujitsu, Mitsubishi, National Semiconductor, Sandia National Laboratories, and ShellCase. In this book, another six WLCSPs from six different new companies are presented.

10.2   EPS/APTOS's WLCSP[28, 29]

In this section, a novel WLCSP is presented.[28, 29] It consists of a copper conductor layer and two low-cost dielectric layers. The bump geometry consists of the SMT-compatible 63wt%Sn-37wt%Pb solder and a copper core on the UBM of the chip. Cross sections of samples are examined for a better understanding of the redistribution, UBM, Cu core, and solder joints. Also, the thermal fatigue life of the solder joint is predicted through a creep analysis and an empirical equation. Finally, the WLCSP solder bumps are subjected to shear test and the WLCSP PCB assembly is subjected to both mechanical shear and thermal cycling tests.

10.2.1   WLCSP redistribution and bumping

Figure 10.1 shows a square chip (9.64 × 9.64 mm) with 144 peripheral-arrayed pads. The pad size is 0.1 × 0.1 mm and the pitch is 0.25 mm. Adding a metal layer on top of the wafer means that the fine-pitch

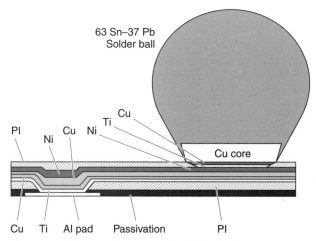

63 Sn–37 Pb
Solder ball

Cu
PI   Cu   Ni   Ti
   Ni

Cu core

Cu   Ti   Al pad   Passivation   PI

Figure 10.2   Structural cross section of the WLCSP.

peripheral-arrayed pads on the chip can be redistributed to a much larger-pitch and area-array pad in the interior of the chip. In this case, the pitch is 0.75 mm and the pad size is 0.3 mm in diameter. Figure 10.2 shows the details of redistribution. It can be seen that the 63wt%Sn-37wt%Pb solder bump is supported by a Cu core, which is connected to the redistributed Cu-Ni pad through the Cu-Ti UBM. The redistributed metal layer is made of Cu-Ni.

The key steps in the wafer-bumping process for the present WLCSP, shown in Fig. 10.2, are briefly discussed as follows. First of all, the 8-in. (200 mm) wafers are ultrasonically cleaned. The remaining steps are as follows:

Step 1: Spin the polyimide on the wafer and cure for an hour. This will form a polyimide layer 4 to 5 μm thick.

Step 2: Apply photoresist and mask, then use photolithography techniques (align and expose) to open vias on the aluminum pads.

Step 3: Etch the desired vias.

Step 4: Strip off photoresist.

Step 5: Sputter Ti and Cu over the entire wafer.

Step 6: Apply photoresist and mask, then use photolithography techniques to open the redistribution trace locations.

Step 7: Electroplate Cu.

Step 8: Electroplate Ni.

Step 9: Repeat step 4.

Step 10: Etch off the Ti-Cu.

Step 11: Repeat step 1.

Step 12: Apply photoresist and mask, then use photolithography techniques to open vias for the desired bump pads and cover the redistribution traces.

Step 13: Repeat step 3.

Step 14: Repeat step 4.

Step 15: Repeat step 5.

Step 16: Apply photoresist and mask, then use photolithography techniques to open the vias on the bump pads to expose the areas with UBM.

Step 17: Electroplate Cu core.

Step 18: Electroplate 63Sn-37Pb eutectic solder.

Step 19: Repeat step 4.

Step 20: Repeat step 10.

Step 21: Reflow the eutectic solder.

A typical cross section of the WLCSP bump is shown in Fig. 10.3, and a typical cross section of the WLCSP PCB assembly is shown in Fig. 10.4.

10.2.2 WLCSP solder bump height

The solder bump height and Cu core height of the WLCSP are measured and the results are shown in Fig. 10.5. It can be seen that the Cu

Figure 10.3 WLCSP solder bump with copper core.

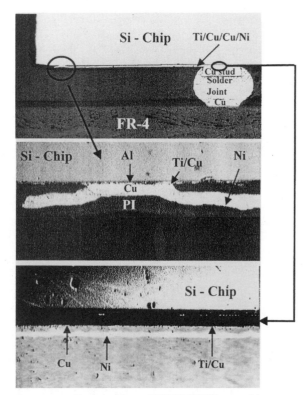

Figure 10.4   Cross sections of WLCSP PCB assembly.

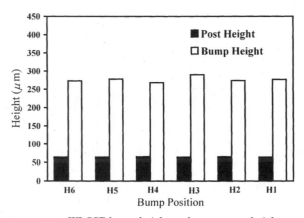

Figure 10.5   WLCSP bump height and copper core height.

core height and the solder bump height are all very uniform. (In this book, the solders on the chip before joining to the PCB are called *solder bumps.* After the solder bumps have been reflowed on the PCB, they are called *solder joints.*)

10.2.3 WLCSP solder bump strength

The solder bumps of the WLCSP are subjected to shear test under the following conditions: (1) shear blade speed is 100 µm/s; (2) the tip of shear blade is 100 µm from the chip surface. The results are shown in Fig. 10.6. It can be seen that the averaged solder bump shear strength is 404 gf, which is many times higher than that of the conventional flip chip solder bumps (~50 gf). It is noted that the failure location is in the solder bump (not at the UBM) and that the fracture surface is shear dominated, as shown in Fig. 10.7.

10.2.4 PCB assembly of WLCSP

It is very easy to assemble the WLCSP on PCB. After the 63Sn-37Pb solder-bumped WLCSP is aligned with the PCB with a look-up camera and a look-down camera, the chip is placed face down on the PCB with a very minimal force. After the chip is placed, it is put on the conveyor belt of a reflow oven with a maximum on-PCB temperature of 220°C. A typical cross section of the WLCSP PCB assembly is shown in Fig. 10.4. The details of the redistribution layer and the UBM are also shown in Fig. 10.2. It can be seen that the Ti-Cu on the Al pad is supporting the redistribution trace (Cu-Ni), which is protected by the polyimide. At

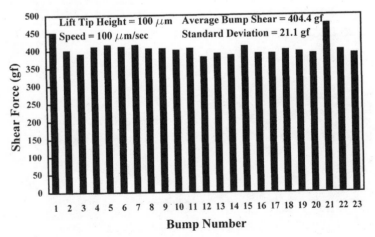

Figure 10.6 Shear strength of WLCSP solder bumps.

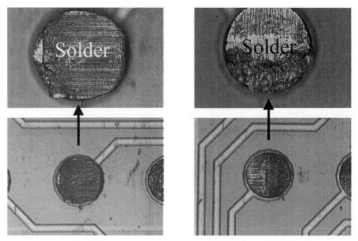

Figure 10.7   Shear fracture surface of WLCSP solder bumps.

the solder joint, the Cu core on the Ni-Cu pad is connected by the Cu-Ti UBM.

10.2.5   Finite element modeling of WLCSP assemblies

Detailed dimensions of the WLCSP assembly under consideration in this study are shown in Fig. 10.8. A commercial finite element code, ANSYS (version 5.5), is employed. A two-dimensional model is established using eight-node plane strain elements. It should be noted that

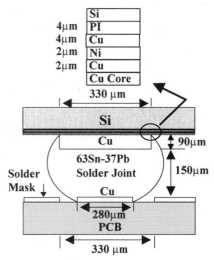

Figure 10.8 Typical structure of the WLCSP solder joint.

all detailed assembly structures such as the chip, Al pads, polyimide, passivation, UBM, Ni layer, Cu layer, solder joint, Cu core, Cu pads, and solder mask are modeled in the finite element analysis. Besides, due to the symmetry in the assembly structure, only half of the diagonal cross section is considered.

10.2.6 Time-temperature-dependent creep analysis

The material properties used in the computational modeling are shown in Table 10.1. Since the goal of this section is to investigate the effects of solder joint height on the solder joint reliability, except that the eutectic solder (63Sn-37Pb) is a time/temperature-dependent creep material, all other constituents are considered as linear elastic materials.

The Garofalo-Arrhenius steady-state creep is generally expressed by (see for example, Eq. 1-188 through 1-191 of Lau[40]):

$$\frac{d\gamma}{dt} = C\left(\frac{G}{\theta}\right)\left[\sinh\left(\omega\,\frac{\tau}{G}\right)\right]^n \exp\left(\frac{-Q}{k\theta}\right)$$

where γ is the steady-state shear creep strain, $d\gamma/dt$ is the steady-state shear creep strain rate, t is the time, C is a material constant, G is the temperature-dependent shear modulus, θ is the absolute temperature (K), ω defines the stress level at which the power law stress dependence breaks down, τ is the shear stress, n is the stress exponent, Q is the activation energy for a specific diffusion mechanism, and k is Boltzmann's constant. For 60wt%Sn-40wt%Pb solder, the material constants have been experimentally determined by Darveaux and Banerji (Eq. 1-44 of Lau[41]) with a single hyperbolic sine function. These values are shown in Fig. 10.9.

The temperature loading imposed on the WLCSP assembly is shown in Fig. 10.10. It can be seen that for each cycle (60 min) the tempera-

TABLE 10.1   Material Properties of the WLCSP PCB Assemblies

Material properties	Young's modulus (GPa)	Poisson's ratio (ν)	Thermal expansion coefficient (α) ppm/°C
FR-4 substrate	22	0.28	18.5
Copper	76	0.35	17
63Sn-37Pb solder	See note	0.4	21
Polyimide	8.3	0.33	3
Silicon chip	131	0.3	2.8
Solder mask	6.9	0.35	19
Ni	20.5	0.3	12.3

Note: Young's modulus and stress-strain relations of solder are temperature-dependent.

60Sn40Pb SOLDER

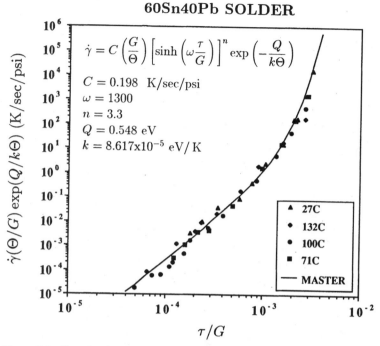

$$\dot{\gamma} = C\left(\frac{G}{\Theta}\right)\left[\sinh\left(\omega\frac{\tau}{G}\right)\right]^{n}\exp\left(-\frac{Q}{k\Theta}\right)$$

$C = 0.198 \ \text{K/sec/psi}$
$\omega = 1300$
$n = 3.3$
$Q = 0.548 \ \text{eV}$
$k = 8.617\text{x}10^{-5} \ \text{eV/K}$

▲ 27C
◆ 132C
● 100C
■ 71C
— MASTER

y-axis: $\dot{\gamma}(\Theta/G)\exp(Q/k\Theta)$ (K/sec/psi)

x-axis: τ/G

Figure 10.9 Constitutive equation of eutectic solder under creep deformation.

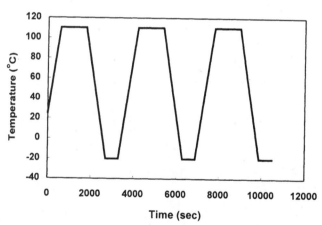

y-axis: Temperature (°C)

x-axis: Time (sec)

Figure 10.10 Temperature profile for modeling and thermal cycling test.

ture condition is between −20 and +110°C with 15 min ramp, 20 min hold at hot, and 10 min hold at cold. Three full cycles are executed.

A typical deformation of the WLCSP assembly is shown in Fig. 10.11. It can be seen that the maximum relative creep deformation (dominated by shear) is at the corner solder joint. This is due to the thermal expansion mismatch between the chip and the PCB. The location of the maximum shear stress and shear creep strain is in the corner joint at the interface between the lower left corner of the Cu core and the solder joint (see Figs. 10.12 and 10.13).

For time-dependent analysis, it is important to study the responses for multiple cycles until the hysteresis loops become stabilized. The shear stress and shear creep strain at the maximum location (corner) are shown (solid lines) in Figs. 10.14 and 10.15, respectively. Figure 10.16 shows the maximum shear stress and shear creep strain hysteresis loops (solid line) for multiple cycles. It can be seen that the creep shear strain is quite stabilized after the first cycle.

Figure 10.17 shows the time history of creep strain energy density at the maximum location. The averaged creep strain energy density per

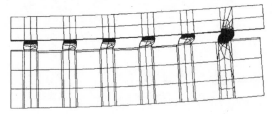

Figure 10.11 Deformed shape of the WLCSP assembly.

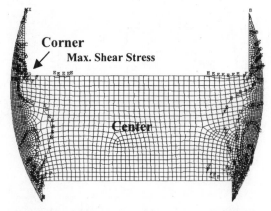

Figure 10.12 Maximum von Mises stress distribution in the corner solder joint.

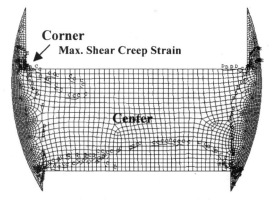

Figure 10.13 Maximum shear strain distribution in the corner solder joint.

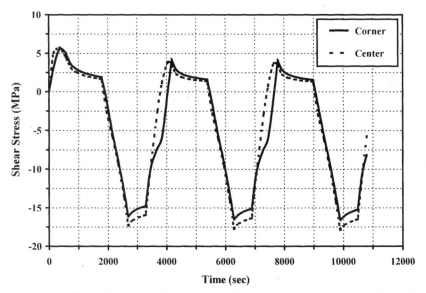

Figure 10.14 Time-dependent shear stress at the maximum and center locations of the corner solder joint.

cycle (ΔW) at the corner solder joint can be obtained by averaging the creep strain energy density of the last two cycles, which is 0.57 N/mm^2 = 82.5 psi. Similarly, the shear stress, shear creep strain, shear stress and shear creep strain hysteresis loops, and the creep strain energy density at the center of the solder joint, are shown (dotted lines) in Figs. 10.14, 10.15, 10.16, and 10.17, respectively. It can be seen that, due to stress concentration at the corner of the Cu core, the shear creep strain at the

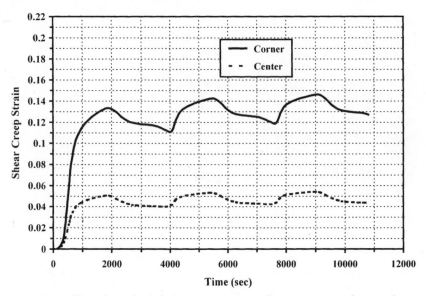

Figure 10.15 Time-dependent shear creep strain at the maximum and center locations of the corner solder joint.

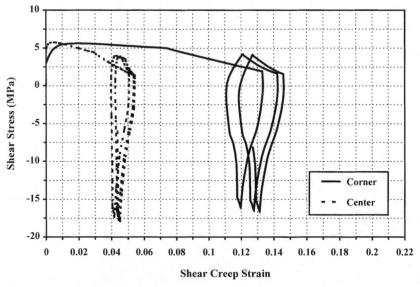

Figure 10.16 Hysteresis loops of the shear stress and creep shear strain at the maximum and center locations of the corner solder joint.

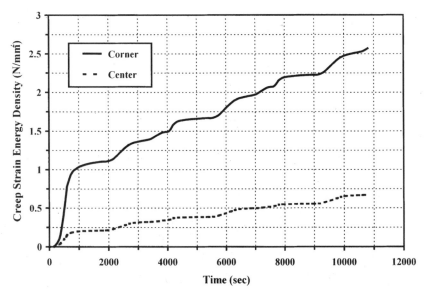

Figure 10.17 Time-dependent creep strain energy density at the maximum and center locations of the corner solder joint.

corner of solder joint is about 3 times that at the center. The averaged creep strain energy density per cycle (ΔW) at the center of the corner solder joint is 0.17 N/mm² = 25.1 psi.

10.2.7 Life prediction for WLCSP corner solder joint

Once we have ΔW, the thermal fatigue crack initiation life (N_o) can be estimated from (Darveaux) Eq. 13.35 of Lau,[27] i.e.,

$$N_o = 7860\Delta W^{-1} = 95 \text{ cycles at the corner}$$

$$N_o = 7860\Delta W^{-1} = 313 \text{ cycles at the center}$$

and the thermal fatigue crack propagation life (N), based on the linear fatigue crack growth rate theory, can be estimated by (Darveaux) Eq. 13.36 of Lau,[27] i.e.,

$$da/dN = 4.96 \times 10^{-8}\Delta W^{1.13}$$

or

$$N = N_o + (a_f - a_o)/(4.96 \times 10^{-8}\Delta W^{1.13})$$

where a is the crack length of the solder joint; a_o is the initial crack length, which is assumed to be zero; and a_f is the final crack length. It can be seen that in order to determine N we need to choose an a_f. For example, if $a_f = 0.364$ mm (solder cracks through near the bottom of the Cu core), then $N = 95 + 1973 = 2068$ cycles. On the other hand, if $a_f = 0.390$ mm (solder cracks through the middle of the solder joint), then $N = 313 + 8111 = 8420$ cycles.

It should be pointed out that, due to a numerical integration scheme error in the old version of the finite element code ANSYS, Darveaux' thermal fatigue life prediction equations may involve errors as high as 16 percent.[42] In a private conversation in early July 1999, Darveaux said that a new version of the equations will be published in the very near future.

10.2.8 Shear test of WLCSP on board

The WLCSP PCB assemblies are subjected to shear tests also. A typical load displacement curve is shown in Fig. 10.18. It can be seen that the maximum force is 17 kgf for 45 solder joints, i.e., 380 gf per solder joint. This is very close to the shear strength of solder bumps measured in Sec. 10.2. Again, the failure mode is in the solder (dominated by shear fracture) and not at the UBM (Fig. 10.19).

10.2.9 Thermal cycling of WLCSP on board

WLCSP PCB assemblies have been subjected to a thermal cycling test with the temperature profile given in Fig. 10.10. At the time of this writing, all the solder joints have survived for more than 1800 cycles without any failure. The test is still ongoing.

10.2.10 Summary

A cost-effective and reliable WLCSP has been investigated in this study. It consists of a Cu-Ni conductor layer and a couple of polyimide

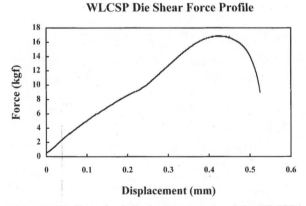

WLCSP Die Shear Force Profile

Figure 10.18 Shear load displacement curve of the WLCSP PCB assembly.

Figure 10.19   Shear fracture surfaces of the WLCSP solder joints.

dielectric layers. The solder bump geometry consists of the SMT-compatible eutectic solder and a copper core on the redistributed Ni-Cu supporting pad with the Ti-Cu as the UBM.

The thermal fatigue life of the corner solder joint of the WLCSP assembly has been predicted by a time/temperature-dependent creep analysis and the empirical equation given by Darveaux. It is found that the thermal fatigue life of the corner solder joint of the WLCSP is more than 2000 cycles (60-min cycle between −20 and +110°C with 15 min ramp, 20 min hold at hot, and 10 min hold at cold). This is adequate for most operating conditions.

The key process steps for WLCSP wafer bumping have been briefly presented. Also, the bump height and bump strength of the WLCSP have been measured. The average bump shear strength is 404 gf. This value is much larger than that (shear strength ~50 gf) for the conventional flip chip solder bumps.

The WLCSP PCB assemblies have been subjected to the shear test. The shear strength per solder joint is 380 gf, which is close to that measured at the WLCSP solder bumps. The failure mode is in the solder joint and is dominated by shear.

10.3   Amkor/Anam's wsCSP™[30, 31, 39]

Amkor Technology Inc. and Anam Semiconductor Inc. have teamed up in the development of a wafer-level package called Wafer Scale Chip

Scale Package (wsCSP), which utilizes typical back-end processes and is reliable for most applications. The detailed assembly process of wsCSP is provided in Refs. 30 and 31. In this section, the overall package design and assembly process are briefly discussed, followed by package- and board-level reliability results.

10.3.1   wsCSP design and assembly flow[30, 31, 39]

The wsCSP utilizes a flexible polyimide circuit tape as the redistribution layer. The circuit tape manufacturing takes place in state-of-the-art PCB or wide-web continuous tape production facilities. The absence of tape windows or vias during the circuit production allows for a high-yield process. The circuit tape is received in the wsCSP assembly line as 6- or 8-in. square panels.

The preparation of the circuit tape for the die attach or wafer lamination process consists of three steps. First, visual inspection of the tape identifies and records possible reject locations. Then, the die attach film adhesive is laminated to the circuit tape by a hot rolling process. This laminate is then subjected to an automatic window punch process that cuts holes at the precise locations of the die bond pads. Since the interconnection method between die and substrate is wire bonding, these windows are cut into the circuit tape to allow access to the die bond pads during the wire-bonding process.

In the next process step, all chips on the wafer are simultaneously attached to the substrate. The circuit tape is aligned with the wafer via a split-beam camera. Then the wafer and the adhesive side of the substrate laminate are brought into contact under pressure and temperature. The adhesive is allowed to cure fully.

The interconnection between chip and substrate is formed through wire bonding. The first bond is placed on the die bond pad and the second is placed on the tape substrate. Since wire bonding is an individual rather than a true wafer-level process, wafer mapping is used to skip the reject locations and to speed up this process.

A stencil printing process is used to encapsulate the wire bonds. The window areas are filled and the wires are covered with encapsulant. A degasing step prior to curing assures that no voids remain in the encapsulant.

The current ball attach process picks up solder balls with a vacuum tool, dips the balls in flux, and places them on the copper pads one device row at a time. This method is chosen to limit the tooling cost. As true wafer-level ball attach methods become available, a higher-efficiency method should be adopted.

The reflow and cleaning processes are in-line. A belt furnace and belt cleaner are used to reflow the solder balls and clean off the flux residue. The wafers are then accumulated in wafer cassettes and are ready for wafer-level electrical test or wafer mounting for the singulation process.

A wafer probe station can be used to perform final electrical testing. Depending on the contact technology and device function, multiple devices can be tested in parallel and a very efficient process can be achieved. The test results are logged in the wafer map.

A state-of-the-art wafer mounting and dicing process is used to singulate the wafer into individual wsCSP units. Wafer-map-compatible pick and place equipment combined with a laser marking head allows separation of good and reject units, marking of units with a customer logo, and placement of the finished packages in shipping trays.

The described assembly flow (Fig. 10.20a) has been successfully proven in the Amkor/Anam pilot line on both 150- and 200-mm wafers. The resulting typical wsCSP package cross section is shown in Fig. 10.20b. Figure 10.21 shows both center-row and edge-row die bond pad designs.

10.3.2 wsCSP package-level reliability

The package-level reliability is evaluated on daisy-chain units of nominal 9×11-mm die and package size. The center row design is selected to closely simulate an actual memory device. The 54 I/Os are arranged in 6 rows of 9 balls each. The pitch between balls in one row is 1.00 mm and the pitch between rows is 0.80 mm. Two center rows are depopulated where the die bond pads are located. The die passivation is silicon nitride.

Table 10.2 lists the package-level tests conducted and their results. The table shows that all tests are successfully passed and the package is qualified as level 2 package.

TABLE 10.2   Package-Level Qualification Test Results

Test	Number failed/ number tested
Level 2: Moisture resistance test	
5 temperature cycles (−55 to +125°C) +	
24 h dry bake at 125°C +	0/22
168 h of 85/60 TH soak +	
3 reflow cycles (FCR) to 240°C	
Temperature cycle condition C (−65 to +150°C): 1000 cycles	0/76
Temperature cycle condition B (−55 to +125°C): 2000 cycles	0/76
Temperature humidity with level 2 preconditioning (85°C/85% RH), 1000 h	0/76
High-temperature storage at 150°C with level 2 preconditioning, 1000 h	0/76
Pressure cooker (121°C/2 atm/100% RH)	
Without preconditioning, 504 h	0/45
With level 2 preconditioning, 168 h	0/45

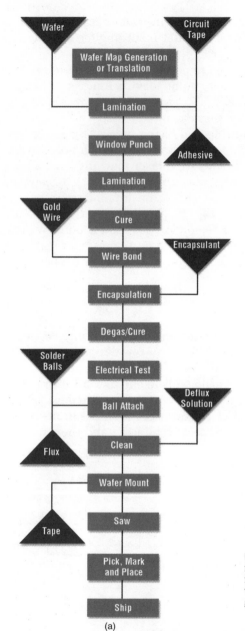

Figure 10.20 (*a*) wsCSP assembly flow chart showing the process from receiving the customer's wafers through shipping the singulated devices.

(a)

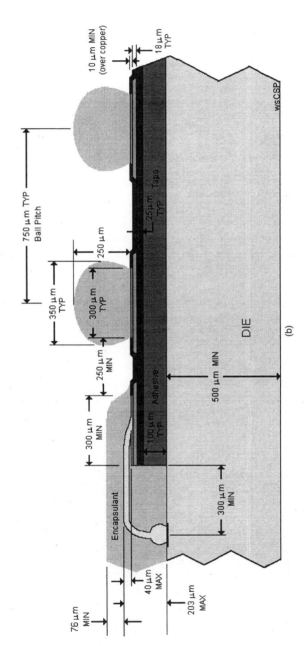

Figure 10.20 (b) Cross-sectional view of a typical wsCSP.

Figure 10.21   Examples of wsCSP.

10.3.3   wsCSP on board reliability tests

Test vehicle design and assembly.   The 54-pin wsCSP package just described is also used for board-level reliability testing. For these tests, wafers of three thicknesses and solder balls of two sizes are selected as listed in Table 10.3.

A four-layer FR-4 PCB with organic solderability preservatives (OSP) surface finish is designed as the test vehicle. The test boards have 15 device locations with designed pad diameters of 0.3 mm. The pads are non-solder mask defined (NSMD), with solder mask openings of 0.45 mm.

Prior to the PCB assembly, all parts are baked for 4 h at 125°C. Boards of two different thicknesses (0.5 and 1.6 mm) are fabricated and the parts are mounted using standard SMT process. The boards are first printed with Indium SMQ92 no-clean type 3 solder paste using a 127-μm-thick (5 mils) laser-cut stencil. The stencil has trapezoidal walls with square aperture openings of 0.30 and 0.35 mm, respectively, at the top and bottom surfaces.

One extra board with parts is used for measuring and adjusting the reflow profile for the assemblies. The board temperature is measured using five thermocouples, one at each corner location and one at the cen-

TABLE 10.3   Package Test Matrix for 54-Pin wsCSP

Package ID	Ball size (mm)	Wafer thickness (mm)	Ball pitch (mm)	Package height (mm)	Package pad diameter (mm)
P1	0.33	0.35	0.8 × 1.0	0.79	0.30
P2	0.33	0.50	0.8 × 1.0	0.94	0.30
P3	0.45	0.50	0.8 × 1.0	1.07	0.30
P4	0.33	0.625*	0.8 × 1.0	1.08	0.30

* No back grind.

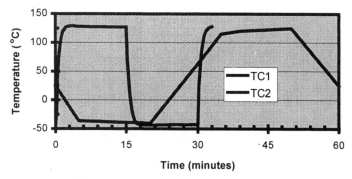

Figure 10.22 Temperature profiles (measured on PCB) for tests.

ter component location of the board. The reflow temperature profile has a peak temperature of 213°C and the time above liquidus is about 60 s.

A total of 18 boards 0.5 mm thick are assembled first and are x-rayed after reflow, which reveals no shorts or bridges. Later, 12 more boards 1.6 mm thick are also assembled without any assembly defects.

Reliability testing. The reliability testing is done in two phases. In the first, 18 boards 0.5 mm thick are tested using two different test conditions (Fig. 10.22):

- **TC1:** −40 to +125°C, single zone, 1-h temperature cycling with 15 min of ramps and dwells

- **TC2:** −40 to +125°C, dual zone, 30-min temperature cycling with 2- to 3-min ramps and 12 to 13 min of dwells at each temperature extreme

In the second phase of testing, the assemblies are tested again under TC2 conditions but with a PCB thickness of 1.6 mm (62 mils). The test matrix for the first phase is shown in Table 10.4.

TABLE 10.4 Test Matrix for 0.5- and 1.6-mm-Thick PCBs

Package ID	Test condition	PCB thickness (mm)		Sample size
P1	TC2	0.5	1.6	45 Each
P2	TC2	0.5	1.6	45 Each
P3	TC2	0.5	1.6	45 Each
P4	TC2	0.5	1.6	45 Each
P2	TC1	0.5	—	45
P3	TC1	0.5	—	45

Results and failure analysis.   Table 10.5 lists the failure data gathered on the test matrix listed in Table 10.4. The tests on sets 5 through 10 are being continued until a failure rate of at least 50 percent is achieved for all data sets. It should be noted that no failures have been observed for parts on test using TC1 conditions.

0.5-mm-thick boards.   The solder joint fatigue failures observed so far are summarized in Table 10.5. Notice that the sample size for sets 1 and 2 is reduced from 45 to 38 and 30, respectively. This is because all 3 boards of set 1 and 1 board of set 2 fell off the racks (at around 1600 cycles) during the test. This resulted in premature failures on 7 parts of Set 1 spread across 3 boards and all 15 parts on 1 board of Set 2. These parts were removed from the test and testing was continued for the remaining parts. However, the data for set 1 is questionable from a solder fatigue standpoint due to this incident, as all three boards and parts accumulated some mechanical damage.

Comparing the effects of wafer thickness, the data clearly shows improved reliability with thinner dies. Comparing the cycles to first failures, reliability improvements of about 1.75 and 2.5 times are observed when the wafer thickness is reduced from 0.625 to 0.50 and 0.35 mm, respectively. Again, for 0.35-mm-thick wafer, the reliability improvement can be potentially much higher if the boards are not damaged during testing. This is because thinner dies lead to less bending stiffness and thus increase the compliance of the assembly, which means less stress and strain in the solder joints.

1.6-mm-thick boards.   Comparing the cycles to first failures in Table 10.5, the PCB-level reliability declines by anywhere from 20 to 60 percent when the same parts are mounted on the thicker board. Most notable are the reductions in life for 0.35- and 0.50-mm-thick wafers.

TABLE 10.5   Summary of Solder Fatigue Failures of the 54-Pin wsCSP

Set ID	Package ID	Test condition	PCB thickness (mm)	Cycles completed and status (7/20/99)	Number failed/ number tested	First failure (cycles)	Mean life (cycles)
1	P1	TC2	0.5	3300—Stopped	10/38	2640	3163*
2	P2	TC2	0.5	3300—Stopped	29/30	1873	2385
3	P3	TC2	0.5	3300—Stopped	43/45	1934	2780
4	P4	TC2	0.5	3300—Stopped	45/45	1065	1418
5	P2	TC1	0.5	1500	0/45	N/A	N/A
6	P3	TC1	0.5	1500	0/45	N/A	N/A
7	P1	TC2	1.6	1325	7/45	1052	1244*
8	P2	TC2	1.6	1325	8/45	1146	1380*
9	P3	TC2	1.6	1325	0/45	N/A	N/A
10	P4	TC2	1.6	1325	10/45	844	1395*

* Extrapolated mean life as failure rate is less than 50 percent.

In fact, as opposed to thin boards, the data from 1.6-mm-thick boards indicates no significant difference in fatigue life, especially when the extrapolated mean life is compared. In other words, the overall assembly stiffness is governed more by the PCB thickness for these packages, and even a 44 percent reduction in die thickness is not enough to mitigate this effect. The fatigue life improvement for thin boards is, however, expected, since thinner boards are more compliant in bending. This compliance reduces the effect of the board in the overall assembly stiffness, and the die thickness becomes more important.

The Weibull life distributions for the failed data are shown in Fig. 10.23 for the 54-pin wsCSP on thin and thick PCBs. Failure analysis of the 0.5-mm-thick PCB shows that failure occurs at the furthermost corner solder joints, with cracks near the package side initiated earlier than near the PCB side. Two representative cross sections are shown in Fig. 10.24.

10.3.4 Summary

The wsCSP is a low-cost wafer-level package. It has been qualified by Amkor/Anam as a level 2 package. The wsCSP on board solder joint reliability has been shown by thermal cycling tests to be superior to most of the other CSP packages for a given chip size, and meets or exceeds most reliability requirements for PC, laptop, and handheld electronics.[30, 31, 39] Also, the electrical performance of the wsCSP is better than that of the μBGA and TSOP (Table 10.6).

10.4 Hyundai's Omega-CSP[32]

10.4.1 Design of omega-CSP

A wafer-level CSP called omega-CSP is being developed by Hyundai MicroElectronics Inc. To assure the solder joint reliability, a stress buffer layer (SBL) that has relatively low tensile modulus compared to benzo-

TABLE 10.6   Electrical Stimulation Results of wsCSP and Other Packages

Inductance	L11			L12		
	Minimum	Maximum	Average	Minimum	Maximum	Average
μBGA	0.893	2.264	1.5785	0.408	0.555	0.4815
wsCSP	0.983	2.354	1.6685	0.278	0.425	0.3515
TSOP	3.76	5.32	4.54	1.625	2.716	2.1705

Capacitance	C11			C12		
	Minimum	Maximum	Average	Minimum	Maximum	Average
μBGA	0.363	0.5296	0.4463	0.1172	0.123	0.1201
wsCSP	0.223	0.3896	0.3063	0.0772	0.083	0.0801
TSOP	0.603	0.784	0.6935	0.208	0.409	0.3085

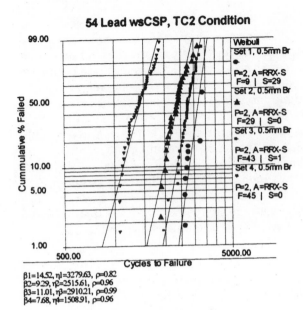

β1=14.52, η1=3279.63, ρ=0.82
β2=9.29, η2=2515.61, ρ=0.96
β3=11.01, η3=2910.21, ρ=0.99
β4=7.68, η4=1508.91, ρ=0.96

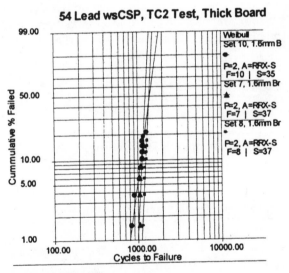

β1=7.36, η1=1488.21, ρ=0.96
β2=17.97, η2=1281.88, ρ=0.83
β3=18.40, η3=1421.22, ρ=0.95

Figure 10.23 Weibull life distributions of the 54-pin wsCSP on thin (top) and thick (bottom) boards.

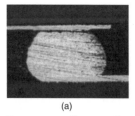 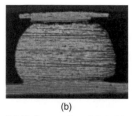

(a) (b)

Figure 10.24 Cross sections of failed samples. (*a*) 0.33-mm ball, 0.35-mm wafer. (*b*) 0.45-mm ball, 0.5-mm wafer.

cyclobutene (BCB) and polyimide (PI) is used as passivation layer between the silicon chip and the runners in the omega-CSP. Also, large solder balls are used to further enhance the PCB level reliability. Figure 10.25 shows a schematic drawing of omega-CSP.

10.4.2 Materials for omega-CSP

The SBL material used in the omega-CSP consists of a modified polyimide prepreg with N-methylpyrollidone (NMP) solvent. The solids content accounts for 60 to 70 percent of the weight. The cured SBL yields a low elastic modulus, good moisture resistance, and low dielectric constant. The structure of the runner metal used in omega-CSP is the three-layered system of Ti-Ni-Cu.

CYCLOTENE™ 4024 is used as solder mask for omega-CSP. The BCB is I-line-sensitive, photosensitive polymer and is derived from B-staged bisbenzocyclobutene chemistry.

10.4.3 Processing of omega-CSP

A schematic representation of the wafer-level omega-CSP process is shown in Fig. 10.26.

SBL process. First, the SBL resin is spin-coated on the wafer and soft-baked at 120°C for 3 min. Then a layer of photoresist is applied, patterned, and developed. Finally, the layer of soft-baked SBL is etched away over bond pads. It is noted that some cracks, as shown in Fig. 10.27, might occur when oxygen plasma is applied on the SBL in order

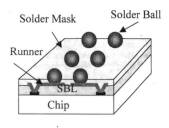

Figure 10.25 Schematic of the omega-CSP.

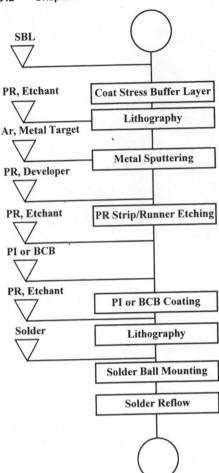

Figure 10.26 Omega-CSP packaging process flow.

to enhance the wetability between the SBL and photoresist. However, this problem can be solved by coating an adhesion promoter. After photoresist is stripped, the SBL is cured at 250°C for 30 min.

Metal sputtering. The sputtered metals consist of three layers. The first layer is Ti, followed by Ni and then by Cu. Ti forms a strong adhesion

Figure 10.27 SBL cracking due to oxygen plasma process.

to the SBL as well as to the Al bond pad. Ni is used as a diffusion barrier for solder, and Cu is used as a seed layer for Cu electroplating. It is noted that cracks or wrinkles, as shown in Fig. 10.28, could occur because of the residual stresses from the sputtering process.

Patterning and electroplating. A layer of photoresist is applied, patterned, and developed. The runner is thickened by Cu electroplating. Then the photoresist is stripped and the seed-layer metals are etched. Runner thickness is a very important parameter of omega-CSP: too thick a runner will reduce the package compliance and lead to cracks in the solder joints, but too thin a runner will lead to adhesive failures in the SBL. Figure 10.29 shows the failure images after solder ball shear test.

Solder mask coating. BCB is selected as a solder mask and the solder bond pads are opened by a photolithography method.

Solder ball formation. Solder balls are formed using pick and place and reflow process.

10.4.4   Reliability of omega-CSP on board

The solder joint reliability of omega-CSP on PCB has been simulated with the Solder Reliability Solution (SRS) software and measured with macroscopic moire interferometry by Hyundai MicroElectronics Inc. It was found that in order to obtain the PCB level reliability for the omega-CSP, the SBL must be applied between the chip and the solder.[32]

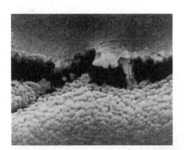

Figure 10.28 Cracking in sputtered metal on SBL due to residual stresses.

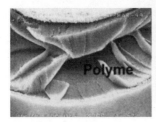

Figure 10.29 Runner failed images after ball shear test.

10.5   FormFactor's WLCSP[33]

FormFactor's WLCSP is based on the company's MicroSpring™ contact On Silicon Technology (MOST™), which provides chip-level interconnects on the wafer. The overall back-end flow, which uses the MOST wafers, wafer-level burn-in, high-speed test, and assembly, is called the wafer on wafer (WOW™) process. In this section, the design, materials, process, reliability, and applications of FormFactor's WLCSP are briefly discussed.

10.5.1   MicroSpring

The MicroSpring contact is a primary element of MOST, as shown in Figs. 10.30 and 10.31. It employs a simple approach using conventional wire-bonding tools to provide the foundation for the MicroSpring contacts (Fig. 10.32). Since there is no leadframe, no die attach, and no molding, it has a significantly lower cost picture than the conventional

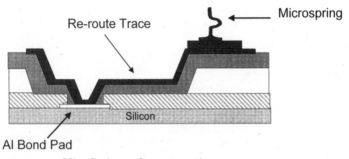

Figure 10.30   MicroSpring wafer cross section.

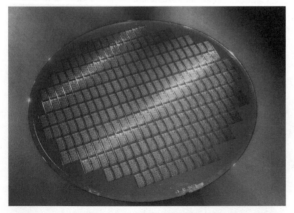

Figure 10.31   Full wafer covered with MicroSpring contacts.

Figure 10.32 MicroSpring contacts by wire bonder.

CSPs. Also, due to the compliance of the MicroSpring contacts, unlike the solder-bumped flip chip on board, underfill is not needed. With the WOW process, the WLCSP can be tested at high speed and burned in at elevated temperatures. The MicroSprings are made of gold wire plated with nickel alloy and formed into an S shape.

10.5.2 MicroSpring flip chip on board

The MicroSpring-contacted flip chip is very easily soldered to a PCB, as shown in Fig. 10.33. Usually, the 63Sn-37Pb solder paste is printed on the pads or vias on the PCB by a stencil. Then the MicroSpring-contacted flip chip is picked and placed by a machine onto the PCB. After SMT reflow, good-looking solder joints are formed between the MicroSprings and the PCB as shown in Figs. 10.33 and 10.34.

The MicroSprings can also be mounted on the PCB as a socket to form a system integration platform. This "socketed" construction allows pitch transformation and ease of rework. Also, it allows avoidance of the import tax, thus lowering motherboard costs.

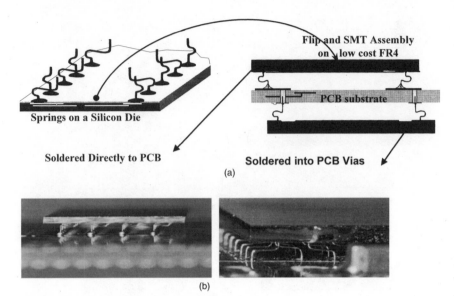

Figure 10.33   MicroSpring assembly on PCB methods.

10.5.3 Reliability of MicroSpring flip chip on board

The MicroSpring-contacted WLCSP and the MicroSpring-contacted flip chip on PCB assemblies have been tested. Some of the test results are shown in Table 10.7 with the temperature profiles shown in Figs. 10.35 and 10.36. It can be seen that both package level and PCB level pass the qualification tests.

10.5.4 Applications of MicroSpring flip chip on board

Although MicroSprings are ideal for many applications such as flash memory, smart/memory cards, PCMCIA cards, and RF identification, the

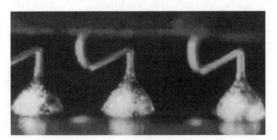

Figure 10.34   Close-up of soldered MOST contacts.

TABLE 10.7   Summary of the Qualification Test Results of wsCSP

Test	Conditions	Result (passed)
Module level		Cycles
Thermal shock	Condition C (Liquid to liquid, −55 to 125°C)	3000
Thermal cycle	Condition B (Air to air, −65 to 150°C)	5000
Package level		Hours
HTOL	125°C, 7 V	1000
LTOL	−10°C, 8 V	1000
Reflow	240°C, 10 sec	3000
THS	65°C, 95% RH	1000

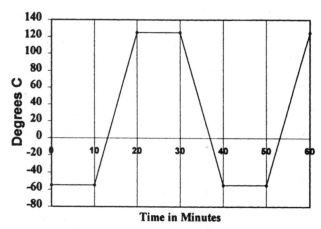

Figure 10.35   Thermal cycling test condition.

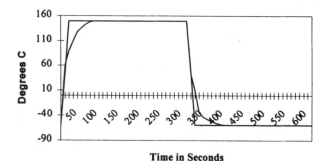

Figure 10.36   Thermal shock test condition.

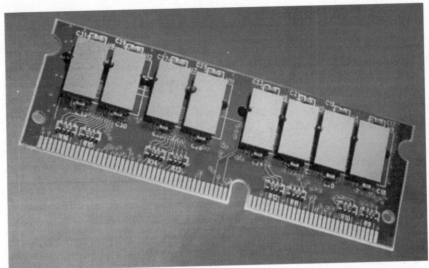

Figure 10.37 Rambus's RIMM module with MicroSpring technology (demonstration only).

optimum product to demonstrate the advantages of the MicroSpring is the DRAM module, specifically the Rambus RIMM module.[33] Figure 10.37 shows the demonstration module with the MicroSpring-contacted flip chip on board technology. The Rambus RIMM module with μBGA package will be discussed further in Chap. 12.

10.6   Tessera's WAVE™[34–37]

Tessera has developed the Wide Area Vertical Expansion (WAVE) technology, which promises to provide a relatively easy method for cost-effectively interconnecting ICs still on the wafer. Based on concepts that allow for the mass assembly and production of compliant packages on the wafer, in this section this important new technology will be discussed in terms of the process and reliability and the solution to die shrink.

10.6.1   Uniqueness of WAVE

Unlike most of the WLCSPs, which process on top of the wafer directly with either solder bumps, wire springs, or beam leads, WAVE is based on the concept shown in Fig. 10.38. It can be seen that the packages and the IC wafer are processed separately and that each is sample-probed for quality before assembly. Good wafers and good packages are joined in a single process that forms all the interconnects simultaneously. Because it uses parallel fabrication, the IC packaging assembly

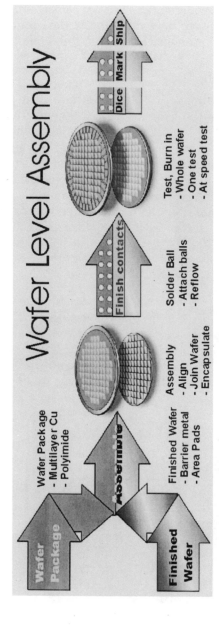

Figure 10.38 Tessera's wafer-level packaging assembly concept.

is manufacturable at high yields. After the wafer is packaged, the solder balls are mounted a whole wafer at a time. Full functional testing is done on the whole wafer only once after packaging. Burn-in and test are also done directly on the wafer. Finally, the complete packaged IC is diced into finished packages at the end of the line.

10.6.2   Design of WAVE

One significant advantage to such processing is the potential to integrate the function of the package more fully with the function of the chip. Thus, power and ground distributions and global wire route or critical clocks could be accommodated for in relatively thick and wide, highly conductive copper instead of very thin and narrow and more resistive aluminum. Such an approach could reduce pin counts on the package while either increasing performance or reducing power requirements, or potentially both. Such a structure defies definition according to current electronics industry lexicon, and thus the term *tile* is offered here (Fig. 10.39).[34–37]

10.6.3   Processing of WAVE

The process flow is very elegant, requiring only a few simple steps as shown in Figs. 10.40 through 10.43. First, the wafer and a pellicle with flex-circuit-based interconnection layers attached are aligned and joined using, for example, the transient liquid-phase metallurgical joining method (Fig. 10.40). The specially processed electrical links are then lifted from the flex base layer to a height of approxi-

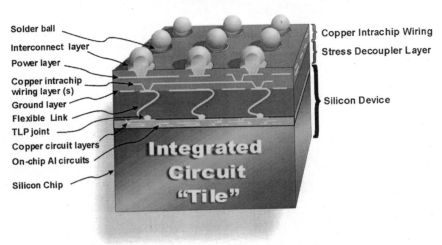

Figure 10.39   By designing silicon with an eye to how it will be packaged, it should be possible to create very high-performance compliant packages directly on the wafer.

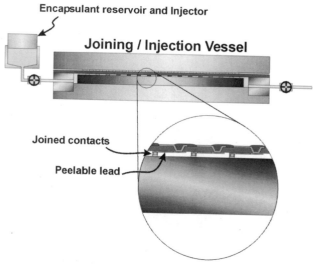

Figure 10.40   Align and join.

mately 100 to 125 µm (Fig. 10.41). A low-modulus encapsulant is in-
jected between the silicon wafer and the flexible film and cured (Fig.
10.42). Finally, solder balls or copper-core balls are mounted on the
whole wafer. After testing and burn-in, the final packages are then
singulated into individual parts. Figure 10.44 shows an x-ray photo of
a WAVE package, indicating the springlike leads encapsulated in the
compliant material.

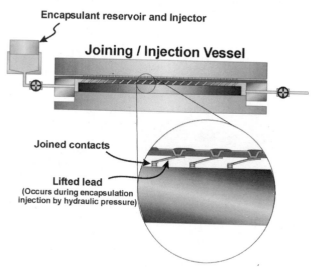

Figure 10.41   Vertical expansion of the electrical links.

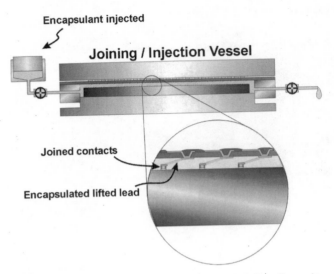

Figure 10.42   Injection of low-modulus encapsulant into the cavity.

10.6.4 Reliability of WAVE

A set of 12.7-mm-square and 458 I/O WAVE packages with fully populated solder joints on an 0.5-mm pitch was subjected to a thermal
cycling test. The test conditions are (A) −65 to +150°C and (B) −40 to
+125°C. The test results show that the solder joints pass 1600 cycles

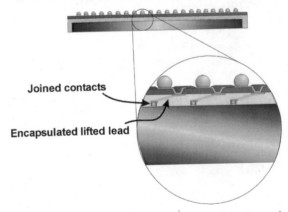

Figure 10.43   Attachment of solder balls, test, burn-in, and singulation of parts.

Top View
(clear encapsulant)

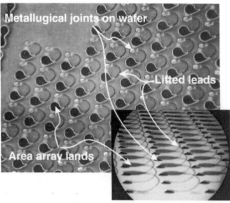

Edge X-ray view

Figure 10.44 X-ray images of the encapsulated leads that connect the die pads (bottom extremity of the lead) and solder ball pads (top extremity).

for condition A and 3000 cycles for condition B. The reason for such good results is the springlike leads, which increase the compliance of the package.

10.6.5 WAVE's solution to die shrink

To solve the die shrink problem, instead of connecting the wafer to the flexible material as discussed earlier, Tessera attaches the singulated die to the flex, and then the rest of the process is preserved. Figure 10.45 shows the process steps for the WAVE Virtual Wafer Level technology for resolving the die shrink issue. Figure 10.46 shows a final package of a SRAM device using Tessera's technology.

10.7 Oxford's WLCSP[38]

A novel yet simple processing and wafer-level packaging method for a 4×4 Resonant Cavity LED (RCLED) chip with individual array has been developed by Oxford University. Figure 10.47 shows examples of optical transmitter systems. An array of emitters is attached to a subcarrier, and an optical component such as an array of fibers guides light from the emitters. Emitters such as RCLEDs and vertical-cavity surface-emitting lasers are attractive for these types of applications, because 2-D arrays can be fabricated, thus achieving paralleled optoelectronic interconnects. Wavelengths in the infrared region are advan-

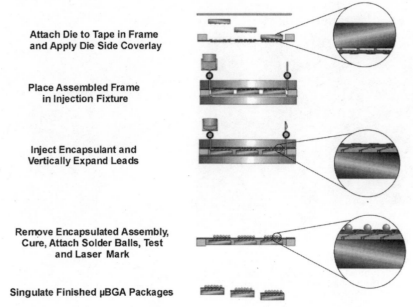

Attach Die to Tape in Frame
and Apply Die Side Coverlay

Place Assembled Frame
in Injection Fixture

Inject Encapsulant and
Vertically Expand Leads

Remove Encapsulated Assembly,
Cure, Attach Solder Balls, Test
and Laser Mark

Singulate Finished µBGA Packages

Figure 10.45   Tessera WAVE Virtual Wafer Level process steps for die shrink.

tageous for the integration of large 2-D emitter arrays with active (such as CMOS driver) components, because at these wavelengths the semiconductor is transparent, so that devices emit through the substrate and the conventional flip chip technology can be applied.

One of the most important advantages of solder-bumped flip chip for optoelectronics products is the self-alignment between the optics (such

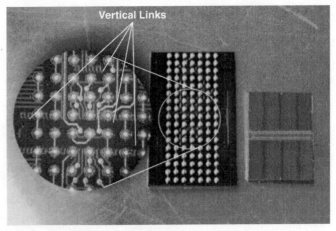

Figure 10.46   A SRAM die packaged using WAVE technology.

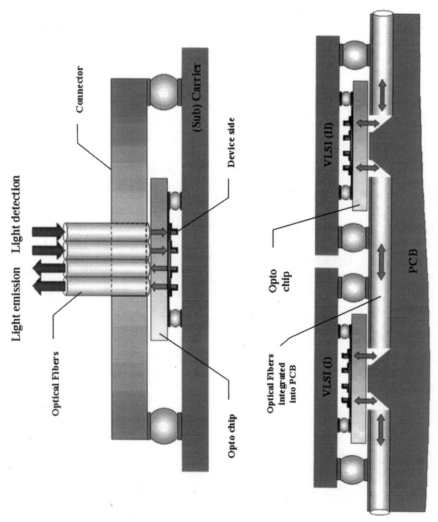

Light emission   Light detection

Connector

Optical Fibers

Opto chip

(Sub) Carrier

Device side

VLSI (I)

VLSI (II)

Opto
chip

Optical Fibers
integrated
into PCB

PCB

Figure 10.47   Optical interconnects with bottom emission and self alignment.

355

as the optical fibers) and the optoelectronics during joining. These often require the micrometer tolerance solder-bumped flip chip joining can produce.

10.7.1   Device design

The emitter device is designed as a WLCSP component. The WLCSP technique is based on processes compatible with device fabrication. The emitters are flip chipped to a substrate that redistributes the connectors to the peripheral bond pads. Figure 10.48 shows a schematic of the device and a cross section through the chip. The basic structure is as follows; a top metal (Pd-Au) layer; three InGaAs16% multiple quantum wells constituting the middle layer (for emission wavelength of 980 nm) surrounded by quarter-wave AlGaAs33% spacer layers, followed by a base of an eight-period distributed Bragg reflector (DBR). Mesas are first defined by wet chemical etching using 8:1:1 (H_2SO_4:H_2O_2:H_2O) solution for 1 min. This gives about 1.5-μm etching.

10.7.2   Device fabrication

Figure 10.49, a through g, shows the fabrication flow of RCLED arrays. The surface is planarized with polyimide to allow metallic tracks running over the surface from the device to the bond pads. For planarization, a photosensitive polyimide is used since it involves a small number of processing steps. This polyimide, which is negatively working, has to be exposed under ultraviolet (UV) light for 70 s (energy is set to 3.5 mW/cm^2), using a Karl Suss (MJB3) mask aligner with 200-W source (Fig. 10.49b). The substrate n-type contact is formed by thermally evaporating Ni-Au-Ge (150 nm) and patterning using a lift-off process. The contact is annealed at 365°C for 15 s to diffuse in the germanium and form the contact.

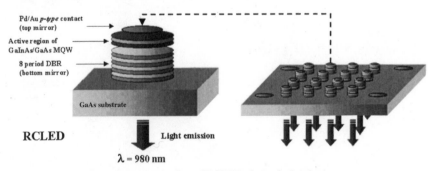

Pd/Au *p-type* contact
(top mirror)

Active region of
GaInAs/GaAs MQW

8 period DBR
(bottom mirror)

GaAs substrate

RCLED   Light emission

λ = 980 nm

Figure 10.48   A schematic cross section of RCLED through the device.

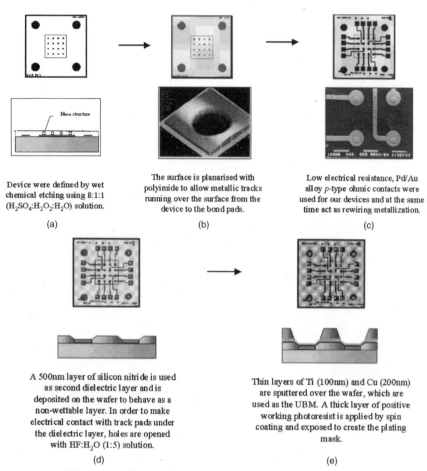

Device were defined by wet chemical etching using 8:1:1 ($H_2SO_4:H_2O_2:H_2O$) solution.

(a)

The surface is planarised with polyimide to allow metallic tracks running over the surface from the device to the bond pads.

(b)

Low electrical resistance, Pd/Au alloy p-type ohmic contacts were used for our devices and at the same time act as rewiring metallization.

(c)

A 500nm layer of silicon nitride is used as second dielectric layer and is deposited on the wafer to behave as a non-wettable layer. In order to make electrical contact with track pads under the dielectric layer, holes are opened with $HF:H_2O$ (1:5) solution.

(d)

Thin layers of Ti (100nm) and Cu (200nm) are sputtered over the wafer, which are used as the UBM. A thick layer of positive working photoresist is applied by spin coating and exposed to create the plating mask.

(e)

Figure 10.49 The process diagram for the fabrication of WLCSP for RCLED arrays.

10.7.3 Processing of WLCSP for optoelectronic devices

Low-electrical-resistance, nonannealed (Pd-Au alloy p-type) contacts are used for the devices and at the same time act as rewiring metallization (Fig. 10.49c). The alloy is thermally evaporated over 5 nm of Pd and 300 nm of Au, respectively, and lifted off the polyimide coating to give low-resistance p-type contacts and tracks. The Pd-Au gives the lowest electrical resistance. This is probably due to the dispersing of the native oxides of GaAs by palladium.

A 500-nm layer of silicon nitride is used as the second dielectric layer and is deposited on the wafer to behave as a nonwetable layer. In order to make electrical contact with track pads under the dielectric layer, holes are opened with $HF:H_2O$ (1:5) solution (Fig. 10.49d).

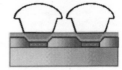

At first a 10 μm high Cu stud is electroplated inside the photomask. The thick copper stud is used as a stem for the solder. Then the (eutectic) solder is electroplated on top of the Cu stud. As a result of electroplating, a mushroom shaped solder bump is formed during the process.

(f)

The final step is to reflow. As a result, a spherical solder bump is formed.

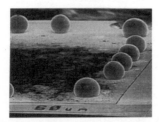

(g)

Figure 10.49   *(Cont.)*

Thin layers of Ti (100 nm) and Cu (200 nm) are sputtered over the wafer and used as UBM. Ti serves as a diffusion barrier for the gold pads; the sputtered Cu layer is used as the plating base. A thick positive working photoresist is applied by spin coating and exposed to create the plating mask (Fig. 10.49e). The sputtered thin layer acts as the electrode during the electroplating. First a 10-μm-high Cu stud is elec-

Figure 10.50 SEM photo of chip with redistributed connectors to the peripheral bond pads (top), and the test vehicle flip chipped to a silicon substrate (bottom).

troplated inside the photomask. The thick copper stud is used as a stem for the solder. Then, the 63Sn-37Pb (eutectic) solder is electroplated on top of the Cu stud. As a result of electroplating, a mushroom-shaped solder bump is formed during the process (Fig. 10.49f). The final step is to reflow the solder into a nice-looking bump, as shown in Fig. 10.49g. Figure 10.50 shows the WLCSP with redistributed connectors to the peripheral bond pads, and a WLCSP on a silicon substrate.

Acknowledgments

The author would like to thank K. L. Jim, G. E. Faulkner, D. C. O'Brien, and D. J. Edwards of Oxford University; T. DiStefano, J. Fjelstad, B. Haba, M. Beroz, and J. Smith of Tessera; J. Healy of FormFactor; I. Park, I. Kang, J. Kim, H. Kim, J. Cho, and D. Choi of Hyundai; P. Hoffman, A. Syed, C. Lee, L. Smith, V. DiCaprio and M. Liebhard of Amkor/Anam; R. Lee, C. Chang, and C. Chen of EPS; and T. Chung of APTOS for sharing their useful information with the industry.

References

1. Young, J., "The MSMT Package for Integrated Circuits," *Micro SMT Application Note,* ChipScale, Inc., San Jose, CA, 1994.
2. Young, J., "Wafer Level Processing: Working Smarter," *Chip Scale Review,* **1**(1): 28–31, 1997.

3. Marcoux, P., "Chip Scale Packaging Meets SMT and MCM Needs," *Proceedings of NEPCON-West*, pp. 228–233, Anaheim, CA, 1995.
4. Young, J., "Designing Wafer Level Chip Scale Packages to JEDEC and EIAJ Outlines," *Proceedings of NEPCON-West*, pp. 1519–1523, Anaheim, CA, 1997.
5. Marcoux, P., "Encapsulant Selection for Micro SMT Wafer Level Chip Scale Packaging," *Proceedings of NEPCON-West*, pp. 360–363, Anaheim, CA, 1997.
6. Young, "Chip Scale Packaging Provides Known Good Die," *Proceedings of NEPCON-West*, pp. 52–59, Anaheim, CA, 1995.
7. Kohl, J. E., C. W. Eichelberger, S. K. Phillips, and M. E. Rickley, "Low Cost Chip Scale Packaging and Interconnect Technology," *Proceedings of the CSP Symposium SMI '97*, pp. 37–43, San Jose, CA, 1997.
8. Elenius, P., "FC2SP-(Flip Chip-Chip Size Package)," *Proceedings of NEPCON-West*, pp. 1524–1527, Anaheim, CA, 1997.
9. Elenius, P., "Flip Chip Bumping for IC Packaging Contractors," *Proceedings of NEPCON-West*, pp. 1403–1407, Anaheim, CA, 1997.
10. Elenius, P., and H. Yang, "The Ultra CSP Wafer-Scale Package," *High-Density Interconnect*, **1**(6): 36–40, October 1998.
11. Hou, M., "Super CSP: The Wafer Level Package," *Proceedings of SEMICON-West*, pp. F1–10, San Jose, CA, July 1998.
12. Hou, M., "Wafer Level Packaging for CSPs," *Semiconductor Int.*, **21**(8): 305–308, July 1998.
13. Yasunaga, M., S. Baba, M. Matsuo, H. Matsushima, S. Nakao, and T. Tachikawa, "Chip Scale Package: A Lightly Dressed LSI Chip," *Proceedings of the IEEE/CPMT International Electronics Manufacturing Technology Symposium*, pp. 169–176, La Jolla, CA, 1994.
14. Baba, S., Y. Tomita, M. Matsuo, H. Matsushima, N. Ueda, and O. Nakagawa, "Molded Chip Scale Package for High Pin Count," *Proceedings of the 46th ECTC*, pp. 1251–1257, Orlando, FL, 1996.
15. Nguyen, L., N. Kelkar, T. Kao, A. Prabhu, and H. Takiar, "Wafer Level Chip Scale Packaging-Solder Joint Reliability," *Proceedings of IMAPS International Symposium on Microelectronics*, San Diego, CA, November 1998.
16. *SRC Cindas Database*, Semiconductor Research Corp., Raleigh, NC.
17. Lau, J. H., "Solder Joint Reliability of Flip Chip and Plastic Ball Grid Array Assemblies Under Thermal, Mechanical, and Vibrational Conditions," *IEEE Trans. CPMT*, part B, **19**(4): 728–735, November 1996.
18. Chanchani, R., K. Treece, and P. Dressendorfer, "MINI Ball Grid Array (mBGA) Technology," *Proceedings of NEPCON-West*, pp. 938–945, Anaheim, CA, 1995.
19. Chanchani, R., K. Treece, and P. Dressendorfer, "A New Mini Ball Grid Array (mBGA) Multichip Module Technology," *Int. J. Microcircuits and Electronic Packaging*, **18**(3): 185–192, 1995.
20. Chanchani, R., K. Treece, and P. Dressendorfer, "Mini Ball Grid Array (mBGA) Assembly on MCM-L Boards" *Proceedings of the 47th ECTC*, pp. 656–663, San Jose, CA, 1997.
21. Badihi, A., and E. Por, "ShellCase—A True Miniature Integrated Circuit Package," *Proceedings of the International Flip Chip, Gall Grid Array, Advanced Packaging Symposium*, pp. 244–252, San Jose, CA, February 1995.
22. Badihi, A., N. Schlomovich, M. de-la-Vega, N. Darligano, and Z. Baron, "SlimCase—An Ultra Thin Chip Size Integrated Circuit Package," *Proceedings of the ICEMCM '96*, pp. 234–238, Denver, CO, April 1996.
23. Zilber, G., M. de-la-Vega, N. Schlomovich, N. Karligano, U. Tropp, and A. Badihi, "Shell-PACK—A Thin Chip Size Integrated Circuit Package," *Proceedings of SEMICON West*, San Jose, CA, June 1996.
24. Zilber, G., H. Gershetman, N. Schlomovich, N. Natan, N. Darligano, U. Tropp, and A. Badihi, "Shell-BGA—A Thin Chip Size Integrated Circuit Package," *Proceedings of the ISHM Workshop on CSP*, Whistler, BC, Canada, August 1996.
25. Badihi, A., "Thin-Film, Related Processes Used for Unique, Ultra-Thin CSP," *Chip Scale Review*, 32–33, 1997.
26. Lau, J. H., and S. W. Ricky Lee, *Chip Scale Package, Design, Materials, Process, Reliability, and Reliability*, McGraw-Hill, New York, 1999.

27. Lau, J. H., *Ball Grid Array Technology*, McGraw-Hill, New York, 1995.
28. Lau, J. H., T. Chung, R. Lee, C. Chang, and C. Chen, "A Novel and Reliable Wafer-Level Chip Scale Package (WLCSP)," *Proceedings of the Chip Scale International Conference*, SEMI, pp. H1–8, September 1999.
29. Lau, J. H., R. Lee, C. Chang, and C. Chen, "Solder Joint Reliability of Wafer Level Chip Scale Packages (WLCSP): A Time-Temperature-Dependent Creep Analysis," ASME Paper No. 99-IMECE/EEP-5, International Mechanical Engineering Congress and Exposition, November 1999.
30. Hoffman, P., "Amkor/Anam wsCSP™," *Nikkei Microdevices Wafer Level CSP Seminar*, Tokyo, Japan, 1998.
31. Syed, A., C. Lee, and M. Liebhard, "Package and Board Level Reliability of wsCSP™ and The Effect of Design Variables," *Proceedings of the Chip Scale International Conference*, SEMI, pp. J1–9, September 1999.
32. Park, I., I. Kang, J. Kim, H. Kim, J. Cho, and D. Choi, "Wafer-Level CSP (Omega-CSP)," *Proceedings of the Chip Scale International Conference*, SEMI, pp. H1–8, September 1999.
33. Healy, J., The Impact of Microsprings on Wafer Level Packaging and Back-End Processing," *Proceedings of the HDI EXPO*, pp. 17–36, August 1999.
34. DiStefano, T. H., "Wafer-Level Fabrication of IC Packages," *Chip Scale Review*, 20–27, May 1997.
35. Fjelstad, J., "W.A.V.E.™ Technology for Wafer Level Packaging of ICs," *Proceedings of the 2nd EPTC Conference*, December 1998, Singapore.
36. Fjelstad, J., "Wafer-Level CSPs Using Flexible Film Interposers," *HDI*, 22–26, March 1999.
37. Haba, B., M. Beroz, and J. Smith, "Wafer Level Packaging: Limitations and Solutions," *Proceedings of the Chip Scale International Conference*, SEMI, pp. E1–8, September 1999.
38. Jim, K. L., G. E. Faulkner, D. C. O'Brien, D. J. Edwards, and J. H. Lau, "Fabrication of Wafer Level Chip Scale Packaging for Optoelectronic Devices," *Proceedings of IEEE/EIA ECTC*, pp. 1145–1147, June 1999.
39. Dicaprio, V., M. Liebhard, and L. Smith, "The Evolution of a New Wafer-Level Chip-Size Package," *Chip Scale Review*, 31–35, May 1999.
40. Lau, J. H., *Thermal Stress and Strain in Microelectronics Packaging*, Van Nostrand Reinhold, New York, 1993.
41. Lau, J. H., *Chip On Board Technologies for Multichip Modules*, Van Nostrand Reinhold, New York, 1994.
42. Anderson, T., I. Guven, E. Madenci, and G. Gustafsson, "The Necessity of Reexamining Previous Life Prediction Analyses of Solder Joints in Electronic Packages," *Proceedings of the IEEE Electronic Components and Technology Conference*, pp. 656–663, June 1999.

Chapter

11

Solder-Bumped Flip Chip on Micro Via-in-Pad Substrates

11.1 Introduction

As discussed in Chap. 4, FR-4 epoxy/glass PCB or BT substrates with buildup and microvia/via-in-pad (VIP) are becoming available, although the prices are still high because of the low manufacturing yield and small market volume. Also, microvias will save substrate drilling area and increase routing space, which are the basic requirements for solder-bumped flip chip technology. Combined with VIP construction, microvias will save even more.[1-10] However, the microvia and VIP are very delicate.[7] Can they survive (1) after the solder bumps are reflowed on them; (2) after they are tightened up to the chip by the underfill encapsulant; and (3) during normal operating conditions? In order to answer these questions, the effects of underfill on the deformations of the buildup surface laminate layer on the well-known SLC structure have been determined by Kunthong and Han.[11] Their useful results are presented in this chapter. However, a new low-cost micro-VIP substrate for supporting the solder-bumped flip chip in a CSP configuration is presented first.[12]

11.2 Flip Chip on Micro-VIP Substrate in a CSP[12]

In this section, the design, analysis, assembly, and modeling of a novel micro-VIP substrate for housing a solder-bumped flip chip in a CSP format are presented. Because of the special design, the substrate consists of a single core of organic material and two metal layers of copper and is manufactured with the conventional PCB process at very low cost. Furthermore, the vias are drilled with laser, thus allowing very small hole sizes (0.15 to 0.1 mm) to be achieved.

The proposed substrate is used to support a functional 32-pin, low-power, high-speed static random access memory (SRAM). The assembled micro-VIP CSP package is soldered onto a PCB. The nonlinear time/temperature-dependent finite element method is used to perform the thermal stress analyses as well as the solder joint life prediction.

11.2.1 IC wafer for the 32-pin SRAM device

The functional 32-pin SRAM, shown in Figs. 11.1a and b, is designed and manufactured at very high yield and low cost by United Microelectronics Corporation (UMC) on an 8-in. wafer. The major function of this SRAM chip is for very high-speed and low-power applications. The major characteristics of the chip for designing the Cu micro-VIP are listed as follows:

- Chip sizes are 5.334×3.662 mm.
- Pad sizes are 0.075×0.075 mm.
- Pad pitch is 0.195 mm minimum.

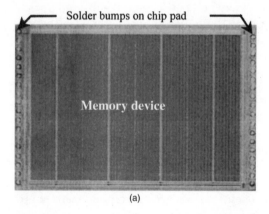

(a)

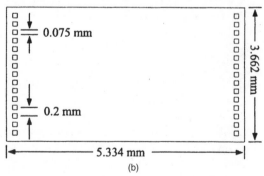

(b)

Figure 11.1 (a) SRAM chip with solder bumps. (b) Dimensions of the chip.

- Chip thickness is 0.675 mm.
- Chip pads are distributed on two shorter sides.
- There are two pads for ground and two pads for power.

11.2.2 Micro-VIP substrate

A novel VIP substrate has been designed for the 32-pin SRAM. Figures 11.2a and b show the top and bottom sides of the substrate, which is an 0.164-mm-thick organic material with a high glass transition temperature. It can be seen from the top side that the traces from the peripheral pads are redistributed (fanning) inward and connected to the copper pads on the bottom side of the package substrate through the microvias. The diameter of the microvias is less than 0.15 mm and the diameter of the

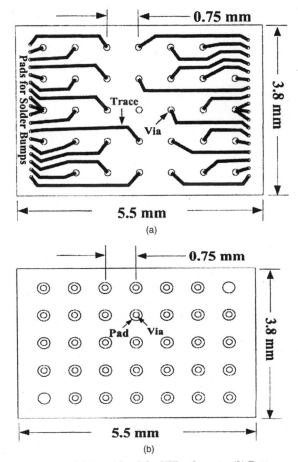

Figure 11.2 (a) Top side of the VIP substrate. (b) Bottom side of the VIP substrate.

copper pads is 0.32 mm. The dimensions of the VIP substrate are 5.5 × 3.8 × 0.164 mm. The pitch of the VIP is 0.75 mm.

The micro-VIP substrate designed in this study is manufactured by the key process steps shown in Fig. 11.3. The raw material is the BT HL832 and the core thickness is 0.1 mm. The major process steps are as follows.

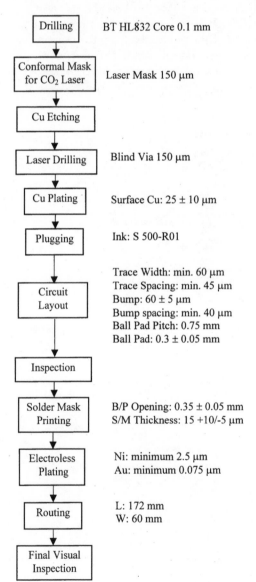

Drilling	BT HL832 Core 0.1 mm
Conformal Mask for CO$_2$ Laser	Laser Mask 150 µm
Cu Etching	
Laser Drilling	Blind Via 150 µm
Cu Plating	Surface Cu: 25 ± 10 µm
Plugging	Ink: S 500-R01
Circuit Layout	Trace Width: min. 60 µm Trace Spacing: min. 45 µm Bump: 60 ± 5 µm Bump spacing: min. 40 µm Ball Pad Pitch: 0.75 mm Ball Pad: 0.3 ± 0.05 mm
Inspection	
Solder Mask Printing	B/P Opening: 0.35 ± 0.05 mm S/M Thickness: 15 +10/-5 µm
Electroless Plating	Ni: minimum 2.5 µm Au: minimum 0.075 µm
Routing	L: 172 mm W: 60 mm
Final Visual Inspection	

Figure 11.3 Key process steps for making the VIP substrate.

Step 1: Use mechanical drill to make holes for handling and referencing.

Step 2: Apply a 30-μm-thick dry-film photoresist on the Cu and open a 150-μm-diameter circle for the microvias.

Step 3: Pre-etch the copper foil at positions where microvias are to be formed.

Step 4: Use a CO_2 laser to drill holes on the BT substrate.

Step 5: Electroplate copper to 25 ± 5 μm thick.

Step 6: Plug in nonconductive ink (S 500-R01).

Step 7: Apply photoresist and circuit layout mask, then use photolithography techniques to open the locations of pad, trace, and Cu ring.

Step 8: Perform inspection.

Step 9: Print and cure solder mask to 15 +10/–5 μm thick.

Step 10: Perform electroless plating.

Step 11: Route the large panel to 172 × 60-mm panels.

Step 12: Perform final inspection.

Figures 11.4*a* and *b* show the top and the bottom sides of the VIP substrate for the 32-pin solder-bumped flip chip SRAM. It can be seen that they are very simple (no dog-bone pads) and low in cost. Figures 11.4*c* and *d* show cross sections of the micro-VIP substrate. Figures 11.4*e* and *f* show the x-ray photos of the microvia-VIPs.

11.2.3 Solder-bumped flip chip on micro-VIP substrate

The assembly process of VIP substrate with the 32-pin SRAM is very similar to that of the NuCSP reported in Lau et al.,[2] except that in this case solder balls are mounted on the VIP substrate. Figure 11.5, *a* through *c,* shows the cross sections of the CSP with the present VIP substrate. Figure 11.5*a* shows the cross section along the chip pads. Figure 11.5*b* shows the cross section along the VIP, and Fig. 11.5*c* shows the schematics. Figures 11.5*d* and *e* show the cross section of the solder-bumped flip chip CSP with micro-VIP substrate. Figure 11.5*f* shows the x-ray photo, indicating the perfect alignment of the micro-VIP CSP assembly. Detailed dimensions of the micro-VIP will be shown in the finite element modeling section.

11.2.4 PCB assembly of the micro-VIP CSP

Figure 11.6 shows the FR-4 epoxy-glass PCB used in this study. It can be seen that there are two different kinds of solder mask layouts on the

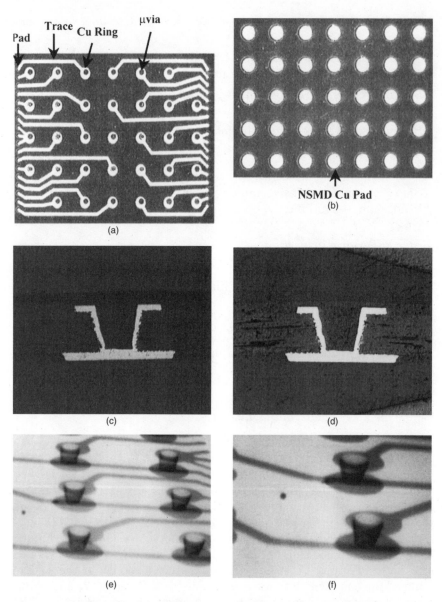

Figure 11.4 (*a*) Top side of the VIP substrate. (*b*) Bottom side of the VIP substrate. (*c*) Cross section of VIP. (*d*) Cross section of VIP. (*e*) X-ray photo of VIP. (*f*) Close-up of VIP.

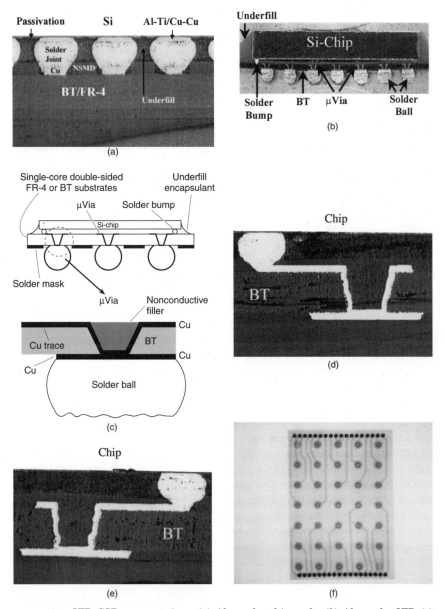

Figure 11.5 VIP CSP cross sections. (*a*) Along the chip pads. (*b*) Along the VIP. (*c*) Schematic of the VIP solder joint. (*d*) Cross section of CSP with VIP. (*e*) Cross section of VIP CSP. (*f*) X-ray photo of the VIP CSP assembly.

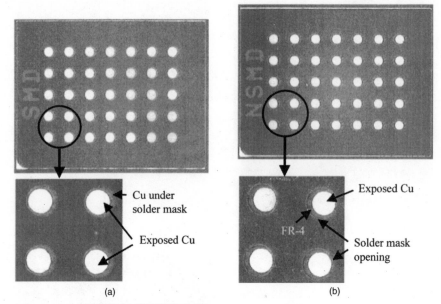

Figure 11.6 (a) SMD PCB. (b) NSMD PCB.

copper pads. These are the solder mask defined (SMD) copper pads and the non-solder mask defined (NSMD) copper pads. The assembly process of the VIP CSP on the PCB is very similar to that of the conventional no-clean SMT.[3]

Figures 11.7a through e show the cross sections of the micro-VIP CSP PCB assembly. Figures 11.7b and d are with SMD copper pads on the PCB, while Figs. 11.7c and e are with NSMD copper pads on the PCB. (The solder balls before reflow in Figs. 11.7c through e are larger than those shown in Fig. 11.7b.) In both cases, it is very easy to perform the PCB assembly with very high yield. Figure 11.8 shows the sections of micro-VIP CSP PCB assemblies with both solder joint on the chip and solder joint on the PCB. It can be seen that, since there are no dogbone pads (because of the VIP) and with microvias on the substrate, space saving is obvious.

11.2.5   Elastoplastic analysis of the micro-VIP

Detailed dimensions of the VIP solder joint are shown in Fig. 11.9. The commercial finite element code ANSYS (version 5.5) is employed in this study. A two-dimensional model is established using the eight-node plane strain elements. It should be noted that all detailed assembly structures such as the chip, underfill, solder mask, BT substrate, Cu VIP, solder joint, Cu pads, and FR-4 PCB are modeled in the finite element analysis. Besides, due to the symmetry in the assembly structure, only half of the cross-section is considered.

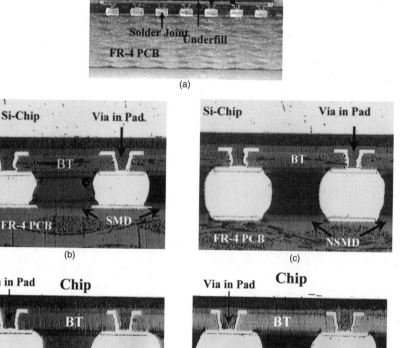

Figure 11.7 (*a*) Cross section of the VIP CSP PCB assembly. (*b*) With SMD copper pads. (*c*) With NSMD copper pads. (*d*) With SMD copper pads. (*e*) With NSMD copper pads.

The material properties used in the computational modeling are shown in Table 11.1. The copper VIP is considered to be an elastoplastic material (yield stress = 54 MPa, yield strain = 0.0007, Young's modulus = 76 GPa, slope of plastic curve = 0.1). The eutectic solder (63Sn-37Pb) is assumed to be a temperature-dependent elastoplastic material, as shown in Figs. 7.16 and 7.17. All other constituents are considered to be linear elastic materials. The temperature conditions for the study in this section are shown in Fig. 11.10 for the case from 25 to 110°C.

A typical deformation of the micro-VIP CSP PCB assembly is shown in Fig. 11.11. It can be seen that the maximum relative deformation (dominated by shear) is at the corner micro-VIP solder joint. This is due to the thermal expansion mismatch between the chip-BT substrate, the PCB, and a rising temperature.

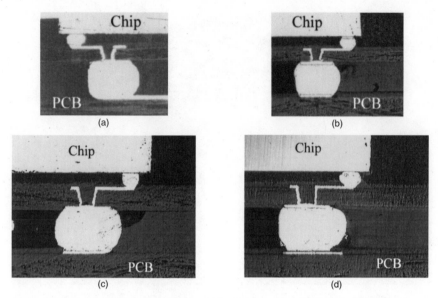

Figure 11.8 Cross sections showing the solder bump, copper VIP, and solder joint.

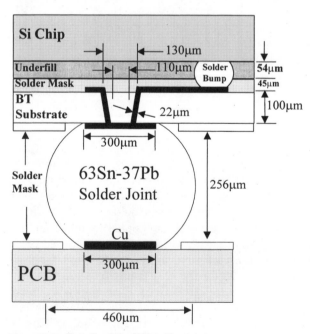

Figure 11.9 Details of the VIP solder joint.

TABLE 11.1   Material Properties of the VIP CSP PBC Assembly

Material properties	Young's modulus (GPa)	Poisson's ratio (ν)	Thermal expansion coefficient (α) ppm/°C
FR-4 substrate	22	0.28	18.5
Copper	76	0.35	17
63Sn-37Pb solder	See note	0.4	21
Underfill	6	0.35	30
Silicon chip	131	0.3	2.8
Solder mask	6.9	0.35	19
Microvia filler	7	0.3	35
BT	26	0.39	15

Note: Young's modulus, as well as stress-strain relationships of solder, are temperature dependent.

The maximum von Mises stress and accumulated equivalent plastic strain range at the corner VIP solder joint and their contour distributions are shown in Figs. 11.12 and 11.13, respectively. It can be seen that the maximum values in the corner VIP solder joint are located at the corner interface between the lower left corner of the Cu VIP and the solder joint. However, these values are very small compared with those of flip chip on board assembly without underfill or WLCSP. This is because the thermal expansion mismatch between the FR-4 PCB and the BT substrate is very small.

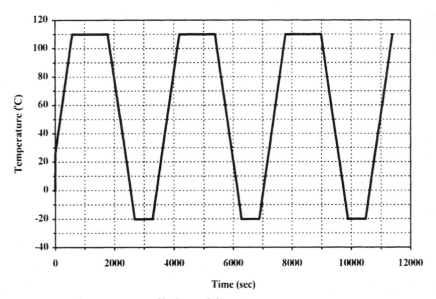

Figure 11.10   Temperature profile for modeling.

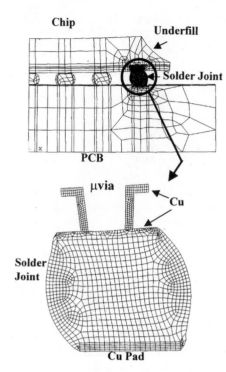

Chip

Underfill

Solder Joint

PCB

μvia

Cu

Solder Joint

Figure 11.11   Deformed shape of the VIP CSP PCB assembly.

Cu Pad

The von Mises stress contour distribution in the Cu VIP is shown in Fig. 11.14. It can be seen that the maximum von Mises stress occurred at the lower left inner corner, and is equal to 74.8 MPa. This is much smaller than the strength of Cu, which is in the range of 200 MPa.

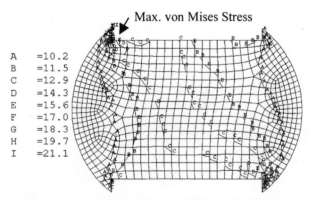

Max. von Mises Stress

A	=10.2
B	=11.5
C	=12.9
D	=14.3
E	=15.6
F	=17.0
G	=18.3
H	=19.7
I	=21.1

Figure 11.12   Maximum von Mises stress in the corner VIP solder joint.

A =.0017
B =.0023
C =.0028
D =.0033
E =.0038
F =.0043
G =.0048
H =.0053
I =.0059

Max. Equivalent
Plastic Strain

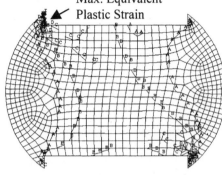

Figure 11.13 Maximum equivalent plastic strain in the corner VIP solder joint.

11.2.6 Solder joint reliability of the micro-VIP CSP assembly

The Garofalo-Arrhenius steady-state creep is generally expressed by[4]

$$\frac{d\gamma}{dt} = C \left(\frac{G}{\theta} \right) \left[\sinh \left(\omega \frac{\tau}{G} \right) \right]^n \exp \left(\frac{-Q}{k\theta} \right)$$

where γ is the steady-state creep shear strain, $d\gamma/dt$ is the steady-state creep shear strain rate, t is the time, C is a material constant, G is the temperature-dependent shear modulus, θ is the absolute temperature (K), ω defines the stress level at which the power law stress dependence

A =60.1
B =62.0
C =63.8
D =65.6
E =67.5
F =69.3
G =71.2
H =73.0
I =74.8

Max. von
Mises
Stress

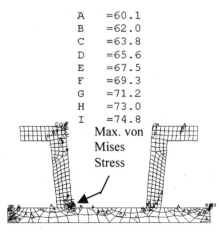

Figure 11.14 Von Mises stress in the corner copper VIP.

breaks down, τ is the shear stress, n is the stress exponent, Q is the activation energy for a specific diffusion mechanism, and k is Boltzmann's constant. For 60wt%Sn-40wt%Pb solder, the material constants have been experimentally determined by Darveaux and Banerji (cited in Lau[4]) with a single hyperbolic sine function. These values are shown in Fig. 11.15.

The temperature loading imposed on the VIP CSP PCB assembly is shown in Fig. 11.10. It can be seen that for each cycle (60 min) the temperature condition is between –20 and +110°C with 15 min ramp, 20 min hold at hot, and 10 min hold at cold. Three full cycles are executed.

The deformed shape is very similar to that of the elastoplastic analysis. Also, the location of the maximum shear stress and creep shear strain hysteresis responses is in the corner joint at the corner interface between the lower left corner of the Cu VIP and the solder joint.

For time-dependent analysis, it is important to study the responses for multiple cycles till the hysteresis loops stabilize. The shear stress and shear creep strain of three cycles at the maximum location are shown in Figs. 11.16 and 11.17, respectively. Figure 11.18 shows the shear stress and shear creep strain hysteresis loops for multiple cycles. It can be seen that the creep shear strain is quite stabilized after the first cycle. Figure 11.19 shows the time history of creep strain energy density at the VIP solder joint's critical location for three cycles. The average creep strain energy density per cycle (ΔW) can be obtained by averaging the creep strain energy density of the last two cycles, which is 0.189 N/mm^2 = 27.4 psi.

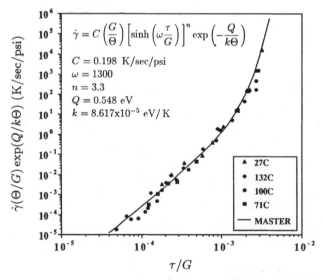

Figure 11.15   Creep constitutive equation for the 60Sn-40Pb solder.

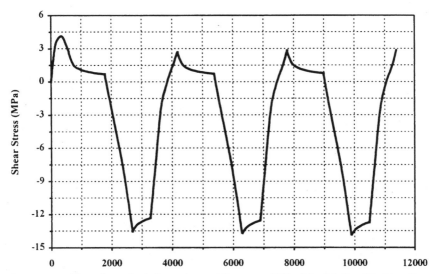

Figure 11.16   Time-dependent shear stress at the corner VIP solder joint's critical location.

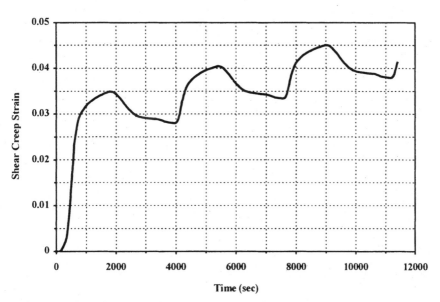

Time (sec)

Figure 11.17   Time-dependent shear creep strain at the corner VIP solder joint's critical location.

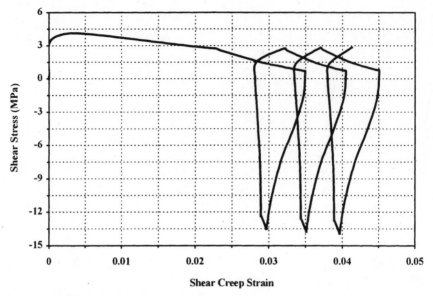

Figure 11.18   Hysteresis loops of the shear stress and creep shear strain at the corner VIP solder joint's critical location.

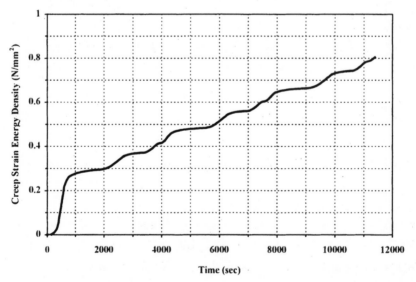

Figure 11.19   Time-dependent creep strain energy density at the corner VIP solder joint's critical location.

Once we have ΔW, the thermal fatigue crack initiation life (N_o) can be estimated from (Darveaux) Eq. 13.35 of Lau,[4] i.e.,

$$N_o = 7860\Delta W^{-1} = 287 \text{ cycles}$$

and the thermal fatigue crack propagation life (N) based on the linear fatigue crack growth rate theory can be estimated by (Darveaux) Eq. 13.36 of Lau,[4] i.e.,

$$da / dN = 4.96 \times 10^{-8}\Delta W^{1.13}$$

or

$$N = N_o + (a_f - a_o)/(4.96 \times 10^{-8}\Delta W^{1.13})$$

where a is the crack length of the solder joint; a_o is the initial crack length, which is assumed to be zero; and a_f is the final crack length. It can be seen that in order to determine N we need to choose an a_f. For example, if $a_f = 0.347$ mm (solder cracks through near the bottom of the Cu VIP), then $N = 6842$ cycles.

On the other hand, if $a_f = 0.433$ mm (solder cracks through the diagonal of VIP solder joint, i.e., from the corner interface between the lower left corner of the Cu VIP and the solder joint to the lower right corner interface between the upper right corner of the Cu pad on the PCB and the solder joint), then $N = 8421$ cycles.

Again, it should be pointed out that due to a numerical integration scheme error in the old version of the finite element code ANSYS, Darveaux' thermal fatigue life prediction equations may involve errors as high as 16 percent.[5]

11.2.7 Summary

A simple micro-VIP substrate for solder-bumped flip chip applications has been presented. It is a single-core, two-sided structure with microvias-in-pads where solder balls are attached. Dog-bone pad structures are not needed and thus there is more space for routing.

The application of the VIP substrate is demonstrated by housing a SRAM device in a CSP format. It is found that the solder-bumped SRAM chip is very easy to assemble on the VIP substrate, and the solder balls are very easy to mount on the copper VIP. Also, the VIP CSP is very easy to assemble on the SMD and NSMD PCB.

The thermal stress in the copper VIP is calculated by a nonlinear finite element method. It is found that the maximum von Mises stress in the copper VIP is much less than the strength of the copper. Thus, under the thermal cycling condition (25 to 110°C), the copper VIP should be reliable in most operating conditions.

The thermal fatigue life of the VIP solder joint is predicted by a creep analysis and by Darveaux's empirical equation. It is found that the average creep strain energy density is very small at the corner solder joint, and thus the joint can last for a long time.

11.3   Effects of Underfill on the Deformations of SLC Substrates[11]

The SLC structure developed by IBM has been discussed in Chap. 4. In this section, the thermal deformations of some key elements of the buildup structure, such as the microvia, the solder mask, and the photosensitive dielectric layer, will be discussed. Most of the results are based on Kunthong and Han.[11]

11.3.1   Problem definition

Figure 11.20 shows a cross section of IBM's SLC structure, which is used to support solder-bumped flip chips with underfill encapsulants. Figures 11.21 a and b show, respectively, the schematic diagrams of a $10 \times 10 \times 0.7$-mm flip chip assembly on the 1.5-mm-thick SLC substrate before and after specimen preparation. It can be seen from the insert that the thickness of the surface laminar layer, including the solder mask, is about 190 µm. The specimen is subjected to a temperature drop of 70°C. A detailed procedure for specimen preparation and isothermal loading for microscopic moire experiments is out of the scope of this section and can be found in Refs. 13–16.

11.3.2   Experimental results: fringe patterns

Figure 11.22 shows the micrographs of the region of interest, which includes the C4 solder bump, microvia, solder mask, and photosensitive dielectric layer. The U and V displacement fields are shown in Fig. 11.23a for the bare substrate and Fig. 11.23b for the C4 flip chip assembly. It should be noted that the zigzag nature of fringes in the core is not caused by optical noise; instead, it represents the hetero-

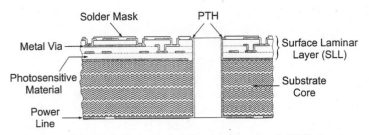

Figure 11.20   Schematic diagram of surface laminar circuit (SLC) structure.

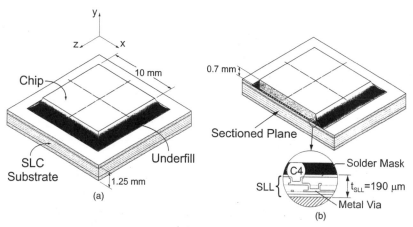

Figure 11.21 Schematic diagram of the flip chip assembly on the SLC substrate (a) before and (b) after specimen preparation.

geneous nature of the multiple epoxy/glass laminate substrate. The region in the core containing the closely spaced fringes represents the epoxy-rich area between the glass fiber bundles, while the sparse fringes indicate the lower TCE of the glass fiber.[11]

One of the advantages of underfill is that it produces a substantial bending of the flip chip assembly and reduces the relative shear deformation of the solder joints. This bending results in a rigid body rotation, which must be nullified by rotating the specimen with respect to the optical system until the shear gradients in U and V of the chip become essentially zero. The resultant patterns of the flip chip assembly after cancellation of the rigid body rotation are shown in Fig. 11.23c. The higher-sensitivity microscopic U and V displacement fields of the region of interest, marked by the dashed box in Fig. 11.22, are

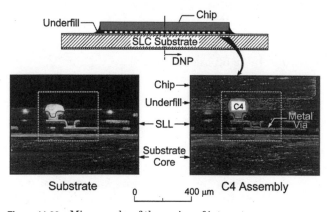

Figure 11.22 Micrographs of the region of interest.

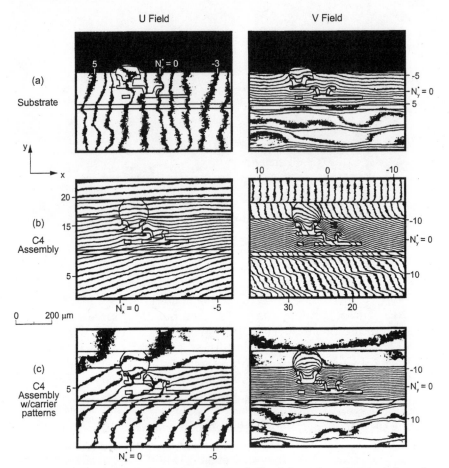

Figure 11.23 Microscopic U and V displacement fields of the region of interest shown in Fig. 11.22. (*a*) Bare substrate. (*b*) Flip chip assembly. (*c*) Flip chip assembly after canceling rigid-body rotation. The contour interval is 104 nm per fringe.

shown in Fig. 11.24*a* for bare substrate and Fig. 11.24*b* for C4 flip chip assembly. The contour interval is 52 nm per fringe.

11.3.3 Global deformation of surface laminar layer

Figure 11.25*a* shows the deformed shape of the surface laminar layer due to the TCE mismatch of the silicon chip, solder joints, underfill encapsulant, and organic SLC, as well as the difference in Young's modulus between these key elements. Due to a temperature drop of 70°C, it can be seen that the surface laminar layer in the bare substrate as well as in the C4 flip chip assembly is contracted. The average shear strain of the surface laminar layer in a C4 flip chip assembly is shown in Fig.

U Field   V Field

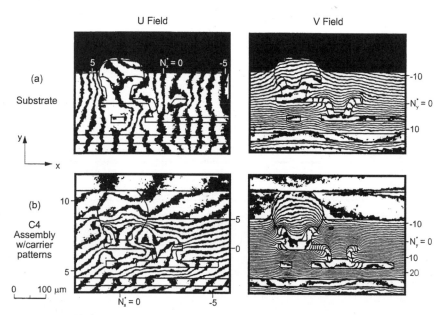

Figure 11.24 Higher-sensitivity microscopic U and V displacement fields of the region, marked by the dashed box in Fig. 11.22. (*a*) Bare substrate. (*b*) Flip chip assembly. The contour interval is 52 nm per fringe.

11.25*b*. It can be seen that the shear strain increases as the distance to neutral point (DNP) increases; the maximum value is about 0.4 percent for a 10×10-mm chip subjected to a temperature change of 70°C.

11.3.4 Local deformation: photosensitive dielectric layer

The normal and shear strain distributions through the thickness of the photosensitive dielectric layer (AA′) are shown in Fig. 11.26. It can be seen that the characteristics of deformation are: (1) $\varepsilon_x > \varepsilon_y > \gamma_{xy}$ for bare substrate; (2) $|\varepsilon_y| > \varepsilon_x > \gamma_{xy}$ for C4 flip chip assembly; (3) ε_x in C4 flip chip assembly is larger than that in bare substrate; (4) $|\varepsilon_y|$ in the C4 assembly is larger than that in bare substrate; and (5) γ_{xy} in C4 assembly is larger than that in bare substrate.

11.3.5 Local deformation: solder mask

The average normal and shear strain distributions along the solder mask are shown in Fig. 11.27. It can be seen that the maximum magnitude of ε_x is about 0.38 percent, slightly less than half of the tensile strength of the solder mask. The magnitudes of ε_y and γ_{xy} are small for the bare substrate. However, they change significantly for the C4 flip chip assembly; peak values are 0.67 percent for ε_y and 0.48 percent for γ_{xy}.

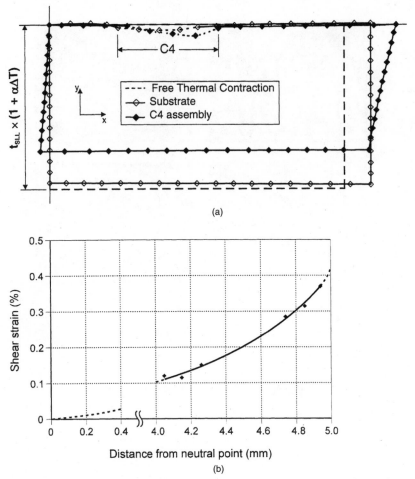

Figure 11.25 (*a*) Deformed shape of the surface laminar layer. (*b*) Average shear strain distribution as a function of DNP.

11.3.6 Local deformation: microvia

The deformed shape of portion CC′ in the flip chip assembly is evaluated from the fringe patterns in Fig. 11.24*b*, and the results are plotted in Fig. 11.28*a*. It can be seen that the microvia segments move in the positive *y* direction (upward) relative to the center portion. This movement is due to the high TCE of the photosensitive material (~70 ppm/°C) and produces a shear strain in the segments. This shear strain distribution is shown in Fig. 11.28*b*. It can be seen that the maximum shear strain (0.38 percent) occurs near the end of the right segment.

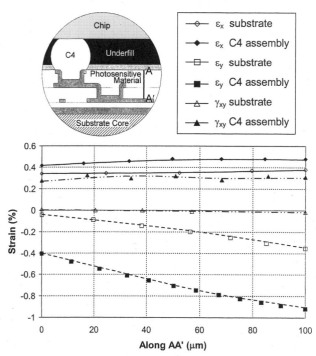

Figure 11.26   Stress-induced strain distributions through the thickness of the photosensitive dielectric layer (AA').

11.3.7   Summary

Figure 11.29 shows the effect of TCE mismatch between the chip and the substrate and modulus difference of the flip chip assembly. It can be seen that (1) without underfill, a large relative deformation is acting at the corner solder joint; (2) with underfill, the chip and substrate deform as a single unit and thus produce substantial bending but very little relative deformation at the corner solder joint; and (3) with the buildup SLC structure, there is a relative deformation acting on the surface laminar layer similar to that in point 1 but on a smaller scale. Thus, when chosing underfill materials for large solder-bumped flip chips on board with buildup layer applications, not only solder joint reliability but also system reliability factors, such as the deformation of other microstructures, should be considered.

Acknowledgments

The author would like to thank P. Kunthong of Clemson University; B. Han of the University of Maryland; C. Chang, C. Chen, and R. Lee of

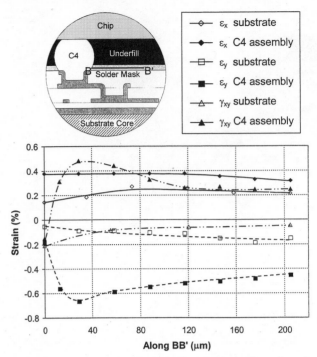

Figure 11.27   Stress-induced strain distributions along the solder mask (BB′).

EPS; and T. Chen, D. Cheng, T. Tseng, D. Lin, and E. Lao of WWEI and UMTC for sharing their useful information with the industry.

References

1. Coombs, C. F., Jr. *Printed Circuits Handbook*, McGraw-Hill, New York, 1996.
2. Lau, J. H., C. Chang, T. Chen, D. Cheng, and E. Lao, "A Low-Cost Solder-Bumped Chip Scale Package—NuCSP," *Circuit World*, **24**(3): 11–25, April 1998.
3. Lau, J. H., and Y. H. Pao, *Solder Joint Reliability of BGA, CSP, Flip Chip, and Fine Pitch SMT Assemblies*, McGraw-Hill, New York, 1997.
4. Lau, J. H., *Ball Grid Array Technology*, McGraw-Hill, New York, 1995.
5. Reboredo, L., "Microvias: A Challenge for Wet Processes," *IPC EXPO 99*, pp. S12-1-1–S12-1-4.
6. Schmidt, W., "High Performance Microvia PWB and MCM Applications," *IPC EXPO 99*, pp. S17-5-1–S17-5-8.
7. Lau, J. H., and C. Chang, "An Overview of Microvia Technologies," *Circuit World* **26**(2): January 2000,
8. Felten, J. J. and S. A. Padlewski, "Electrically Conductive Via Plug Material for PWB Applications," *IPC EXPO '97*, pp. S6-6-1–S6-6-8, March 1997.
9. McDermott, B. J. and S. Tryzbiak, "The Practical Application of Photo-Defined Micro-Via Technology," *SMI Proceedings*, pp. 199–207, San Jose, CA, September 1997.

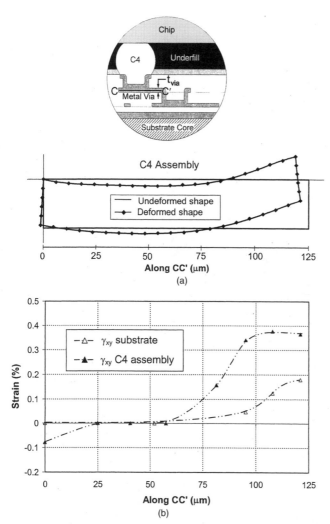

Figure 11.28 (a) Deformed shape of the microvia segment CC′ in the flip chip assembly. (b) Shear strain distribution along CC′.

10. Itagaki, M., K. Amami, Y. Tomura, S. Yuhaku, Y. Ishimaru, Y. Bessho, K. Eda, and T. Ishida, "Packaging Properties of ALIVH-CSP using SBB Flip-Chip Bonding Technology," *IEEE Trans. Advanced Packaging*, **22**(3): 366–371, August 1999.
11. Kunthong, P., and B. Han, "Effect of Underfill on Deformations of Surface Laminar Circuit in Flip Chip Package," ASME Paper No. 99-IMECE/EEP-8, November 1999.
12. Lau, J., C. Chang, C. Chen, C. Ouyang, and R. Lee, "Via-In-Pad (VIP) Substrates for Solder Bumped Flip Chip Applications," *Proceedings of the SMTA International Conference*, pp. 128–136, September 1999.
13. Post, D., B. Han, and P. Ifju, *High Sensitivity Moire Experimental Analysis for Mechanics and Materials*, Springer-Verlag, New York, 1994.

COOLING

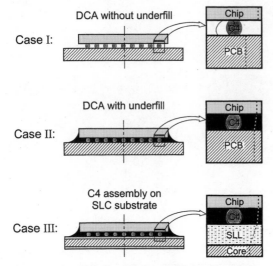

Figure 11.29 Effects of the TCE and modulus between the chip, underfill, and substrate.

14. Han, B., "Deformation Mechanism of Two-Phase Solder Column Interconnections under High Accelerated Thermal Cycling Condition: An Experimental Study," *J. Electronic Packaging, Trans. ASME,* 119: 189–196, 1997.
15. Han, B., "High Sensitivity Moire Interferometry for Micromechanics Studies," *Optical Eng.,* **31**(7): 1517–1526, 1992.
16. Han, B., and D. Post, "Immersion Interferometer for Microscopic Moire Interferometry," *Exp. Mechanics,* **32**(1): 38–41, 1992.

12

PCB Manufacturing, Testing, and Assembly of RIMMs™

12.1 Introduction

The Rambus™-in-line memory modules (RIMMs™) are general-purpose high-performance memory subsystems that have been endorsed by Intel as a new type of memory for its high-speed chipsets. They are suitable for use in a broad range of applications, such as computer memory, personal computers, servers, and workstations, where high bandwidth and low latency are required.

The RIMMs come in several densities: (1) the 64-MB module consists of eight 64-Mbit Direct Rambus DRAM (DRDRAM) devices, and (2) the 128-MB module consists of sixteen 64-Mbit DRDRAM devices. (It should be noted that 128- and 144-Mbit DRDRAMs are also available.) These are extremely high-speed CMOS DRAMs organized as 4-Mbit words by 16 or 18 bits. The Rambus Signaling Level (RSL) technology permits transfer rates of 600, 700, or 800 MHz while using conventional system and board design technologies. DRDRAM devices are capable of sustained data transfers at up to 1.25 ns per 2 bytes (10 ns per 16 bytes). Also, the architecture of DRDRAM allows the highest sustained bandwidth for multiple simultaneous randomly addressed memory transactions. The separate control and data buses with independent row and column control yield over 95 percent bus efficiency. The DRDRAM's multiple banks support up to four simultaneous transactions.

The 64-Mbit DRDRAM is packaged with a μBGA® package[1] with either 74 balls (for the edge-bonded device) or 54 balls (for the center-bonded device). The ball pitch is 0.75 mm in the x and y directions for the edge-bonded device and 0.8 mm in the x direction and 1 mm in the y direction for the center-bonded device, as shown in Fig. 12.1. For both devices, the package total thickness is between 0.6 mm (minimum) and 1.2 mm (maximum). The ball diameter for both devices is between 0.33 mm (minimum) and 0.5 mm (maximum). In this chapter, the focus is on the 74-ball μBGA package.

The PCB of the RIMMs (with the 74-ball μBGA package) is schematically shown in Fig. 12.2. It can be seen that it has 184 pins (92 pins on each side) at a 1-mm spacing. Table 12.1 lists the pin names and numbers for the RIMMs. The module dimensions are 133.35 × 34.93 × 1.27 mm.

In this chapter, the PCB manufacturing and testing of the RIMMs are presented. Also, information is provided on how to assemble the μBGA CSP on both sides of the RIMMs reliably.

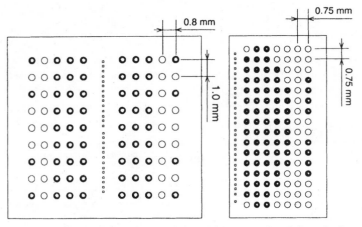

Figure 12.1   Solder ball patterns of the μBGA, center-bonded, and edge-bonded devices.

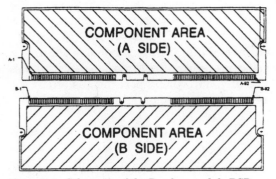

Figure 12.2   Schematic of the Rambus module PCB.

TABLE 12.1   Pin Numbers and Names for RIMMs

Pin	Pin name	Pin	Pin name	Pin	Pin name	Pin	Pin name
A1	Gnd	B1	Gnd	A47	NC	B47	NC
A2	LDQA8	B2	LDQA7	A48	NC	B48	NC
A3	Gnd	B3	Gnd	A49	NC	B49	NC
A4	LDQA6	B4	LDQA5	A50	NC	B50	NC
A5	Gnd	B5	Gnd	A51	Vref	B51	Vref
A6	LDQA4	B6	LDQA3	A52	Gnd	B52	Gnd
A7	Gnd	B7	Gnd	A53	SCL	B53	SA0
A8	LDQA2	B8	LDQA1	A54	Vdd	B54	Vdd
A9	Gnd	B9	Gnd	A55	SDA	B55	SA1
A10	LDQA0	B10	LCFM	A56	SVdd	B56	SVdd
A11	Gnd	B11	Gnd	A57	SWP	B57	SA2
A12	LCTMN	B12	LCFMN	A58	Vdd	B58	Vdd
A13	Gnd	B13	Gnd	A59	RSCK	B59	RCMD
A14	LCTM	B14	NC	A60	Gnd	B60	Gnd
A15	Gnd	B15	Gnd	A61	RDQB7	B61	RDQB8
A16	NC	B16	LROW2	A62	Gnd	B62	Gnd
A17	Gnd	B17	Gnd	A63	RDQB5	B63	RDQB6
A18	LROW1	B18	LROW0	A64	Gnd	B64	Gnd
A19	Gnd	B19	Gnd	A65	RDQB3	B65	RDQB4
A20	LCOL4	B20	LCOL3	A66	Gnd	B66	Gnd
A21	Gnd	B21	Gnd	A67	RDQB1	B67	RDQB2
A22	LCOL2	B22	LCOL1	A68	Gnd	B68	Gnd
A23	Gnd	B23	Gnd	A69	RCOL0	B69	RDQB0
A24	LCOL0	B24	LDQB0	A70	Gnd	B70	Gnd
A25	Gnd	B25	Gnd	A71	RCOL2	B71	RCOL1
A26	LDQB1	B26	LDQB2	A72	Gnd	B72	Gnd
A27	Gnd	B27	Gnd	A73	RCOL4	B73	RCOL3
A28	LDQB3	B28	LDQB4	A74	Gnd	B74	Gnd
A29	Gnd	B29	Gnd	A75	RROW1	B75	RROW0
A30	LDQB5	B30	LDQB6	A76	Gnd	B76	Gnd
A31	Gnd	B31	Gnd	A77	NC	B77	RROW2
A32	LDQB7	B32	LDQB8	A78	Gnd	B78	Gnd
A33	Gnd	B33	Gnd	A79	RCTM	B79	NC
A34	LSCK	B34	LCMD	A80	Gnd	B80	Gnd
A35	Vcmos	B35	Vcmos	A81	RCTMN	B81	RCFMN
A36	SOUT	B36	SIN	A82	Gnd	B82	Gnd
A37	Vcmos	B37	Vcmos	A83	RDQA0	B83	RCFM
A38	NC	B38	NC	A84	Gnd	B84	Gnd
A39	Gnd	B39	Gnd	A85	RDQA2	B85	RDQA1
A40	NC	B40	NC	A86	Gnd	B86	Gnd
A41	Vdd	B41	Vdd	A87	RDQA4	B87	RDQA3
A42	Vdd	B42	Vdd	A88	Gnd	B88	Gnd
A43	NC	B43	NC	A89	RDQA6	B89	RDQA5
A44	NC	B44	NC	A90	Gnd	B90	Gnd
A45	NC	B45	NC	A91	RDQA8	B91	RDQA7
A46	NC	B46	NC	A92	Gnd	B92	Gnd

12.2 PCB Manufacturing and Testing of Rambus Modules[2]

Because of the very strict electrical requirements of the RIMMs, manufacturing and testing of the PCB pose great challenges. The key process steps and the important parameters of each step are discussed. Cross sections of the PCB are examined for a better understanding of the eight-layer structure. Categories of electrical performance such as controlled impedance, propagation delay, crosstalk, and attenuation of the PCB are measured by the time domain reflectometer (TDR).[10] The results are compared against the specifications.

12.2.1 Electrical requirements of Rambus modules

The electrical requirements of the RIMMs are given as follows:

Controlled impedance.

$$Z_1 = 28 \ \Omega \pm 10\%; Z_2 = 56 \ \Omega \pm 10\%; Z_3 = 28 \ \Omega \pm 10\%$$

Same propagation delay for all traces.

Maximum propagation delay is 1.5 ns.

Forward crosstalk requirement (300-ps input rise time 20 to 80 percent).

$$V_{xf}/V_{in} \leq 0.8\%$$

Backward crosstalk requirement (300-ps input rise time 20 to 80 percent).

$$V_{xb}/V_{in} \leq 1.0\%$$

where V_{in} is the voltage of input signal on the active line and V_{xf} and V_{xb} are, respectively, the crosstalk induced at the far end and the near end of the quiet line.

Attenuation limit.

$$V_a/V_{in} \leq 6.25\%$$

where V_a is the attenuation of the input signal propagated along the trace.

12.2.2 Manufacturing of Rambus modules

Since the electrical requirements of the PCB for the RIMMs are very strict, the process window is very small. In order to meet the specifications, special care and attention should be given to each process step. The key manufacturing steps of the PCB for the RIMMs are shown in Fig. 12.3. A brief description of each step follows.

Step 1: Prepare inner layers and dry film. Pay attention to exposure energy and film life. Also, control the laminate thickness and pits and dents.

Step 2: Perform automatic online inspection. Pay attention to circuitry line width and spacing.

Step 3: Apply prepreg. Control the gel time, resin flow and content, and thickness.

Step 4: Laminate all layers together. Take special care with temperature and pressure. Control board thickness and pits and dents.

Step 5: Drill holes using drill data, drill table, and drill template. Hole size and hole location should be within 0.05 mm. Take special care on hole roughness.

Step 6: Electroplate Cu. Pay attention to Cu concentration and current density. Control the back light and Cu thickness.

Step 7: Prepare outer layers and dry film. Devote special attention to exposure energy, film life, and impedance control.

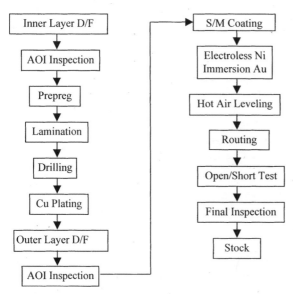

Figure 12.3 Key process steps of the PCB for the Rambus module.

Step 8: Perform automatic online inspection. Watch out for line width and spacing as well as etching factor.

Step 9: Apply solder mask and cure. Pay attention to exposure energy, film life, and curing temperature. Control registration, thickness and cleanliness, adhesion, and via plug.

Step 10: Expose connector pads to Ni-Au plating. Pay attention to chemical analysis and current density. Control Ni and Au thickness.

Step 11: Expose μBGA solder ball pads on the PCB to hot-air leveling (HAL) with Sn-Pb solder or to Entek organic coating. For HAL, pay attention to solder pot temperature, air knife pressure, and Cu contamination. Control Sn-Pb thickness and content, solder balls, solder height, and solder unevenness.

Step 12: Route to smaller panels. Watch for bit life.

Step 13: Perform electrical test for opens/shorts.

Step 14: Perform final inspection. Watch for appearance, cleanliness, flatness, thickness, and hole size. Control impedance.

Step 15: Store the units. Pay attention to storage conditions such as temperature, moisture, and shelf life.

The temperature and pressure are the most important process parameters for laminating multilayer PCB. They affect the thickness of the dielectric layer and thus the controlled impedance. Figure 12.4 shows the lamination temperature and pressure needed for making the eight-layer PCB for the RIMMs. It can be seen that the maximum pressure of

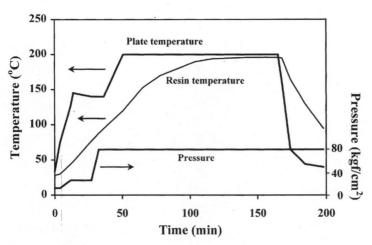

Figure 12.4 Temperature and pressure profiles for PCB lamination.

about 80 kgf/cm² is kept for about 160 min in a vacuum chamber. The maximum temperature at the press plate is about 200°C for about 110 min. At this temperature and duration, the resin temperature can reach 175°C for about 90 min. The whole lamination cycle time is about 200 min.

Figure 12.5 shows the manufactured PCB and the cross sections. It can be seen that an eight-layer PCB is used, with the top and bottom layers for signals (0.5 oz. copper), two inner layers for ground (0.5 oz. copper), two inner layers for power (1 oz. copper), and two inner signal layers (0.5 oz. copper), as shown schematically in Fig. 12.6. Signals are routed with 14-mil (0.355-mm) and 5-mil (0.127-mm) traces.

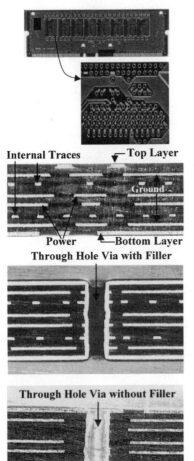

Figure 12.5 Top and cross-sectional views of PCB for the Rambus module.

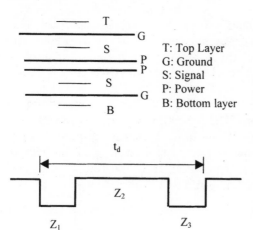

T: Top Layer
G: Ground
S: Signal
P: Power
B: Bottom layer

Figure 12.6 Schematic of the eight-layer PCB and a trace on the PCB.

12.2.3 Electrical measurement of Rambus modules

According to the electrical specifications given by Intel and Rambus, the impedance, propagation delay, crosstalk, and attenuation of the traces (Fig. 12.6) on the PCB for the RIMMs must be measured. In this study, they are measured by the Tektronix Digital Sampling Oscilloscope and Cascade Microtech microprobe. Although the HP network analyzer yields more accurate results (since it works in the frequency domain), the TDR measurement is more than adequate.[10, 11] The electrical measurements are applied to four panels (W, X, Y, and Z), each of which has six double-sided PCBs. The electrical properties to be measured are:

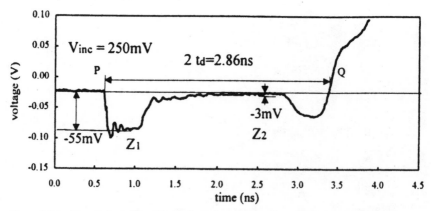

Figure 12.7 A typical profile of the reflected waveform from TDR measurement.

Impedance of trace. A typical profile of the reflected waveform from TDR measurement is shown in Fig. 12.7. From this figure, the impedance (Z_1 and Z_2) can be obtained using the following formula:

$$V_{read} = V_{incident} + V_{reflected}$$

$$Z_i = Z_o \frac{V_{incident} + V_{reflected}}{V_{incident} - V_{reflected}}$$

where Z_o (50 Ω) is the impedance of the cable and $V_{incident}$ (250 mV) is the voltage of the input impulse. The reading and calculation of the impedance are automatically handled by the Tektronix digital sampling oscilloscope. The locations where the impedance data of Z_1 and Z_2 are measured are shown in Fig. 12.7. Z_1 is the impedance of the trace before going into the µBGA package and Z_2 is the impedance at the region of the µBGA package.

Propagation delay. As shown in Fig. 12.7, the time delay is measured form point P to point Q. The time delay measurement can indicate the length of the traces.

Crosstalk of the traces. The typical curves of the forward and backward crosstalk are shown in Figs. 12.8 and 12.9, respectively. These curves are obtained from TDR measurement.

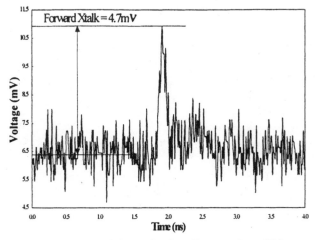

Figure 12.8 A typical forward crosstalk curve from TDR measurement.

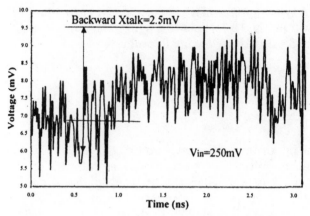

Figure 12.9   A typical backward crosstalk curve from TDR measurement.

Attenuation.   The signals measured at the near end and at the far end of the trace are shown in Fig. 12.10. These curves can be used to determine the attenuation.

12.2.4   Measurement results

Controlled impedance.   Since the impedance Z_3 is the same as Z_1, only Z_1 and Z_2 are shown here. The third PCB of each panel is chosen for measuring the impedance. Figure 12.11 shows the measurement results for Z_1 and Z_2 for many traces on panels W through Z. It can be seen that all the Z_1 values meet the specification (28 Ω ± 10 percent). The average Z_1 is approximately 30 Ω, as shown in Table 12.2.

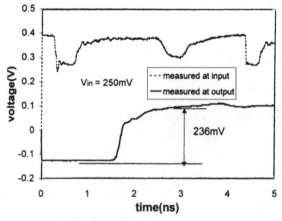

Figure 12.10   Typical signals measured at the near and the far ends of a trace.

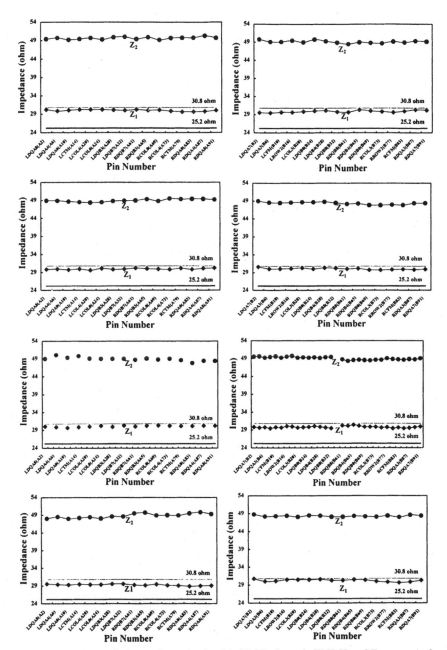

Figure 12.11   Controlled impedance for the third PCB of panels W, X, Y, and Z, respectively (from top to bottom). The left column is for the A side and the right column is for the B side.

TABLE 12.2 Results of Controlled Impedance
and Propagation Delay of the Traces on Panel W

Pin name and pin number	Z_1	Z_2	Time delay
LDQA8(A2)	30.21	49.4	1.43
LDQA4(A6)	29.83	49.7	1.43
LDQA0(A10)	30.02	49.2	1.43
LCTM(A14)	30.21	49.4	1.43
LCOL4(A20)	30.21	49.7	1.43
LCOL0(A24)	30.21	49.3	1.43
LDQB3(A28)	30.12	49.9	1.43
LDQB7(A32)	30.02	50.0	1.43
RDQB7(A61)	30.21	49.4	1.43
RDQB3(A65)	30.02	49.9	1.43
RCOL0(A69)	30.12	49.2	1.43
RCOL4(A73)	29.83	49.7	1.43
RCTM(A79)	29.65	49.8	1.43
RDQA0(A83)	29.65	49.8	1.43
RDQA4(A87)	29.74	50.3	1.43
RDQA8(A91)	30.02	49.7	1.43

The measurement results show that the Z_2 vales are lower than the specification ($Z_2 = 56\ \Omega \pm 10$ percent). It should be noted, however, that this result does not mean that the Z_2 cannot meet the specification, because there are a few vias between layers and short metal lines (28 Ω) connecting the vias on the top layer between the µBGA packages. These vias are capacitive and reduce the characteristic impedance.

In order to verify this point, another two panels (called M and N) are specially made with test coupons. Again, it is found that all Z_2 values from these two panels are very similar to those for panels W through Z and below the specification (56 $\Omega \pm 10$ percent). However, when the impedances of the test coupons on panels M and N are measured, it is found that they all meet the specification (Table 12.3). Thus, it is rea-

TABLE 12.3 Impedance Results of the
Test Coupons on the Specially Made PCB

Panel	Board number	Z_2 (ohm)
M	1	55.3
M	2	55.1
M	3	55.1
M	4	55.0
M	5	55.1
M	6	55.6
N	1	55.6
N	2	56.0
N	3	55.6
N	4	55.7
N	5	56.2
N	6	55.1

sonable to state that the Z_2 of panels W through Z meets the specification. In order to measure the impedance (Z_2) accurately, the manufacturing of coupon lines (test coupons) on the Rambus module boards is highly recommended. These test coupons can be treated as standard traces and can be used to evaluate the consistency of the manufacturing process.

Propagation delay. From the measurement results shown in Table 12.2, it can be seen that the propagation delay values of signals transmitted on all traces on panels W through Z are very close to each other (~1.43 ns). Also, the time delay of these signals meets the specification (maximum time delay is 1.5 ns).

Crosstalk. On the third PCB of each panel, the step impulse (V_{in}) is launched at pin A20 (LCOL4). The forward crosstalk is measured at pin A71 (RCOL2) and the backward crosstalk is measured at pin A22 (LCOL2). The rise time of the signal on the active line is 30 ps (not 300 ps due to equipment limitations). Table 12.4 shows the results of measurements. It looks like they do not meet the specification. However, since (1) the amplitude of crosstalk on the quiet line heavily depends on the rise time of the signal on the active line, and (2) the input rise time in this study is only 10 percent of that recommended by the specification, it is worthwhile to do some simulations to show that all the PCBs do meet the crosstalk specification.

First of all, a simulation of the present structure with a 30-ps rise time input is performed. The results are shown in column 2 of Table 12.5. It can be seen that they are very close to those from measurements. With this confidence, another simulation of the present structure with a 300-ps rise time input is performed; the results are shown in column 4 of Table 12.5. It can be seen that the results are within the specification shown in column 5.

Attenuation. On the third PCB of each panel the attenuation is measured for each trace. The results are shown in Fig. 12.12. It can be seen that except for one trace, all other traces meet the specification.

TABLE 12.4   Crosstalk Measurement Results

Panel	V_{xf}/V_{in} (%)	V_{xb}/V_{in} (%)
W	1.4 ± 0.4	0.9 ± 0.4
X	1.6 ± 0.4	1.0 ± 0.4
Y	2.0 ± 0.4	0.9 ± 0.4
Z	1.5 ± 0.4	1.0 ± 0.4
Average	1.63	0.95

TABLE 12.5   Crosstalk Simulation Results

	Rise time			
	30 ps		300 ps	
	Simulation	Average measurement	Simulation	Specification
V_{xf}/V_{in} (%)	1.68	1.63	0.96	1.0
V_{xb}/V_{in} (%)	1.05	0.95	0.78	0.8

12.2.5   Summary and recommendations

The manufacturing process for PCB for the RIMMs has been presented. Key process steps and the important parameters of each step are also provided. It is found that in order to meet the strict electrical specifications, the PCB thickness and the line width and thickness of the signal traces must be controlled.

Cross sections of PCB for the RIMMs have been examined for a better understanding of the manufactured PCB. It is found that almost all the top and bottom layers for signals, the two inner layers for ground, the two inner layers for power, and the two inner signal layers are quite consistent and properly done.

The controlled impedance, propagation delay, crosstalk, and attenuation of the PCB for the RIMMs have been measured and analyzed by TDR. It is found that (1) $Z_3 = Z_1$, (2) Z_1 and Z_2 meet the requirements of the specifications, (3) the propagation delays (~1.43 ns) of signals transmitted on all traces are almost the same and are less than 1.5 ns, (4) the forward crosstalk ratio (0.96) and the backward crosstalk ratio (0.78) are within the requirements of the specifications (1.0 and 0.8),

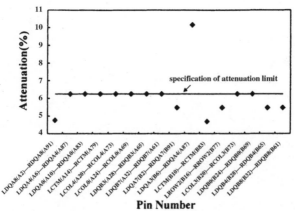

Figure 12.12   Typical attenuation results for both sides of the PCB of panel W.

respectively, and (5) the attenuation ratio of all the traces (except one) on each PCB fulfills the requirement (≤6.25 percent). Also, it should be pointed out that the measurement data are very consistent for all the PCBs, which means that the materials, process, geometry, and quality of the PCBs are quite uniform.

In order to measure the controlled impedance (Z_2) accurately, the manufacturing of coupon lines (test coupons) on the PCB for the RIMMs is highly recommended. These test coupons can be treated as standard traces and can be used to evaluate the consistency of the manufacturing process.

12.3 PCB Assembly of μBGA on Rambus Modules[3]

This section treats the key elements and steps, such as the no-clean solder paste, stencil printing, pick and place, and reflow temperature profile, for assembling the daisy-chained TV-74 Rambus μBGA packages on both sides of the daisy-chained test board. Also, an x-ray imaging system is used to show that no bridging and minimal voiding exist in the solder joints and that there are no broken wires in the μBGA packages during solder reflow. Furthermore, a C-mode scanning acoustic microscope (SAM) is used to show that there are no delaminations or voids in the μBGA packages. Solder joint reliability of the two-sided assembly is evaluated by the shear test and thermal cycling test (−20 to +110°C, 1 h per cycle) and finite element modeling.

12.3.1 Tessera's μBGA component[12]

The surface mount component used in this study is the daisy-chained TV-74 Rambus μBGA, as shown in Figs. 12.13 and 12.14. In general, the unique features of the μBGA package are the ribbonlike flexible leads for chip-level interconnection and the compliant elastomer between the interposer and the chip to relieve the thermal mismatch. The leads are bonded to the Al bond pads on the die one at a time by a thermosonic bonder. The bonding location is not restricted to the perimeter and may be anywhere within the die. Encapsulation is required to protect the bonded leads. The package terminals of the μBGA may be plated bumps, solder balls, or solid-core metal spheres. The array pitch may be 0.5, 0.75, 0.8, or 1.0 mm.

In this study, the dimensions of the daisy-chained TV-74 Rambus μBGA are 12.7 × 8.17 mm. The solder balls are made of 63%wtSn-37wt%Pb solder and are about 0.3 mm in diameter. The bump height is about 0.19 mm above the polyimide tape and the total height of the package is about 0.81 mm. The Young's modulus of solder is 1.5×10^6 psi (10 GPa), the Poisson's ratio is 0.4, and the thermal coefficient of linear expansion is 21×10^{-6}/°C.

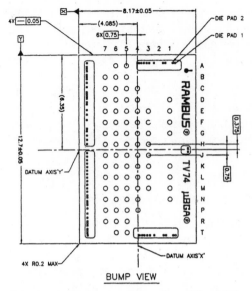

BUMP VIEW

Figure 12.13 Diagrammatic top view of the µBGA.

The solder ball pitch is 0.75 mm. The pads on the bottom surface of the µBGA are interconnected via copper traces in an alternating pattern so as to provide daisy-chained connections when the µBGA is soldered to the PCB. Figure 12.15 shows the top surface of the daisy-chained TV-74 Rambus µBGA.

12.3.2 Test board

Figure 12.16 schematically shows the top and bottom sides of the PCB used in this study. There are eight sites of TV-74 Rambus µBGA on the

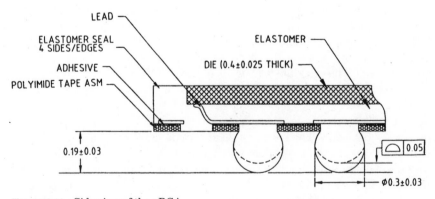

Figure 12.14 Side view of the µBGA.

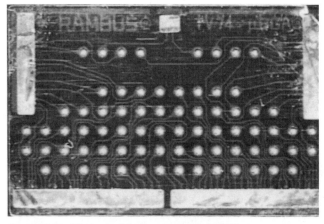

Figure 12.15   Photographic top view of the µBGA.

top side. The pattern is the same on the bottom side. The dimensions of the test board are $133.35 \times 34.93 \times 1.27$ mm. The PCB is made of FR-4 epoxy/glass with etched copper conductors and land patterns (solder pads). The thermal coefficient of expansion of the FR-4 material is 18×10^{-6}/°C in the in-plane direction and 70×10^{-6}/°C in the transverse direction. The Young's modulus of the PCB is 1.6×10^6 psi (11GPa) and the Poisson's ratio is 0.28. Traces are routed on the test board to allow for daisy chaining. For the electroplated copper pads (with Ni-Au) and traces on the PCB, the Young's modulus is 11×10^6 psi (76 GPa), the Poisson's ratio is 0.35, and the thermal coefficient of linear expansion is 17×10^{-6}/°C. A close-up view of the PCB is shown in Fig. 12.17.

12.3.3   Assembly flow chart

The key steps of the two-sided assembly process are shown in Fig. 12.18. It can be seen that the top-side assembly of the TV-74 Rambus µBGA is performed with solder paste, whereas the bottom-side assembly uses flux only. The finished two-sided assemblies will be subjected to x-ray and C-mode SAM inspections, shear test, and thermal cycling test.

12.3.4   Paste, printing, and pick and place

In this study, the no-clean paste TUP NL-005S42B is used. The metal composition is 63 wt% Sn and 37 wt% Pb with a melting point at 183°C and a specific gravity of about 8.4. The metal content is 90 wt%. The powder size is type 3, i.e., more than 90 wt% of the nominal diameters are between 20 and 45 µm, fewer than 1 wt% of the nominal diameters are larger than

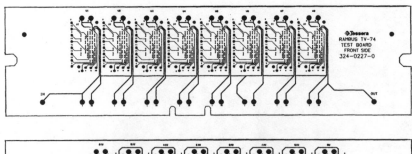

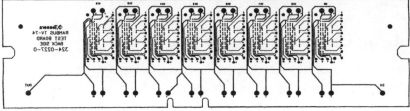

Figure 12.16   Top and bottom sides of the μBGA test board.

45 μm, no more than 10 wt% of the nominal diameters are less than 20 μm, and none of the powder particles are larger than 50 μm. The flux is REM0 [resin with moderate (0%) activity level] type. The viscosity (tested by the Brookfield TF Bar at 5 rpm and 25°C) is 650 to 900 kcps (1 kcps = 1000 centipoise).

The solder paste is printed through a metal stencil with an 80 to 90° rubber squeegee. The squeegee pressure is set to 200 kPa with a squeegee hardness of 90 durometer. The squeegee speed is 25.4 mm/s. The stencil opening to the solder pad is approximately 1:1. The stencil is 0.006 in. (0.15 mm) thick. The printer used in this study is an MPM inline solder stencil system. A fully automated pick and place system (KME) is used to pick and place the TV-74 Rambus μBGA package on the test board.

12.3.5   Solder reflow

A forced-convection oven is used to reflow the components on the two-sided PCB. The reflow temperature profile (Fig. 12.19) is very similar to

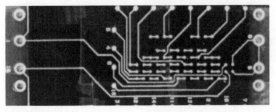

Figure 12.17   Close-up of the μBGA test board.

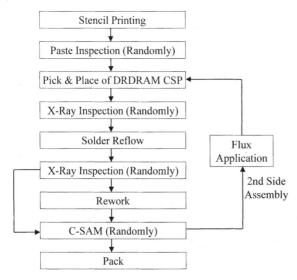

Figure 12.18 Key assembly process steps of μBGA packages on PCB.

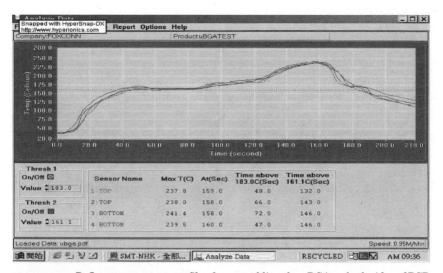

Figure 12.19 Reflow temperature profiles for assembling the μBGA on both sides of PCB.

that of the fine pitch technology. It can be seen that the maximum temperatures of 237 to 241°C are achieved for the TV-74 Rambus μBGA. Also, the time durations above 161 and 183 for the package are, respectively, 132 to 146 and 47 to 66. The belt speed is 0.95 m/min.

12.3.6 Two-sided assembly results

X-ray images of the two-sided assembly are shown in Figs. 12.20 and 12.21. It can be seen directly from 90° (Fig. 12.20) and from a tilted angle (Fig. 12.21) that there is no bridging and only very small voids in the solder joints, and that there is no broken wire in the μBGA package due to solder reflow. Figure 12.22 shows the C- and A-Mode SAM images of the μBGA PCB assembly. It can be seen that there are no voids or delaminations in the μBGA package.

Figure 12.23 shows a typical cross section of the two-sided assembly with the μBGA packages reflowed on the top side with solder paste and on the bottom side with flux only. The typical solder joints with solder paste and with flux, respectively, are shown in Figs. 12.24 and 12.25. They look very similar except for the solder joint heights: the joints with solder paste are 140 to 150 μm high, and the ones with flux are 113 to 118 μm high.

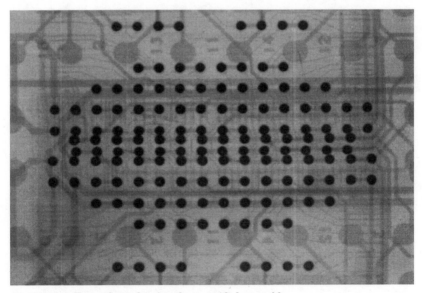

Figure 12.20 X-ray photo showing the two-sided assembly.

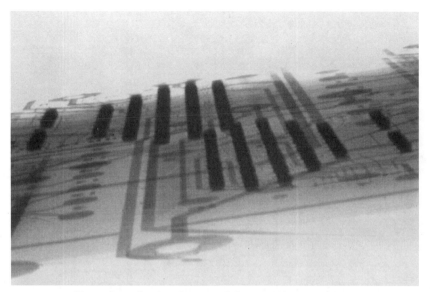

Figure 12.21   X-ray photo showing the two-sided assembly at an angle.

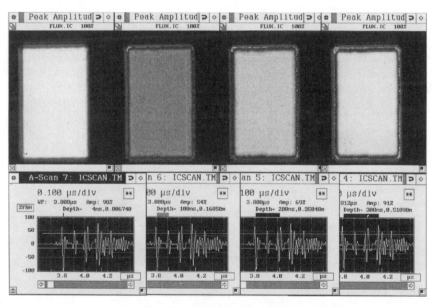

Figure 12.22   C-Mode and A-Mode SAM images of the µBGA assembly.

Figure 12.23   Cross section of the two-sided µBGA assembly.

12.3.7   Shear test and results

The Royce Instruments Model 550 is used to perform the mechanical shear tests. The shear wedge is placed against one edge of the µBGA package 10 mils from the surface of the test board, which is clamped on the stage. A push of the wedge (1 mil/s) is applied to shear the chip/encapsulant/joint away (a destructive test). Typical load displacement curves are shown in Figs. 12.26 and 12.27, respectively, for the case with

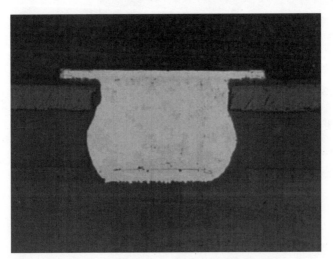

Figure 12.24   Solder joint cross section with solder paste.

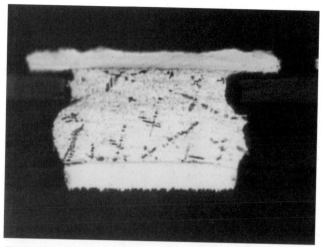

Figure 12.25   Solder joint cross section with flux only.

solder paste and with flux only. It can be seen that the curves are about the same and that the average shear force per solder joint is about 315 gf (Table 12.6). Figures 12.28 and 12.29 show the typical fracture surfaces on the chip and on the PCB, respectively. In general, some of the copper pads are peeled off from the PCB and attached to the tape, and some of the solder joints are peeled off from the tape.

12.3.8   Thermal cycling test and results

All the test boards are grouped and wired together in such a way that the resistance of all the solder joints can be measured. The test boards are tested in an air-to-air thermal cycling chamber. They are subjected

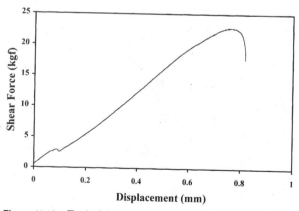

Figure 12.26   Typical load displacement curve for the solder joints with paste.

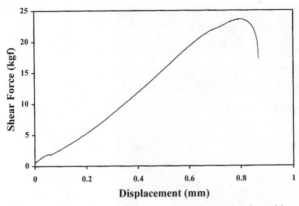

Figure 12.27 Typical load displacement curve for the solder joints with flux only.

to the temperature cycling profile shown in Fig. 12.30.[11–15] It can be seen that the cycle period, ramp-up time, ramp-down time, dwell time at maximum temperature (110°C), and dwell time at minimum temperature (–20°C) are 60, 15, 15, 20, and 10 min, respectively.

A data acquisition system continuously monitored the electrical resistance of the µBGA solder joints and logged the failure times. The failure of a solder joint is defined as total opening (separation) between the solder ball and the PCB (or the tape substrate). The number of cycles to failure is defined as the time of the first solder joint failure in a µBGA package. The sample size is 68. At the time of this writing, all the solder joints have been survived for more than 1800 cycles. The test is ongoing.

Figure 12.28 Fracture surface on the µBGA top surface.

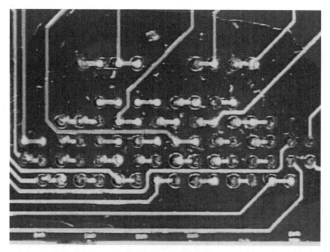

Figure 12.29   Fracture surface on the PCB.

12.3.9   Finite element modeling and results

Figure 12.31 shows the finite element model for the nonlinear temperature-dependent analysis of the assembly. Only one quarter of the assembly is analyzed. The stress-strain relations and Young's modulus of the 63wt%Sn-37wt%Pb are shown in Figs. 7.16 and 7.17. The boundary condition is shown in Fig. 12.30. It is found that the maximum stress and accumulated equivalent plastic strain occurred in the corner solder joint. Also, the maximum accumulated equivalent plastic strain is very small (less than 1 percent) due to the compliant leads and elastomer of the µBGA package. Thus, the solder joints of the assembly should be reliable under most operating conditions.

TABLE 12.6   Die Shear Force of µBGA on the Two-Sided Rambus Test Board

SMT process with flux		SMT process with paste	
Samples	Shear force (Kgf)	Samples	Shear force (Kgf)
F1-3	23.4	S1-3	25.7
F1-4	25	S1-4	24.5
FR1-1	21.6	SR1-1	22.7
FR1-2	24.4	SR1-2	22.8
FR2-1	23.6	SR2-1	22.4
FR2-2	24.7	SR2-2	21.3
Average	23.8	Average	23.2

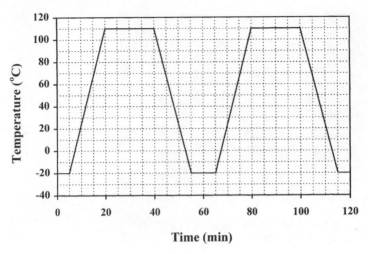

Figure 12.30 Temperature cycling test condition for the two-sided assembly.

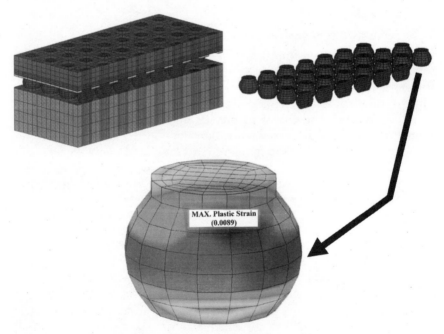

MAX. Plastic Strain
(0.0089)

Figure 12.31 Finite element analysis of quarter of the assembly.

12.3.10   Summary

The TV-74 Rambus μBGAs have been assembled on two sides of PCB with a mass-reflow no-clean process. The cross sections, x-ray images, and C-mode SAM images of the two-sided assemblies have been examined for a better understanding of the present process. It is found that the assembly process of the TV-74 Rambus μBGA packages on PCB is very robust either with solder paste or with flux only.

Temperature-dependent finite element modeling shows that the maximum accumulated plastic strain in the corner solder joint is less than 1 percent, too small to create solder joint reliability issues.

The reliability of the TV-74 Rambus μBGA solder joints has been studied by shear tests and temperature cycling tests. It is found that the average μBGA solder joint shear force is about the same (315 gf) for both pasted and fluxed assemblies. Also, at the time of this writing (1800 cycles), there is no failure yet (−20 to +110°C, 1 h per cycle).

Acknowledgments

The author would like to thank L. Elliott Pflughaupt, Young Gon Kim, and Michael Warner of Tessera; Ricky Lee, Leo Lee, and Meileng Liu of Hong Kong University Science and Technology; Henry Chao and Shang Fu Hung of Foxconn International; Chris Chang and Chih-Chiang Chen of EPS; and Nader Gamini of Rambus for their friendly guidance, effective help, and constructive comments on this project.

References

1. Kim, Y., N. Gamini, and C. Mitchell, "CSP Die Shrink Solution for Memory Devices," *Proceedings of IEEE Electronic and Components Technology Conference,* pp. 627–637, June 1999.
2. Lau, J., H. Chen, C. Chen, C. Chang, M. Lee, D. Cheng, and T. Tseng, "PCB Manufacturing and Testing of the Direct Rambus RIMM Modules," *Proceedings of the Surface Mount International Conference,* pp. 526–533, September 1999.
3. Lau, J. H., C. Chang, C. Chen, H. Chao, and N. Gamini, "Reliable Two-Side Printed Circuit Board Assembly of the Direct Rambus RIMM Modules," *ASME Paper No. 99-IMECE/EEP-3, ASME Winter Annual Meeting,* Tennessee, 1999.
4. "Direct Rambus RIMM Module," www.rambus.com.
5. "Direct Rambus RIMM Continuity Module," www.rambus.com.
6. "Applications for Rambus Interface Technology," www.rambus.com.
7. "Rambus Component Catalog," www.rambus.com.
8. "Direct Rambus RIMM Module Validation Specification," www.rimm.com.
9. "RIMM Module Reference Design," www.intel.com.
10. Tektronix, *TDR Tools in Modeling Interconnects and Packages,* Application Note, Tekronix, Beaverton, OR, 1993.
11. Lau, J. H., C. P. Wong, J. Prince, and W. Nakayama, *Electronic Packaging: Design, Materials, Process, and Reliability,* McGraw-Hill, New York, 1998.

12. Lau, J. H., and S. W. Lee, *Chip Scale Package, Design, Materials, Process, Reliability, and Applications,* McGraw-Hill, New York, 1999.
13. Lau, J. H., and Y.-H. Pao, *Solder Joint Reliability of BGA, CSP, Flip Chip, and Fine Pitch SMT Assemblies,* McGraw-Hill, New York, 1997.
14. Lau, J. H., *Flip Chip Technologies,* McGraw-Hill, New York, 1996.
15. Lau, J. H., *Ball Grid Array Technology,* McGraw-Hill, New York, 1995.

13

Wire Bonding Chip (Face-Up) in PBGA Packages

13.1 Introduction

As mentioned earlier, face-up PBGA packages are very popular today for housing IC devices with more than 250 I/Os at the cost of a penny a pin. However, it should be pointed out that since the PBGA packages are not hermetically sealed,[1–10] they will absorb moisture while they are being made and stored. During the early stage of heating in a reflow oven, cracks may initiate due to the thermal expansion mismatch (TEM) of the package. Growth of the initial cracks (i.e., *popcorning*) may occur during rapid heating at higher temperatures due to a combination of the TEM of the package and the moisture-vaporized pressure acting on the crack surfaces inside the package.

In this chapter, the real-time popcorning of PBGA packages during solder reflow on a PCB is studied using electrical resistance strain gauge measurements. Then, the effects of popcorning on PBGA packages are analyzed using the method of fracture mechanics. Finally, the method for assembling a PBGA on PCB reliably with a large PQFP directly on the opposite side is presented.

13.2 Measurements of Popcorning of PBGA Packages[2]

When does popcorning occur? This is a frequently asked question, and its answer could be useful in designing the PBGA packages and assembling them on the PCB. In this section, the real-time deformation of the PBGA package as shown in Fig. 13.1 during solder reflow on PCB is measured using the electrical resistance strain gauge technique.

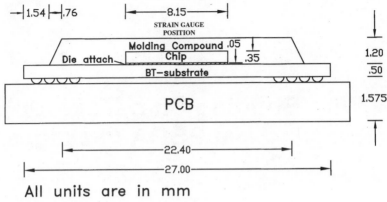

All units are in mm

Figure 13.1 Cross section of a PBGA package.

The PBGA package has four rows of solder balls. Figure 13.2 shows top views of four different layers of the PBGA: the die surface layer (Fig. 13.2a), the die attach material layer (Fig. 13.2b), the solder mask layer (Fig. 13.2c), and the copper layer (Fig. 13.2d). The strain gauge system consists of a three-element rectangular rosette is mounted on the back center of the package (Fig. 13.3) with measurement directions of x, y, and 45° angle as shown in Fig. 13.4, and a data acquisition system to record sufficient data to completely define the strain and stress fields of the PBGA package during solder reflow. From the time-history response curves during reflow, it is possible to identify when popcorning occurs.

The C-mode and A-mode scanning acoustic microscope (SAM) method and cross sections of the packages are utilized to aid in understanding the phenomenon of popcorning of PBGA packages. The tomographic acoustic microimaging (TAMI) technique[11] is then used to perform failure analysis (i.e., to detect delaminations and voids). Unlike the conventional SAM technique, the TAMI technique scans the assembly only once. The output is 30 C-scan images representing thin slices of the assembly. Paging through the pictures rapidly gives the effect of peeling away the assembly layer by layer.

13.2.1 Electrical resistance strain gauge method

The basics. The electrical resistance strain gauge is one of the methods of measuring strains. Ideally, a strain gauge bonded to a specimen would respond only to the applied strain in the specimen and be unaffected by other variables in the environment. In fact, the electrical resistance of the strain gauge varies not only with strain, but with tem-

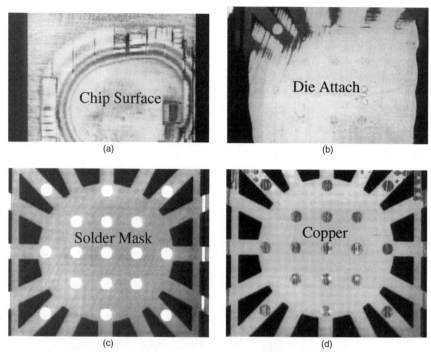

Figure 13.2   Top views of the PBGA package. (*a*) Die surface after grinding out the molding compound. (*b*) Die attach materials (with some area of solder mask) after grinding out the die. (*c*) Solder mask (with some die attach materials remained in the thermal vias) after grinding out the die attach. (*d*) Copper layer (with die attach materials inside the thermal vias) after grinding out the solder mask layer.

perature as well. In addition, the relationship between strain and resistance change (*gauge factor*) itself varies with temperature.

A temperature-induced resistance change is independent of, and unrelated to, the mechanical (stress-induced) strain in the specimen to which the strain gauge is bonded. It is purely due to temperature change, and is thus called the *thermal output* of the gauge.

Figure 13.3   Strain gauge (Fig. 13.4) mounted on top of PBGA package.

Thermal output is caused by two concurrent and algebraically additive effects in the strain gauge installation. First, the electrical resistivity of the grid conductor is somehow temperature dependent; as a result, the gauge resistance varies with temperature. The second contribution to thermal output is due to the differential thermal expansion between the grid conductor and the test specimen to which the gauge is bonded. Thus, the thermal output of the gauge is due to the combined resistance changes from both sources. The net resistance change can be expressed as the sum of resistivity and differential expansion effects as follows:

$$(\Delta R/R_I)_T = [\beta_G + F_G (\alpha_S - \alpha_G)] \Delta T \qquad (13.1)$$

where
R_I = initial reference resistance
ΔR = change in resistance from R_I
$(\Delta R/R_I)_T$ = unit change in resistance from R_I due to thermal output
β_G = thermal coefficient of resistance of the grid conductor
F_G = gauge factor of the strain gauge
α_S = thermal expansion coefficient of speciman
α_G = thermal expansion coefficient of grid conductor
ΔT = temperature change

The thermal output of the gauge is usually expressed in strain units by dividing Eq. 13.1 with the gauge factor setting of the instrument,

$$\varepsilon_T = (\Delta R/R_I)_T/F_I \qquad (13.2)$$

where ε_T is the thermal output in strain units, that is, the strain magnitude registered by a strain indicator (with a gauge factor setting of F_I) when the gauge installation is subjected to a temperature change ΔT under conditions of free thermal expansion for the specimen. F_I is the gauge factor setting of the instrument. In most cases, F_I is set equal to F_G, i.e.,

$$F_I = F_G \qquad (13.3)$$

Many factors affect the thermal output of strain gauges. Some of the more important are test specimen material and shape, grid alloy and lot, gauge series and pattern, transverse sensitivity of the gauge, bonding and encapsulating materials, and installation procedures. If the measurement is a pure mechanical strain (no thermal output engaged in the output), then the result from the strain indicator is a real strain provided the strain gauge and bond adhesive are properly chosen and undergo an appropriate installation procedure. However, if the test is temperature related, then all thermal influences have to be considered. These deviations from the ideal behavior can be important under certain circumstances, and can cause significant errors if not properly accounted for. When the underlying phenomena are thoroughly understood, however, the errors can be controlled. We can rewrite the indicated strains (thermal output) (Eq. 13.2) for the test specimen and reference material, respectively, in view of Eq. 13.1 and 13.3, as follows:

$$\varepsilon_{T(G/S)} = [\beta_G + F_G (\alpha_S - \alpha_G)] \, \Delta T/F_G \qquad (13.4)$$

$$\varepsilon_{T(G/R)} = [\beta_G + F_G (\alpha_R - \alpha_G)] \, \Delta T/F_G \qquad (13.5)$$

where α_R is the thermal expansion coefficient of a standard reference material R, $\varepsilon_{T(G/S)}$ is the thermal output for grid alloy G on the specimen material S, and $\varepsilon_{T(G/R)}$ is the thermal output for grid alloy G on a standard reference material R. Subtracting Eq. 13.5 from Eq. 13.4 and rearranging, we have

$$\alpha_S \Delta T = \alpha_R \Delta T + [\varepsilon_{T(G/S)} - \varepsilon_{T(G/R)}] \qquad (13.6)$$

Therefore, the thermal strain of the specimen $\alpha_S \Delta T$ can be obtained by measuring $\varepsilon_{T(G/S)}$ and $\varepsilon_{T(G/R)}$ and incorporating a well-known reference material. In our case, the reference material is titanium silicate ($\alpha_R = 0.03 \times 10^{-6}/°C$, which is very small).

Strain rosette. Electrical resistance strain gauges are normally employed on the free surface of a specimen to establish the stress at a particular point on this surface. In general it is necessary to measure three strains at a point to completely define either the stress or the strain field.

One of the important rules in the proper interpretation of rosette measurements is the gauge element numbering. *Numbering* refers to the alphabetic sequence in which the gauge elements in a rosette are identified during strain measurement and in the substitution of measured strains into data reduction equations. With any three-element rosette (Figs. 13.3 and 13.4), misinterpretation of the rotational sequence [clockwise (CW) or counterclockwise (CCW)] can lead to incorrect principal strain directions. In case of the rectangular rosette, an improper numbering order will produce completely erroneous principal strain magnitudes and directions. These errors occur when the gauge user's numbering sequence differs from the data reduction equations.

Principal strains and stresses. The three-element rectangular rosette employs gauges placed at the 0, 45, and 90° positions (ε_A, ε_B, and ε_C in Fig. 13.4). The axis of ε_B must be 45° away from that of ε_A, and ε_C must be 90° away from ε_A in the same rotational direction. Thus, by measuring the strains ε_A, ε_B, and ε_C, the cartesian components of strain ε_{xx}, ε_{yy}, and γ_{xy} can be obtained:

$$\varepsilon_{xx} = \varepsilon_A \tag{13.7}$$

$$\varepsilon_{yy} = \varepsilon_C \tag{13.8}$$

$$\gamma_{xy} = 2\varepsilon_B - \varepsilon_A - \varepsilon_C \tag{13.9}$$

Next, the principal strains can be calculated as

$$\varepsilon_{1,2} = \tfrac{1}{2}(\varepsilon_{xx} + \varepsilon_{yy}) \pm \tfrac{1}{2}\sqrt{(\varepsilon_{xx} - \varepsilon_{yy})^2 + (\gamma_{xy})^2} \tag{13.10}$$

and the principal angle ϕ (from the x-axis) is given by

$$\tan 2\phi = \gamma_{xy}/(\varepsilon_A - \varepsilon_C) \tag{13.11}$$

The maximum shear strains in the xy plane are associated with axes at 45° to the directions of the principal strains, and are given by

$$\gamma_{max} = \sqrt{(\varepsilon_{xx} - \varepsilon_{yy})^2 + (\gamma_{xy})^2} \tag{13.12}$$

If the test material is homogeneous in composition, and is isotropic in its mechanical properties, and if the stress-strain relationship is linear, then the principal stresses are:

$$\sigma_{1,2} = E\left[\frac{\varepsilon_{xx} + \varepsilon_{yy}}{2(1 - v)} + \frac{1}{2(1 + v)}\sqrt{(\varepsilon_{xx} - \varepsilon_{yy})^2 + (\gamma_{xy})^2}\right] \tag{13.13}$$

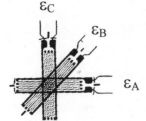

Figure 13.4 The three-element rectangular rosette electrical resistance strain gauge.

13.2.2   Solder reflow of dried PBGAs

Preparation of dried PBGA. The face-up 256-pin PBGA package shown in Figs. 13.1 and 13.2 is used in this study. Ten PBGAs are randomly chosen and put into an oven at 125°C for 24 h to bake out the moisture. The average weight of the baked packages (2.3976 g) is assigned as a reference point for zero moisture absorption (0 wt% change).

Strain gauge installation and data acquisition system. The strain gauge is installed on the PBGA packages and the reference material (ULE™ titanium silicate Code 7971, Corning Glass Company). Thermally produced resistance changes in the leadwires of the strain gauge will generate circuit outputs that are indistinguishable for the thermal outputs being measured. If these differ in any way between the reference and test specimen, the indicated differential expansion data will be in error accordingly. To minimize such effects, leadwire resistance should be kept as low as possible by employing a generous wire size and keeping the leads short. The wiring should also be identical in size, length, and routing for both specimens. The proper gauge installation is a basic requirement for accurate measurement of thermal strain.

1. Prepare the surface.

- Thoroughly degrease the gauging area with GC-6 isopropyl alcohol.
- Remove any scale or oxide material by dry and wet abrasion of the gauging area with 320- and 400-grit silicon-carbide paper, respectively.
- Neutralize the gauging area.

2. Transfer the gauge to the gauging area.

3. Coat both the gauging area and the gauge with the prepared adhesive to bond the gauge and the PBGA.

4. Apply force by spring clamp or dead weight until a clamping pressure of 5 to 20 psi (35 to 135 kN/m²) is reached. Place a soft silicone gum pad over the gauge install ion to evenly distribute the clamping force over the gauge.

5. Cure the adhesive.

6. Protect the gauge by coating it with a thin layer of polyurethane.

7. Wire the gauge.

The M-Bond adhesive 600 is used to bond the strain gauges. The operating temperature range for this material is –269 to +370°C. A

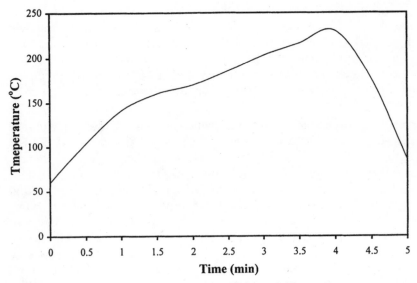

Figure 13.5   Reflow temperature profile of the PBGA on PCB.

curing temperature of 75°C for 4 h is used in our installation. Coatings are applied to protect the gauges from moisture and mechanical damage.

The strain gauges are then connected to our strain gauge measurement and data acquisition system (DAS). The system consists of an analog-to-digital converter, a strain gauge amplifier, a power supply, a GPIB interface board, operating software, and a 486+ computer. The system is able to measure up to 16 channels of signals, to convert data at a maximum rate of 50,000 samples per second, and to measure strains produced by both mechanical and thermal loading. Frequency of data recording and data storage are controlled by Windows-based software.

Measurement and results of dried PBGAs.   The reflow temperature profiles, obtained by mounting a thermal couple on the PCB, are shown in Fig. 13.5 for all the cases considered in this study. The maximum reflow temperature is about 235°C. During reflow, the gauge on the package and the gauge on the reference material will record data and store it in the DAS. Subtracting the gauge data for the package from the gauge data for the reference material and adding the thermal expansions of the reference material will yield the pure PBGA package strains during reflow.

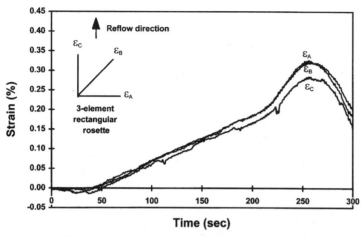

Figure 13.6   Strain components of dried PBGA (24 h at 125°C) during solder reflow.

Figures 13.6 and 13.7 show the time histories of strain measurements for the package only and for the package soldered on the PCB, respectively. It can be seen that the time history of deformation basically follows the reflow temperature profile—the higher the temperatures, the larger the deformations. The response curves are quite smooth. Typical cross sections of the reflowed PBGA packages are shown in Figs. 13.8a and b. It can be seen that there is no delamina-

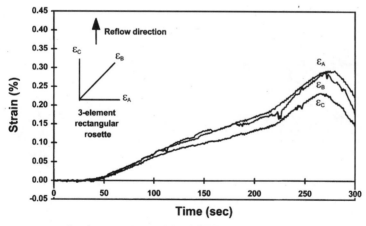

Figure 13.7   Strain components of dried PBGA (24 h at 125°C) on PCB during solder reflow.

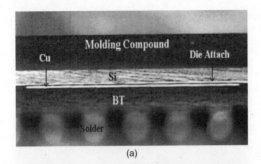

(a)

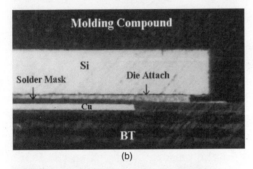

(b)

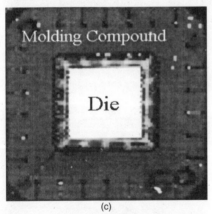

(c)

Figure 13.8   (a) Cross section of dried PBGA package (no delamination). (b) Cross section of dried PBGA package (no crack). (c) C-SAM of dried PBGA package (no delamination; 24 h at 125°C).

tion. This is further verified by C-SAM (Fig. 13.8c). It should be pointed out that there is not much difference between the package soldered on PCB and the one not soldered, except that the PBGA soldered on PCB exhibits smaller strain values. This could be due to the constraints of solder balls reflowed on the PCB.

13.2.3 Solder reflow of moistured PBGAs

Preparation of moistured PBGAs

1. *Controlled moisture content.* Twelve samples are baked for 24 h at 125°C. Their average weight after baking is 2.3976 g. The samples are then put into a humidity chamber at 85°C and 85 percent relative humidity for 168 h. The average moisture absorption is 15.2 mg with a standard deviation of 1.75 mg (Fig. 13.9a). The percent moisture absorption is 0.63 wt%.

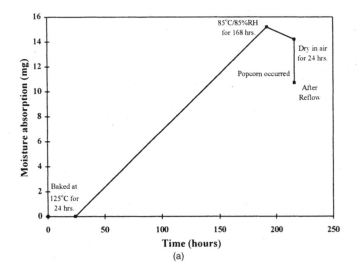

(a)

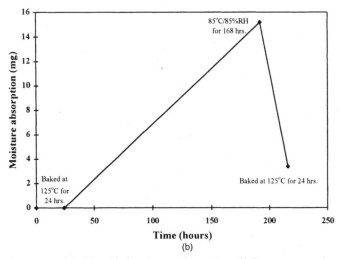

(b)

Figure 13.9 (*a*) Controlled moisture absorption. (*b*) Permanent moisture absorption.

2. *Permanent moisture absorption.* Six of the samples are put into an oven at 125°C for 24 h in an attempt to remove the moisture absorbed during treatment at 85°C/85 percent RH for 168 h. There is 11.8 mg of moisture removed from the samples and 3.4 mg of moisture left inside the PBGA package permanently (Fig. 13.9*b*). This means that baking at 125°C for 24 h cannot remove all the moisture absorbed during treatment at 85°C/85 percent RH for 168 h.

3. *Dry in air and reflow.* The other six PBGA packages are left at room temperature for 24 h before reflow. Very little (~1 mg) moisture is removed from the packages. After reflow (package only), another 3.5 mg of moisture escapes from the packages (Fig. 13.9*a*). This means there is still 10.7 mg of moisture trapped inside the PBGA packages. This trapped moisture could be the driving force on the initial crack surface for popcorning of the package during solder reflow.

Measurement and results of moistured PBGA.   Figures 13.10 and 13.11 show, respectively, the strain components of the moistened (85°C/85 percent RH for 168 h) PBGA package only and of the PBGA soldering on the PCB during solder reflow. It can be seen that there is a jump (discontinuity) in the time-history response curves, indicating that popcorning has occurred. It is interesting to know that the popcorning does not occur at the peak reflow temperature; rather, it occurs earlier.

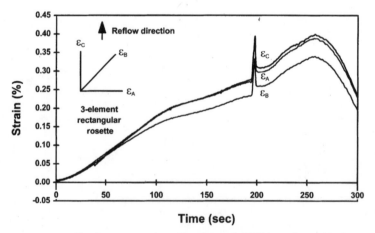

Figure 13.10   Strain components of moistened PBGA package (24 h at 125°C, then 85°C/85% RH for 7 days) during solder reflow.

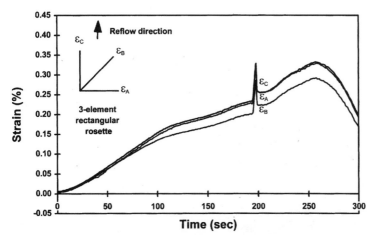

Figure 13.11   Strain components of moistened PBGA package (24 h at 125°C, then 85°C/85% RH for 7 days) on PCB during solder reflow.

The cross sections of these packages (Fig. 13.12a and b) show that there are interfacial delaminations and cracks in the packages. It can be seen that the crack is initiated at the die attach materials near the chip corner and then propagated through the die attach materials, the solder mask, and the interface between the solder mask and the copper layer. There is no crack between the die and the die attach, nor between the solder mask and the BT substrate. The C-SAM image shown in Fig. 13.12c confirms the delaminations. Again, the strain for the PBGA package soldered on the PCB is smaller than that for the PBGA package alone.

Conventional C-mode SAM images of the conditioned (moistened) PBGA on PCB after solder reflow were performed. Figure 13.13 shows these as well as the corresponding A-mode scans. It can be seen that if the A-mode scan is at the delamination area (A-Scan 1), it shows an inverted peak with relatively high amplitude (±30 percent or more), and the corresponding C-mode (C-Scan 1) image is red. On the other hand, if the A-mode scan is not at the delamination area (A-Scan 2), no phase inversion can be observed in the second pulse.

Figure 13.14 shows selected images of the PBGA package generated from a TAMI scan. Figure 13.14a shows the interface of die and die attach material. The points labeled A and B show gap-type defects such as delaminations. The points labeled 1, 2, 3, and 4 are gap-type defects such as voids or disbonds. Figure 13.14b shows the die attach material. At this layer point A becomes a shadow while point B is still

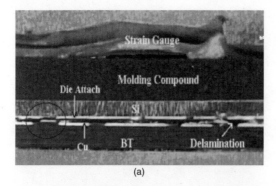

(a)

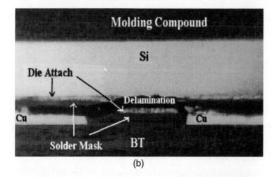

(b)

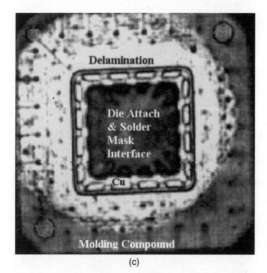

(c)

Figure 13.12 (*a*) Cross section of moistened PBGA package (with delamination). (*b*) Cross section of moistened PBGA package (with cracks). (*c*) C-SAM of the moistened PBGA package (with delamination) (24 h at 125°C, then 85°C/85% RH for 7 days).

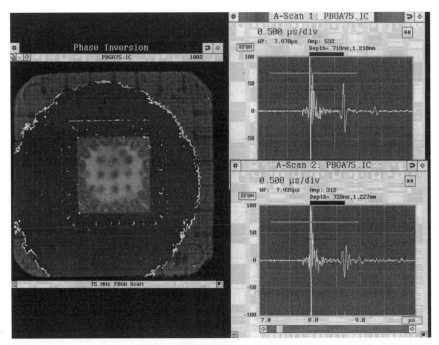

Figure 13.13   Conventional scan images of the moistened PBGA package after solder reflow, C-mode on the left, A-mode on the right.

in bright white form, which means that delamination at A is stopped while delamination at B continues to grow through the die attach material. Figure 13.14c shows the solder mask. At this layer the status of points A and B remains the same as in Fig. 13.14b. The four bright white spots in Fig. 13.14a become shadows, indicating that the disbonds stop at this layer. Figure 13.14d shows the copper layer. It can be seen that the delamination goes through most of this layer. Figure 13.14e shows the acoustic waveform of the A-mode scan with the corresponding locations from Figs. 13.14a through d. The die attach and solder mask contain fewer delaminations because both have lower amplitudes.

Figure 13.15a shows a time-of-flight image 3-D view of the overall PBGA package under the die. It can be seen that delaminations in the 3-D view occur under the die location. The arrow indicates the most likely location for the initiation of delamination. The other spots also represent the delamination growth through the die attach material, solder mask, and copper layer. Figure 13.15b shows a cross-sectional view of the whole PBGA package after popcorning. The arrow showing delamination at the corner corresponds to the arrow in Fig. 13.15a.

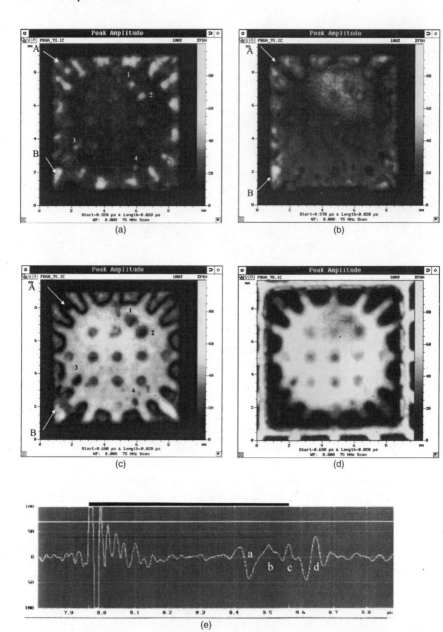

Figure 13.14   Selected images of the PBGA package generated from a TAMI scan. (*a*) The interface of die and die attach. (*b*) The die attach. (*c*) The solder mask. (*d*) The copper layer. (*e*) A typical acoustic waveform of an A-mode scan, showing the corresponding locations from a to d.

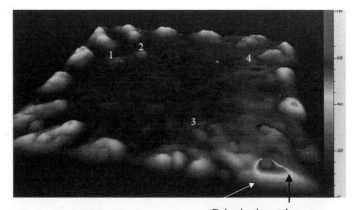

Figure 13.15 (*a*) A time-of-flight image showing a 3-D view of the overall PBGA package under the die. Delamination in this 3-D view is observed under the die location. The arrow points to the most likely location for initiation of delamination. (*b*) A cross-sectional view of the whole PBGA package.

13.2.4 Summary

The real-time popcorning of PBGA packages during solder reflow on PCB has been studied by the electrical resistance strain gauge measurements. Also, detailed equipment setup, strain-gauge installation, and measurement techniques have been provided. Furthermore, the complete time histories of deformations of the packages have been provided. Finally, cross sections and C-SAM images of the packages have been examined for a better understanding of the PBGA package popcorning. Two groups of specimens have been tested, one after 24 h at 125°C and the other after 24 h at 125°C followed by 85°C/85 percent RH for 168 h. Some important results are summarized in the following list.

- The real-time stress/strain measurement is a useful method of identifying when popcorning occurs in PBGA packages during solder reflow.

- Popcorning will not occur for dried ("moisture-free") PBGA packages during solder reflow.

- Popcorning will occur for moistened (85°C/85 percent RH for 168 h) PBGA packages during solder reflow.

- Popcorning occurs not at the peak solder-reflow temperature, but earlier.

- During reflow, the deformations of the PBGA package soldering on the PCB are smaller than those for the PBGA package alone.

- Baking at 24 h at 125°C cannot bake out *all* the moisture absorbed by a PBGA package treated at 85°C/85 percent RH for 168 h.

- TAMI is a very useful technique for performing failure analysis to find delaminations and voids layer by layer in plastic packages.

13.3 Popcorning of PBGA Packages by Fracture Mechanics[5]

Moisture-induced cracking during solder reflow is a critical reliability problem for the plastic ball grid array (PBGA) packages. In this section, the popcorning effect of the PBGA package shown in Fig. 13.16 is analyzed using the method of fracture mechanics. Three specific problems are studied: (1) crack initiation of the uncracked die attach of the

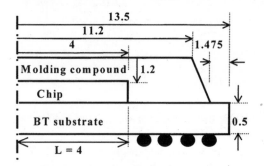

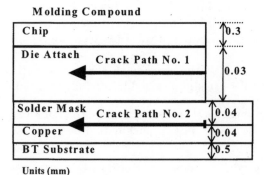

Units (mm)

Figure 13.16 Schematic diagrams of the PBGA package with two different crack paths.

TABLE 13.1   Material Properties of the PBGA Assembly

Material Property	E (MPa)	ν	α (ppm/°C)
BT substrate	26,000	0.39	15
Eutectic solder	10,000	0.4	21
FR-4 PCB	22,000	0.28	18
Silicon die	131,000	0.3	2.8
Molding compound	16,000	0.25	15
Die attach	As shown in Figure 13.17	0.3	59 (<53°C) 195 (≥53°C)
Solder mask	6,870	0.35	19
Copper	76,000	0.34	17

package, (2) crack growth in the die attach, and (3) crack growth at the interface between the solder mask and copper. Two different methods, namely crack tip opening displacement (CTOD) and virtual crack closure technique (VCCT), are used to determine the crack tip parameters such as the strain energy release rate, stress intensity factors, and phase angle for different crack lengths and temperatures.

13.3.1   Crack initiation due to thermal expansion mismatch

In this section the stress and strain distributions of the uncracked PBGA during the reflow process are calculated. The 2-D plane strain element, which has eight nodal points, each with 2 degrees of freedom, is used. Because of the symmetry, only one half of the package is analyzed. Table 13.1 shows the material properties used in the present simulation. The Young's modulus of the die attach is temperature dependent, as shown in Fig. 13.17. The experimental shear strength of the die attach as a function of temperature is shown in Fig. 13.18.

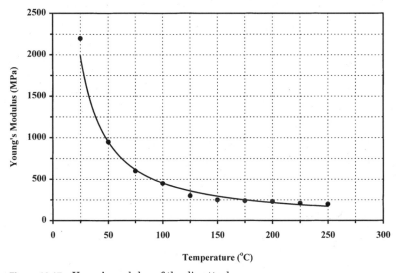

Figure 13.17   Young's modulus of the die attach.

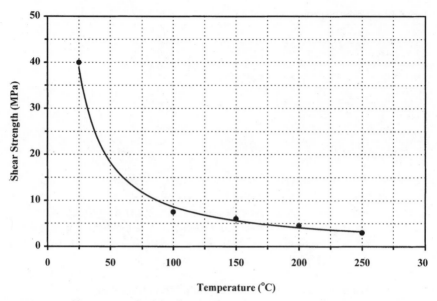

Figure 13.18 Shear strength of the die attach.

The simulation results are used to predict the possible die attach crack initiation inside the PBGA package due to thermal expansion mismatch. During reflow, a thermal couple attached to the PCB is used to measure the time-history temperature profile as shown in Fig. 13.19, which is the thermal loading of the present calculation.

The initiation of cracks has always been a very complicated phenomenon. The initial crack length is also difficult to define. In this study, it is assumed that the initial crack length is the distance from the corner of the chip to the position where the shear stress is equal to

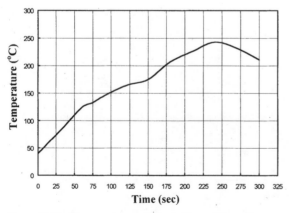

Figure 13.19 Temperature profile during solder reflow.

the shear strength, as shown in Fig. 13.18 for the die attach. Figure 13.20 shows the shear stress distributions of the die attach from chip center to chip corner for different temperatures (x is measured from the chip center and L = 4 mm is half the length of the chip). It can be seen that the higher the temperatures, the larger the initial crack lengths. For example, at 180°C, the initial crack length is about 0.95 mm from the chip corner.

13.3.2 Popcorning due to thermal expansion mismatch and pressure

In this study, two different sets of loading are imposed on the cracked structure. The first is the thermal stress loading that is due to the thermal expansion mismatch of the package. The second is the steam pressure loading. During solder reflow in surface mounting, the moisture absorbed in the plastic package is vaporized and exerts a pressure on the internal crack surfaces of the package. The induced vapor pressure coupled with the thermal expansion mismatch could result in a package failure that is often referred to as *popcorning*. Many researchers have proposed various equations for estimating the moisture-induced vapor pressure in the IC packages. In this study, the vapor pressure is estimated following the equation[10]

$$P(T) = C \cdot \exp\left\{4640\left(\frac{1}{T_i} - \frac{1}{T_f}\right)\right\} \qquad (13.14)$$

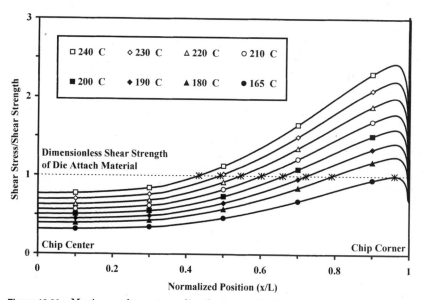

Figure 13.20   Maximum shear stress distribution in the die attach at various temperatures.

where C is a constant (0.485 atm), $T_i = (273 + 85)$ K, and T_f (K) is the reflow temperature. The moisture vapor pressure $P(T)$ as a function of temperature is shown in Fig. 13.21. Even though it is difficult to distinguish the resultant stresses, it is reasonable to assume that the vapor pressure contributes to the opening mode of the crack, while the thermal expansion mismatch contributes mainly to the shearing mode of the crack.

13.3.3 Fracture mechanics methods

In order to understand the fracture behavior of popcorning during solder reflow, finite element analyses of the package with initial crack lengths for different temperatures obtained previously were performed using a commercial code called ANSYS.

In the present study, eight-node plane strain elements are used to model the PBGA package. The cracks in the die attach adhesive and along the interface between the solder mask and the copper are simulated by an array of double nodes. In addition to uniform temperature loading, normal traction is applied to the crack surfaces to account for the steam pressure originating from the absorbed moisture in the package. It should be noted that, due to the thermal mismatch between materials and the applied pressure, both types of cracks have mixed-mode fracture characteristics. Furthermore, since the crack surfaces are no longer traction free, the conventional J-integral formula may

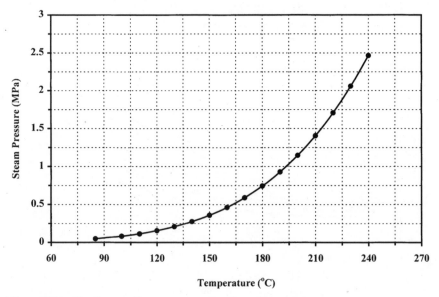

Figure 13.21 The "popcorn" pressure as a function of temperature.

not be valid. In order to evaluate the fracture parameters at the crack tip, two other approaches are employed.

The first method is to calculate the stress intensity factor (SIF) from the crack tip opening displacement (CTOD).[12–14] Details of this method have been discussed in Sec. 8.3. In this section another computational fracture mechanics approach, the virtual crack closure technique (VCCT), is used as well for comparison. With this method, a single element at the crack tip is artificially closed as illustrated in Fig. 13.22. Consequently, the strain energy release rate may be evaluated by calculating the work per unit length required to close the crack at the tip. It should be noted that, since eight-node elements are used in this study, the crack closure procedure involves two pairs of nodal points. In

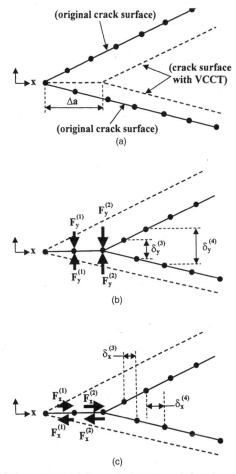

Figure 13.22   Schematic diagram of the virtual crack closure technique (VCCT).

principle, the work required to close the crack should be calculated from the product of nodal forces and nodal displacements of the same pair of nodes. However, in order to reduce the cost of computation, an assumption of self-similarity at the crack tip is introduced by Rybicki and Kanninen.[15] Therefore, the formula for evaluating the strain energy release rate using the modified VCCT can be written as

$$G = \frac{W}{\Delta a} = \frac{1}{2\Delta a} \; [F_x^{(1)}\delta_x^{(3)} + F_y^{(1)}\delta_y^{(3)} + F_x^{(2)}\delta_x^{(4)} + F_y^{(2)}\delta_y^{(4)}] \qquad (13.15)$$

where F_x and F_y denote the nodal forces in the x- and y-directions, respectively, and all symbols in the equation are defined in Fig. 13.22. In this study the strain energy release rates derived from both methods are evaluated and compared. The good agreement verifies the accuracy of results obtained in the present finite element analysis with fracture mechanics.

13.3.4   Fracture mechanics results

Figure 13.23 shows the strain energy release rate (at the crack tip inside the die attach, Fig. 13.16) calculated with the CTOD and

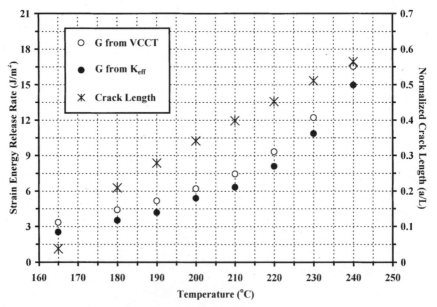

Figure 13.23   Strain energy release rate for different crack lengths and temperatures using the VCCT and CTOD methods (crack in the die attach).

VCCT methods for different temperatures and crack lengths determined in Fig. 13.20. It can be seen that the maximum difference between these two approaches to fracture mechanics is less than 9 percent. Also, the strain energy release rate increases with the crack length. Thus, the cracks will not arrest and popcorning could occur.

Figure 13.24 shows the stress intensity factors and the phase angles (at the crack tip inside the die attach) for different temperatures and crack lengths determined in Fig. 13.20. It can be seen that at lower temperatures, the fracture is dominated by the shear mode ($K_{II} > K_I$ and $\psi \sim 90°$) due to thermal expansion mismatch. As the temperature and crack length increase, K_{II} becomes smaller and K_I becomes larger, and $\psi < 45°$. This is due to more normal forces acting on the larger crack surface to open the crack. Figure 13.25 shows the effective stress intensity factor at the crack tip inside the die attach for different crack lengths and for temperatures up to 240°C. It is interesting to note that connecting the stress intensity factors at different initial crack lengths yields a conservative stress intensity factor–versus–various crack length curve (dotted line in Fig. 13.25).

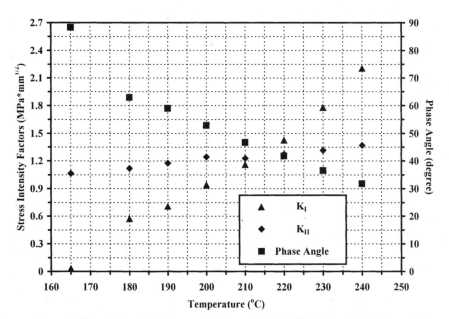

Figure 13.24 Stress intensity factor and phase angle for different crack lengths and discrete temperatures using the CTOD method (crack in the die attach).

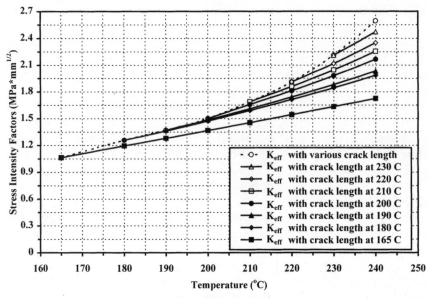

Figure 13.25   Stress intensity factors for different crack lengths and continuous temperatures.

13.3.5   Crack growth in the middle of the die attach

In the previous calculations, the initial crack length is assumed by comparing the maximum shear stress distribution with the shear strength of the die attach. This is a rough estimation. In reality, the initiation of the crack is difficult to determine. Hence, for a more regular approach, the strain energy release rate, phase angle, and stress intensity factor should be calculated using different crack lengths.

Figures 13.26 through 13.29 show the calculated strain energy release rate (at the crack tip in the middle of the die attach, Fig. 13.16) as a function of the solder reflow temperature for crack lengths of 0.4, 1.2, 2, and 2.8 mm, respectively. Again, the maximum difference between the results achieved via the VCCT and CTOD methods is no more than 9 percent. Also, the differences are smaller for larger crack lengths and lower temperatures. It can be seen from Figs. 13.26 through 13.29 that the strain energy release rates are larger for larger crack lengths. Also, for all the crack lengths, the strain energy release rates are larger for higher temperatures. Thus, the crack in the middle of the die attach will not arrest and popcorning could occur.

Figures 13.30 through 13.33 show the stress intensity factors (K_I, K_{II}) and phase angles (ψ) at the crack tip in the middle of the die attach as a function of the solder reflow temperature for crack lengths of 0.4,

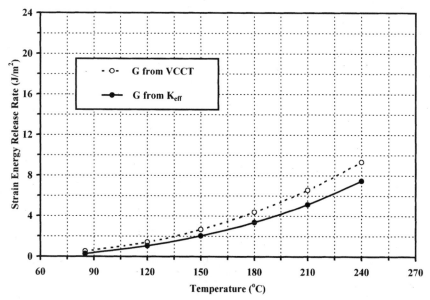

Figure 13.26   Strain energy release rate with $a/L = 0.1$ (crack in the die attach) using VCCT and CTOD methods.

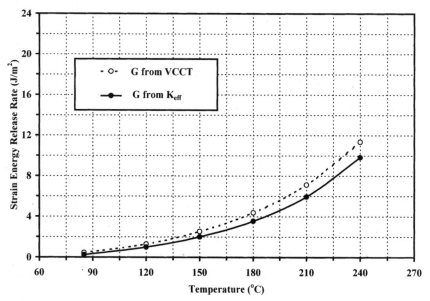

Figure 13.27   Strain energy release rate with $a/L = 0.3$ (crack in the die attach) using VCCT and CTOD methods.

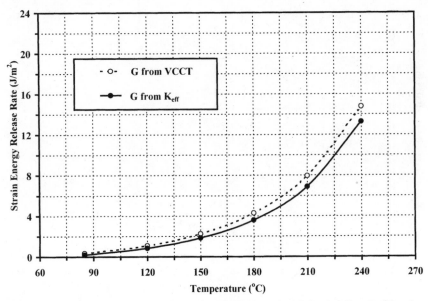

Figure 13.28 Strain energy release rate with $a/L = 0.5$ (crack in the die attach) using VCCT and CTOD methods.

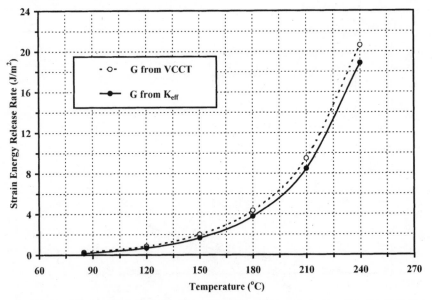

Figure 13.29 Strain energy release rate with $a/L = 0.7$ (crack in the die attach) using VCCT and CTOD methods.

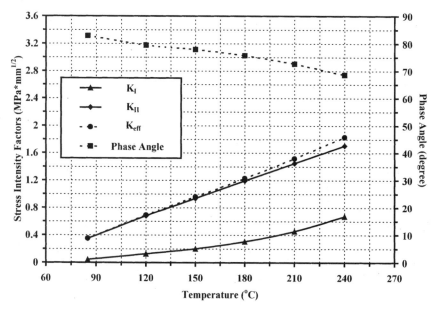

Figure 13.30 Stress intensity factors and phase angle with $a/L = 0.1$ (crack in the die attach) using the CTOD method.

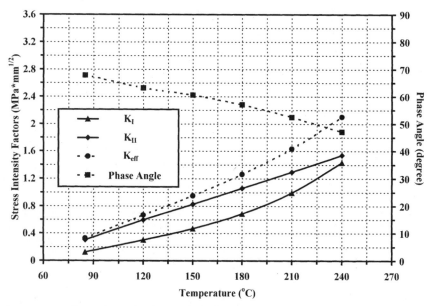

Figure 13.31 Stress intensity factors and phase angle with $a/L = 0.3$ (crack in the die attach) using the CTOD method.

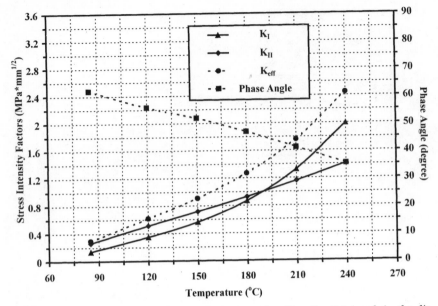

Figure 13.32 Stress intensity factors and phase angle with $a/L = 0.5$ (crack in the die attach) using the CTOD method.

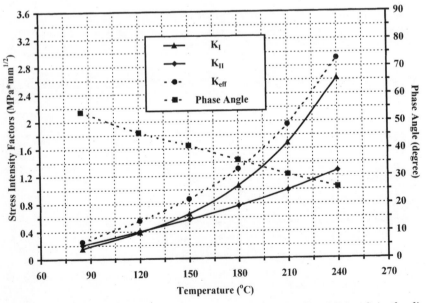

Figure 13.33 Stress intensity factors and phase angle with $a/L = 0.7$ (crack in the die attach) using the CTOD method.

1.2, 2, and 2.8 mm, respectively. It can be seen that (1) for very small crack lengths, $K_{II} \sim K_{eff} > K_I$ and $\psi > 70°$, and thus the fracture is dominated by the shearing mode due mainly to thermal expansion mismatch and small pressure loading; (2) for larger crack lengths, $K_I \sim K_{eff} > K_{II}$ for higher temperatures and $\psi < 25°$, and thus the fracture is dominated by the opening mode due mainly to the moisture-vaporized pressure on the larger crack surface and thermal mismatch; (3) the larger the crack length, the larger the stress intensity factor; and (4) for all crack lengths, the higher the temperature the larger the stress intensity factor.

13.3.6 Crack growth at the interface between the solder mask and copper

Figures 13.34 through 13.37 show the strain energy release rate (at the crack tip of the interface between the solder mask and copper, Fig. 13.16) as a function of the solder reflow temperature for crack lengths of 0.4, 1.2, 2, and 2.8 mm, respectively. For the same crack lengths, Figs. 13.38 through 13.41 show the stress intensity factors and phase angles. It can be seen that the trends of the crack tip parameters are almost the same as in the case where cracks are in the middle of the die attach (Sec. 13.3.5), except that (1) the stress intensity factors are at least 5 times larger; (2) for longer crack lengths and at higher temper-

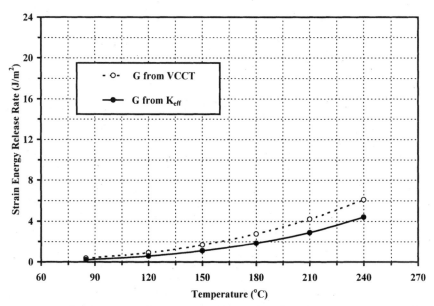

Figure 13.34 Strain energy release rate at $a/L = 0.1$ (crack at the interface between the solder mask and copper) using VCCT and CTOD methods.

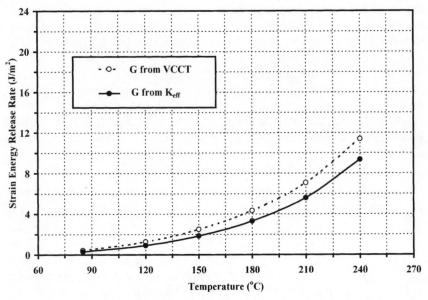

Figure 13.35   Strain energy release rate with $a/L = 0.3$ (crack at the interface between the solder mask and copper) using VCCT and CTOD methods.

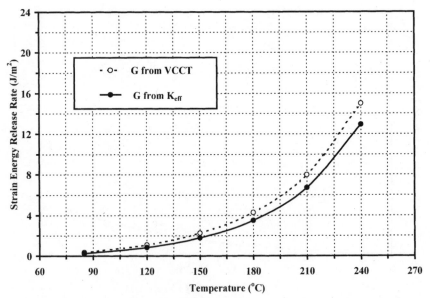

Figure 13.36   Strain energy release rate with $a/L = 0.5$ (crack at the interface between the solder mask and copper) using VCCT and CTOD methods.

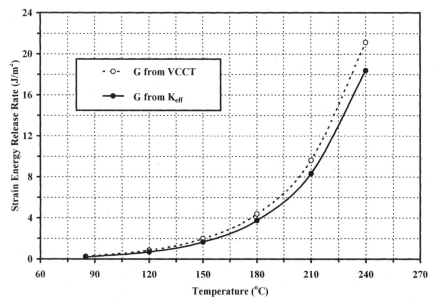

Figure 13.37 Strain energy release rate with $a/L = 0.7$ (crack at the interface between the solder mask and copper) using VCCT and CTOD methods.

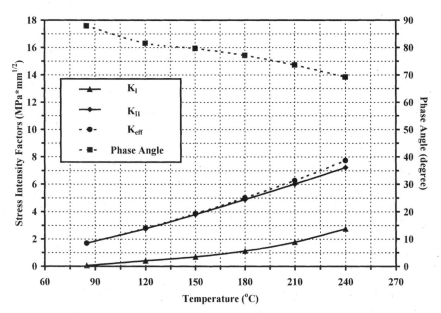

Figure 13.38 Stress intensity factors and phase angle with $a/L = 0.1$ (crack at the interface between the solder mask and copper) using the CTOD method.

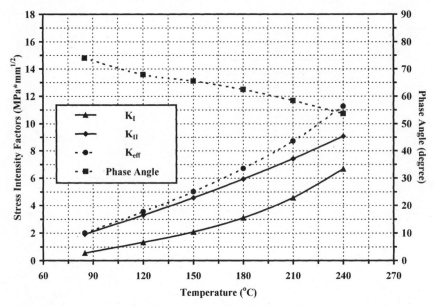

Figure 13.39 Stress intensity factors and phase angle with $a/L = 0.3$ (crack at the interface between the solder mask and copper) using the CTOD method.

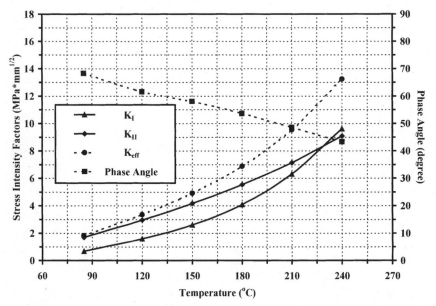

Figure 13.40 Stress intensity factors and phase angle with $a/L = 0.5$ (crack at the interface between the solder mask and copper) using the CTOD method.

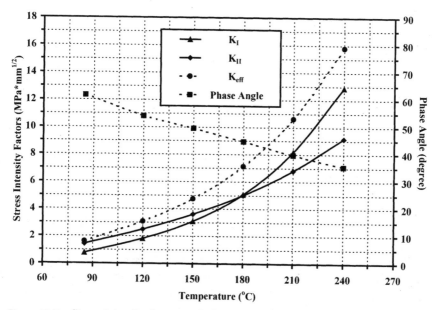

Figure 13.41   Stress intensity factors and phase angle with $a/L = 0.7$ (crack at the interface between the solder mask and copper) using the CTOD method.

atures, even $K_I > K_{II}$, but K_I is not dominant; and (3) even for longer crack lengths but at lower temperatures, K_I is not only not larger than K_{II} but smaller. These effects could be due to the presence of composite materials at the crack tip between the solder mask and copper.

13.3.7   Summary and recommendations

The temperature-dependent popcorning of a PBGA package during solder reflow has been analyzed by fracture mechanics. A simple method has been proposed for estimating the initial crack length of the die attach of the package. Two different paths of crack growth (one in the middle of the die attach and the other in the interface between the solder mask and the copper) are considered. The crack tip parameters, such as the strain energy release rate, the stress intensity factors, and the phase angle, are determined. Some important results are summarized as follows.

- The maximum difference between the CTOD and VCCT methods of calculating the crack tip parameters is less than 9 percent. The differences are even smaller for longer crack lengths and lower temperatures.

- For both crack growth paths, the strain energy release rates at the crack tip are larger for longer crack lengths.

- For both crack growth paths and for all the crack lengths considered, the strain energy release rates at the crack tip are larger for higher temperatures.

- For the crack growth path in the middle of the die attach and for very short crack lengths, the fracture is dominated by the shearing mode. When the crack length increases, the fracture is dominated by the opening mode.

- For the crack growth path in the middle of the die attach, the longer the crack lengths the larger the stress intensity factors. Also, for all the crack lengths considered, the higher the temperatures, the larger the stress intensity factors.

- For the crack growth path in the interface between the solder mask and the copper, the stress intensity factors are more than 5 times larger than for cracks in the middle of the die attach.

- For the crack growth path in the interface between the solder mask and the copper, and for longer crack lengths and at higher temperatures, even $K_I > K_{II}$, but K_I is not dominant. Also, even for longer crack lengths but at lower temperatures, K_I is not only not larger than K_{II} but smaller.

13.4   PCB Assembly of PBGA with Large PQFP Directly on the Opposite Side[4]

For high-density PCB assemblies, it is general practice not to attach two large PBGAs directly on opposite sides of the PCB in view of thermal dissipation, inspection, rework, and solder joint reliability issues.[9] However, it is not uncommon to have a large plastic package, e.g., the 28 × 28-mm PQFP, attached directly on the opposite side of the PCB from a large PBGA package, e.g., the 27 × 27-mm or 35 × 35-mm PBGAs.

In this section, the key elements and steps such as the no-clean solder paste, stencil printing, pick and place, and reflow temperature profile for assembling the PBGA packages on a PCB with the PQFP packages directly on the opposite side are presented. Also, an x-ray imaging system is used to show that no bridging and minimum voiding exist in the solder joints and that there are no broken wires in the PBGA packages. Furthermore, C-mode SAM is used to show there are no delaminations or voids in the PBGA packages. Solder joint reliability of the double-sided assembly is evaluated by the shear test and thermal cycling test (−20 to +110°C, 1 h per cycle).

13.4.1   PBGA and PQFP components

The surface-mount components used in this study are the 225-, 256-, 388-, and 396-pin PBGAs and the 208-pin PQFP, as shown in Fig. 13.42. The components meet the Joint Electronics Devices Engineering

(a) 396PGGA

(b) 225PBGA

(c) 208PQFP

Figure 13.42   Top and bottom views of some of the SMT packages.

Council (JEDEC) package specifications. The dimensions of the PQFP are 28 × 28 × 3.35 mm. The thermal coefficient of linear expansion of the plastic encapsulant is $22 \times 10^{-6}/°C$; the Young's modulus is 2×10^6 psi (14 GPa) and the Poisson's ratio is 0.3. The leadframe material is made of copper; the thermal coefficient of linear expansion is $17 \times 10^{-6}/°C$, the Young's modulus is 17.5×10^6 psi (121 GPa), and the Poisson's ratio is 0.35. The foot length, width, and thickness of the gull-wing-shaped leads are 0.5, 0.15, and 0.115 mm, respectively. These PQFP leadframes are interconnected inside the package in an alternating pattern so as to provide a daisy-chained connection when the PQFP is soldered to the test board.

Unlike the PQFP packages, the 225-, 256-, 388-, and 396-pin PBGAs have an organic substrate with solder balls (instead of the lead frame). The substrate material is bismaleimide triazine resin (BT), made by Mitsubishi Gas Chemical Company. For this material, the glass transition temperature is higher than 200°C, and the thermal coefficient of expansion is 15×10^{-6}/°C in the in-plane direction and 52×10^{-6}/°C in the transverse direction. The flexural strength is 510 MPa and the water absorption is 0.06 percent. The dimensions of the substrates are $27 \times 27 \times 0.56$ mm for the 225- and 256-pin PBGAs and $35 \times 35 \times 0.56$ mm for the 388- and 396-pin PBGAs.

The solder balls are made of 63wt%Sn-37wt%Pb solder and are about 30 mils (760 µm) in diameter. The Young's modulus of the solder is 1.5×10^6 psi (10 GPa), the Poisson's ratio is 0.4, and the thermal coefficient of linear expansion is 21×10^{-6}/°C. The solder ball pitches are 1.27 and 1.5 mm for the PBGAs under investigation. The pads on the bottom surfaces of the PBGAs are interconnected via copper traces in an alternating pattern so as to provide daisy-chained connections when the PBGAs are soldered to the PCB.

13.4.2 Test board

Figure 13.43 schematically shows the top side of the PCB for this study. There are five sites of 208-pin PQFPs, five sites of 225-pin PBGAs with 1.5-mm pitch, four sites of 396-pin PBGAs with 1.5-mm pitch, five sites of 256-pin PBGAs with 1.27-mm pitch, and eight sites of 388-pin PBGAs with 1.27-mm pitch. The pattern is the same on the bottom side. The dimensions of the test board are $320 \times 250 \times 1.57$ mm. The PCB is made of FR-4 epoxy/glass with etched copper conductors and land patterns (solder pads). The thermal coefficient of expansion of the FR-4 material is 18×10^{-6}/°C in the in-plane direction and 70×10^{-6}/°C in the transverse direction. The Young's modulus of the PCB is 1.6×10^6 psi (11 GPa) and the Poisson's ratio is 0.28.

The diameter of the copper pad is 0.635 mm and the diameter of the solder mask opening is 0.889 mm. The width of the copper trace is 0.25 mm. For the electroplated copper pads and traces on the PCB, the Young's modulus is 11×10^6 psi (76 GPa), the Poisson's ratio is 0.35, and the thermal coefficient of linear expansion is 17×10^{-6}/°C. The solder mask height is 0.025 mm (0.066 mm when it is covering the copper traces). Traces are routed on the test board to allow for daisy chaining.[1]

13.4.3 Assembly flow chart

The key steps in fabricating the two-sided assembly are shown in Fig. 13.44. It should be noted that in order to assemble the large PBGAs on PCB reliably with the large PQFPs directly on the opposite side, the PBGAs must be assembled first. Otherwise, the solder paste on the sec-

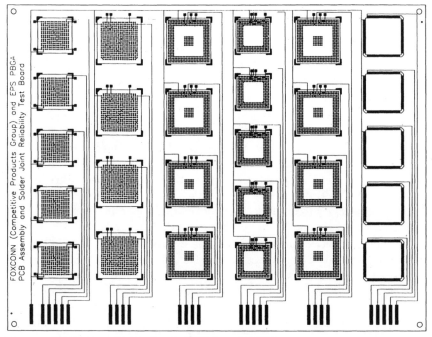

Figure 13.43   Test board with sites for PBGAs and PQFPs.

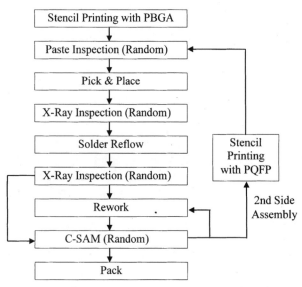

Figure 13.44   Key process steps in assembling PBGAs and PQFPs on both sides of PCB.

ond side of the PCB and solder balls of the PBGAs will not be fully reflowed.

13.4.4 Paste, printing, and pick and place

In this study the no-clean paste TUP NL-005S42B is used. The metal composition is 63 wt% Sn and 37 wt% Pb with a melting point at about 183°C and a specific gravity of about 8.4. The metal content is 90 wt%. The powder size is type 3; i.e., more than 80wt% of the nominal diameters are between 20 and 45 µm, less than 1wt% of the nominal diameters are larger than 45 µm, no more than 10wt% of the nominal diameters are less than 20 µm, and none of the particles are larger than 50 µm. The flux is REM0 [resin with moderate (0%) activity level] type. The viscosity (tested by the Brookfield TF Bar at 5 rpm and 25°C) is 650 to 900 kcps.

The solder paste is printed through a metal stencil with an 80 to 90° rubber squeegee. The squeegee pressure is set to 200 kPa with a squeegee hardness of 90 durometers. The squeegee speed is 25.4 mm/s. The stencil opening to the solder pad is approximately 1:1. The stencil is 0.006 in. (0.15 mm) thick. The printer used in this study is an MPM in-line solder stencil system. A fully automated pick and place system (KME) is used to pick and place the PQFPs and PBGAs on the test board.

13.4.5 Solder reflow

A forced-convection oven is used to reflow the components on the two-sided PCB. The reflow temperature profile (Fig. 13.45) is very similar to that for the fine-pitch technology. It can be seen that maximum temperatures of 221.4 and 213°C, respectively, are achieved for the 27 × 27-mm and 35 × 35-mm PBGAs. Also, the time durations above 150 and 183°C for 27 × 27-mm and 35 × 35-mm PBGAs are, respectively, 157 and 65 s, and 145 and 61 s. The belt speed is 0.95 m/min.

13.4.6 Two-sided assembly results

Figure 13.46 shows a typical cross section of a two-sided assembly with PQFPs reflowed first and PBGAs reflowed second. It can be seen that near the central portion of the PBGA package, the solder paste and solder balls are not fully reflowed. This is due to the presence of the PQFP, which absorbs the heat so that there is not enough heat to totally reflow the solder balls and paste.

It is important to point out that the continuity test results of the two-sided assembly shown in Fig. 13.46 indicate that the assembly is satisfactory. However, the x-ray images (Fig. 13.47) show that it is not! This

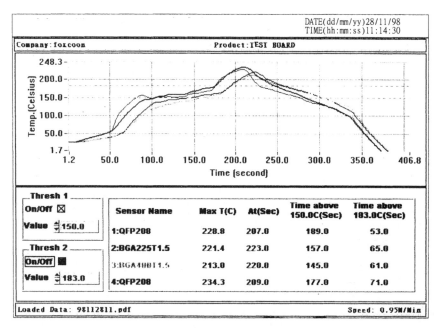

Figure 13.45   Reflow temperature profiles for assembling PBGAs and PQFPs on both sides of PCB.

is further verified by the shear test results (Fig. 13.48). It can be seen from Fig. 13.49 that near the central portion of the PBGA package there are solder balls and paste that have not fully reflowed. Also, on the PCB, there are many locations with solder paste that has not fully reflowed.

Figure 13.49 shows a typical cross section of the two-sided assembly with the PBGAs reflowed first and PQFPs reflowed second. The solder

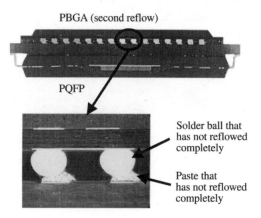

PBGA (second reflow)

PQFP

Solder ball that
has not reflowed
completely

Paste that
has not reflowed
completely

Figure 13.46   Two-sided assembly with PQFPs reflow first and PBGAs reflow second.

208PQFP

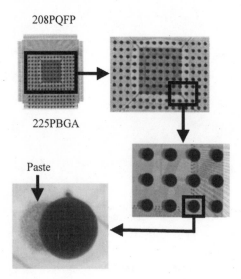

225PBGA

Paste

Figure 13.47   X-ray photos showing the two-sided assembly with less than fully reflowed solder ball and paste.

joints for the PBGAs and PQFPs are shown in Figs. 13.49 and 13.50. It can be seen that they are very nice-looking solder joints.

The x-ray images of the two-sided assembly are shown in Fig. 13.51. Everything looks normal. Figures 13.52 and 13.53 show the shear test results. It can be seen that the failure mode in Fig. 15.52 is very differ-

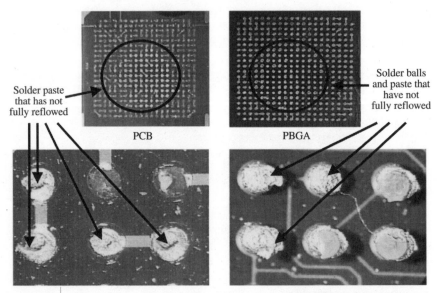

Solder paste that has not fully reflowed

Solder balls and paste that have not fully reflowed

PCB   PBGA

Figure 13.48   Shear test result of the two-sided assembly with PQFPs reflowed first and PBGA reflowed second.

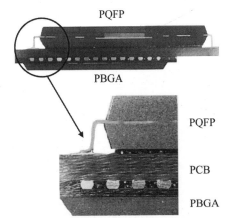

PQFP

PBGA

PQFP

PCB

PBGA

Figure 13.49 Two-sided assembly with PBGAs reflowed first and PQFPs reflowed second.

ent from that of the two-sided assembly with PQFPs reflowed first (Fig. 13.48). In the present case, there are many solder balls and copper pads (from the PCB) on the PBGA package. This means that the solder balls and paste are fully reflowed, and that the bond between the solder and the BT substrate is stronger than that between the solder and the PCB.

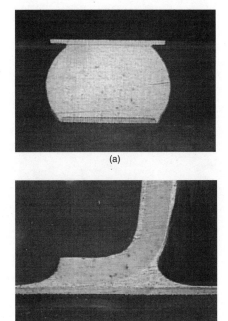

(a)

(b)

Figure 13.50 Cross sections of (a) PBGA solder joint and (b) PQFP solder joint.

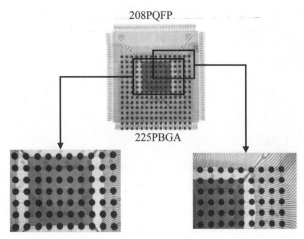

Figure 13.51 X-ray photos showing the two-sided assembly with PBGAs reflowed first and PQFPs reflowed second.

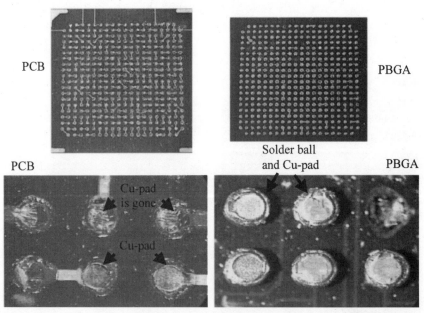

Figure 13.52 Shear test results of the two-sided assembly with PBGAs reflowed first and PQFPs reflowed second.

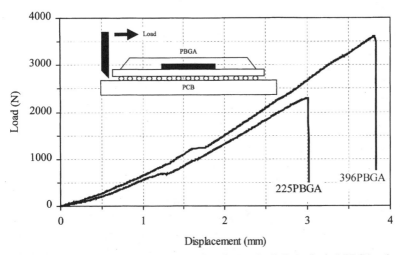

Figure 13.53 Typical load displacement curves for the fully reflowed PBGA solder joints.

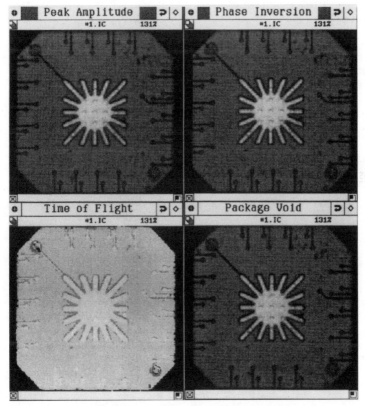

Figure 13.54 C-Mode SAM of a PBGA package on the test board.

The load displacement curves of the 225- and 396-pin PBGAs are shown in Fig. 13.53. It can be seen that the average shear force per solder joint is about 0.9 to 1.0 N. Figure 13.54 shows the C-Mode SAM images of the PBGA-PCB assembly. It can be seen that there are no voids or delaminations in the PBGA package. (All the PBGAs have been baked for 24 h at 125°C before reflow.)

13.4.7 Thermal Cycling Test and Results

All the test boards were grouped and wired together in such a way that the resistance of all the solder joints could be measured. The test boards were tested in an air-to-air thermal cycling chamber, and were subjected to the temperature cycling profile shown in Fig. 13.55. It can be seen that the cycle period, ramp-up time, ramp-down time, dwell time at maximum temperature (110°C), and dwell time at minimum temperature (−20°C) are 60, 15, 15, 20, and 10 min, respectively.

A data acquisition system continuously monitored the electrical resistance of the PBGA solder joints and logged the failure times. The test was run nonstop for more than 6 mo. The failure of a solder joint is defined as total opening (separation) between the solder ball and the PCB (or the BT substrate). The number of cycles to failure is defined as the first solder joint failure in a PBGA package. The sample size is 60.

In this study, reliability of the PBGA solder joints is modeled by the Weibull distribution[9] and the thermal cycling test results are shown in Fig. 13.56. It can be seen that the Weibull slope is 2.54 and the charac-

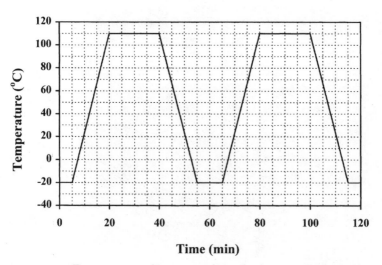

Figure 13.55 Temperature cycling test condition for the two-sided assembly.

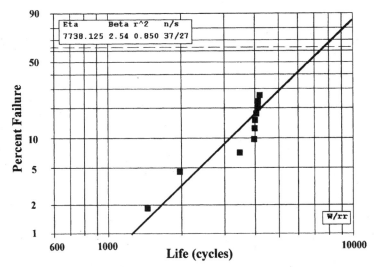

Figure 13.56 Weibull life distribution of the PBGA solder joints in the two-sided assemblies.

teristic life is 7738 cycles. This is an excellent result, and the PBGA solder joints can be used reliably under most operating conditions.[9]

13.4.8 Summary

Large PBGAs and PQFPs have been assembled on two sides of a PCB with a mass-reflow no-clean process. The cross sections, x-ray images, and C-mode SAM images of the two-sided assemblies have been examined for a better understanding of the present process. It was found that, in order to assemble the large PBGAs on PCB reliably with the large PQFPs directly on the opposite side, the PBGAs should be reflowed first.

The reliability of PBGA solder joints has been studied by means of shear tests and temperature cycling tests. It was found that the average PBGA solder joint shear force is 0.9 to 1 N. Also, the Weibull slope is 2.54 and the characteristic life of the PBGA solder joints under thermal cycling conditions (−20 to +110°C, 1 h per cycle) is 7738 cycles.

Acknowledgments

The author would like to thank Henry Chao and Shang Fu Hung of Foxconn International; Tony Chen, Chris Chang, Ray Chen, and C. Chen of EPS; and Ricky Lee, Eric Yan, Xingjia Huang, Leo Lee, and Meileng Liu of Hong Kong University of Science and Technology for their effective help and constructive comments for this chapter.

References

1. Lau, J. H., *Ball Grid Array Technology,* McGraw-Hill, New York, 1995.
2. Lau, J., R. Chen, and C. Chang, "Real-Time Popcorn Analysis of Plastic Ball Grid Array Package During Solder Reflow," *Proceedings of the 23rd IEEE International Electronics Manufacturing Technology Symposium,* pp. 455–463, October 1998.
3. Teo, Y., E. Wong, and T. Lin, "Enhancing Moisture Resistance of PBGA," *IEEE Proceedings of Electronic Components and Technology Conference,* pp. 930–935, May 1998.
4. Lau, J. H., S. W. Lee, and H. Chao, "How to Assemble a Large PBGA on PCB Reliably with a Large PQFP Directly on the Opposite Side," *Proceedings of the IEEE IEMTS,* pp. 448–456, October 1999.
5. Lau, J. H., and S. W. Lee, "Temperature-Dependent Popcorning Analysis of Plastic Ball Grid Array Package During Solder Reflow with Fracture Mechanics Method," *J. Electronic Packaging, Trans. ASME,* March 2000.
6. Kuo, A. Y., W. Chen, L. Nguyen, K. Chen, and G. Slenski, "Popcorning—A Fracture Mechanics Approach," *Proceedings of IEEE Electronic Components and Technology Conference,* pp. 869–874, May 1996.
7. Holalkere, V., S. Mirano, A. Group, A. Kuo, W. Chen, and C. Sumithpibul, "Evaluation of Plastic Package Delamination via Reliability Testing and Fracture Mechanics Approach," *Proceedings of IEEE Electronic Components and Technology Conference,* pp. 430–435, May 1997.
8. Lau, J., J. Miremadi, J. Gleason, R. Haven, S. Ottoboni, and S. Mimura, "No Clean Mass Reflow of Large Over Molded Plastic Pad Array Carriers (OMPAC)," *Proceedings of the 15th IEEE/CHMT International Electronics Manufacturing Technology Symposium,* pp. 63–75, October 1993.
9. Lau, J. H., and Y.-H. Pao, *Solder Joint Reliability of BGA, CSP, Flip Chip, and Fine Pitch SMT Assemblies,* McGraw-Hill, New York, 1997.
10. Shirley, A. G., and J. T. McCullen, "Component Reliability," Tutorial at the 45th IEEE Electronic Components and Technology Conference, 1995.
11. Sigmund, J., and M. Kearney, "TAMI Analysis of Flip Chip Packages," *Adv. Packaging,* pp. 50–54, July/August 1998.
12. Huntchinson, J. W., and Z. Suo, "Mixed Mode Cracking in Layered Materials," *Adv. Appl. Mechanics,* **29:** 64–187, 1992.
13. Rice, J. R., "Elastic Fracture Mechanics Concepts for Interfacial Cracks," *ASME Trans., J. Appl. Mechanics,* **55:** 98–103, 1988.
14. Malyshev, B. M., and R. L. Salganik, "The Strength of Adhesive Joints Using the Theory of Cracks," *Int. J. Fracture Mechanics,* **1:** 114–128, 1965.
15. Rybicki, E., and M. Kanninen, "A Finite Element Calculation of Stress Intensity Factors by a Modified Crack Closure Integral," *Eng. Fracture Mechanics,* **9:** 931–938, 1997.

Wire Bonding Chip (Face-Down) in PBGA Packages

14.1 Introduction

The conventional face-up and overmolded PBGA packages for housing an IC device have been reported in Lau[1] and in Chap. 13 of this book from a perspective of popcorning problems. In general, the thermal performance of the conventional PBGAs is very limited (~3 W). Even the copper-cored face-up PBGAs can only dissipate ~5 W of heat. On the other hand, the face-down PBGAs[1–9] allow the back side of the chip to be attached to a copper heat spreader. This device spreads the heat from the back of the chip over its entire free surface to the air and transfers the heat to the substrate, the solder balls, and then the PCB. Better thermal performance can be achieved by using a thicker heat spreader, adding a heat sink, or both.

One of the notable face-down PBGAs is the SuperBGA,[2,3] which utilizes the concentric power and ground rings surrounding the IC device to facilitate very low interconnect inductance to the power plane and ground plane. The bonding finger ring for I/Os is outside the power and ground rings. In this chapter, a family of face-down and glob-topped (or overmolded) NuBGAs (new and useful BGAs) is presented. These are very similar to the SuperBGA except that the power and ground rings surrounding the IC device are combined into a single ring with many small segments. The flexibility of this design provides programmable power and ground planes and transmission line structures for signal traces.[5]

In this chapter the design concept of NuBGA is described. The electrical and thermal performance of a family of NuBGAs (35 and 40 mm in size, 4- and 5-row solder ball configurations, electroplated and elec-

TABLE 14.1   A NuBGA Package Family

4/4 Line/Space (Ball Pitch at 1.27 mm)						
Body (mm × mm)	Ball/Side	Balls (4 Rows)	I/O	VDD/VSS	Cavity Size (mm × mm)	Die size (mm × mm)
35	26	352	264	80	12.0	10.0
40	30	416	324	84	15.0	13.0
3/3 Line/Space (Ball Pitch at 1.27 mm)						
Body (mm × mm)	Ball/Side	Balls (5 Rows)	I/O	VDD/VSS	Cavity Size (mm × mm)	Die size (mm × mm)
35	26	420	308	112	13.0	10.0
40	30	500	380	120	15.0	13.0

Pitch = 1.27 mm.

troless bonding fingers), as shown in Table 14.1, is studied using both numerical simulations and experimental measurements. Parasitic parameters such as the resistance, inductance, and capacitance are extracted from time domain reflectometer (TDR) and time domain transmission (TDT) measurements. Crosstalk and simultaneous switch output (SSO) noise of NuBGA are also investigated through simulations.

Measures of thermal performance, such as the thermal resistance and temperature distribution, are determined by finite element simulation and wind tunnel measurements. Also, the solder joint responses of NuBGA on PCB are obtained using a nonlinear three-dimensional finite element method.

Finally, another new NuBGA package with a very thin (less than 150 μm thick) substrate and nonuniform heat spreaders is presented. It can be shown that the electrical and thermal performance of the new package is much better than that of packages with thicker substrates and uniform heat spreaders.

14.2   NuBGA Design Concepts

Figure 14.1 shows a schematic of an electronic packaging system. The critical electrical issues of electronic packaging systems are signal integrity and power integrity. The signal integrity is affected by transmission line properties such as delay, reflection, controlled impedance, crosstalk, and SSO noise. The power integrity is affected by the power supply resonance due to core logic and gate switching. The purpose of power (VDD) and ground (VSS) conductors or planes for I/O signals is to lower inductance, provide multiple voltages, isolate SSO noise, and arrange solder ball and bond pad distributions. The purpose of VDD and VSS planes for core logic is to isolate I/O, provide loop inductance

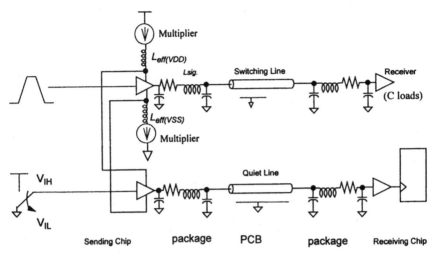

Figure 14.1 Schematic of an electronic packaging system.

to avoid power supply resonance, and arrange bond pad and solder ball distributions.

NuBGA is a single-core (0.55-mm-thick, high–glass transition temperature organic materials), two–metal layer, cavity-down (face-down) package with a heat spreader covering its back surface. Excellent electrical performance on the substrate is achieved by applying the concept of split via connection (SVC) or split wraparound (SWA). With these concepts, programmable power/ground split-ring segments for wire bonding, shorter wire bond length, and controlled-impedance microstripline and coplanar stripline signal traces can be designed. The copper heat spreader is attached to the substrate through a standard PCB prepreg lamination process. Figures 14.2*a* and *b* illustrate a cross section of a standard NuBGA design.

14.2.1 Programmable VDD/VSS SVCs and microstripline and co-planar stripline traces

Figure 14.3 shows the SVC on the top side of a NuBGA package, and Figs. 14.4*a* and *b* illustrate the design concept. Unlike SuperPBGA, NuBGA has only one ring for power and ground wire bonding. This ring is split or segmented to support a pad ratio (signal to power/ground) of at least 4:1. To provide transmission line structures for signal traces on the top side of the substrate, split planes on the bottom side of the substrate are designed to match the configuration of the split ring segments. Split-ring segments and split planes are then connected through vias of the substrate. Signal traces on the top side of the sub-

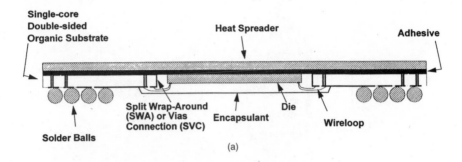

Figure 14.2   (a) Schematic of a standard NuBGA. (b) Cross section of a standard NuBGA.

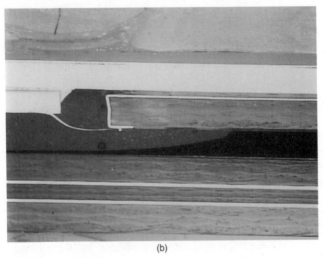

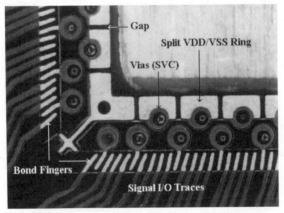

Figure 14.3   Split via connection (SVC) on a NuBGA.

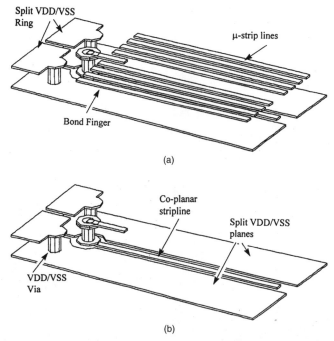

Split VDD/VSS Ring

μ-strip lines

Bond Finger

(a)

Co-planar stripline

Split VDD/VSS planes

VDD/VSS Via

(b)

Figure 14.4 The split power/ground ring segments on the top side of the substrate (*a*), and the split power/ground planes on the bottom side of the substrate (*b*), are connected through vias (SVC). Signal traces are routed in either microstripline (*a*) or coplanar stripline (*b*).

strate are covered by the split power/ground planes on the bottom side of the substrate to form the microstripline structure. Signal traces routed on the bottom side of the substrate are sandwiched by the split power/ground planes to form coplanar stripline structures. The microstripline and coplanar stripline structures provide controlled-impedance transmission line characteristics and smaller crosstalk for signal traces. The transmission line structures also provide stronger coupling between signal traces and power/ground planes, which reduces the effective power/ground inductance and SSO noise for better signal and power loop integrity.

14.2.2 Programmable VDD/VSS SWA and microstripline and coplanar stripline traces

Another way to connect the single split-ring segments on the top side of the substrate and the split VDD/VSS planes on the bottom side of the substrate is to use the side wall of the cavity of the substrate through split wraparound (SWA). Figure 14.5 shows the SWA on the substrate and Figs. 14.6*a* and *b* illustrate the design concept.

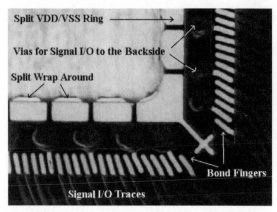

Figure 14.5   Split wraparound (SWA) on a NuBGA.

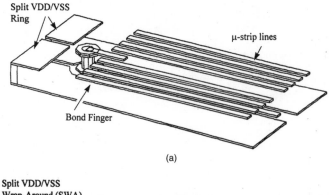

(a)

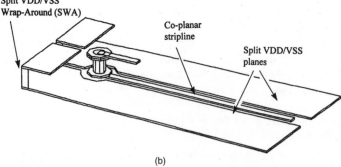

(b)

Figure 14.6   In contrast to SVC, the split-ring segments on the top side (a) of the substrate connected to the split power/ground planes on the bottom side (b) through split wraparound (SWA). Signal traces are routed in either microstripline (a) or coplanar stripline (b).

14.3   NuBGA Design Examples[5, 6]

Layout designs for a 40 × 40-mm NuBGA are illustrated in Figs. 14.7*a* and *b* for a four-row configuration (416 solder balls) and in Figs. 14.8*a* and *b* for a five-row configuration (500 solder balls). In these cases, the bonding fingers are made with electroless Ni-Au. For electroplated Ni-Au bonding fingers, the layouts are basically the same except more complicated.

Figures 14.7*a* and 14.8*a* show the top-side layout of split power/ground ring segments, vias, bonding fingers for the microstripline and coplanar stripline traces, and solder balls. Figures 14.7*b* and 14.8*b* show the bottom-side layout of split VDD/VSS planes and coplanar stripline traces. The layout designs of the 35 × 35-mm NuBGAs have been reported in Lau and Chou.[5] From the preceding discussion, NuBGA designs can be summarized as follows:

14.3.1   Conventional PCB design rules and process

- Package body size: less than 50 mm

- Chip size: less than 20 mm

- Number of solder balls: less than 600

- Solder ball distributions: 4 or 5 rows

- Solder ball pitch: 50 or 40 mils (1.27 or 1 mm)

- Bond finger pitch: 4/4 or 5/3 mils (0.1/0.1 or 0.125/0.075 mm)

- Trace width/space: 0.1/0.1 mm (4-row) or 0.075/0.075 mm (5-row)

- Through-hole via diameters: 0.3 mm (4-row) or 0.25 mm (5-row)

- Through-hole land diameter: 0.56 mm (4-row) or 0.5 mm (5-row)

- Solder ball land diameter: 30 mils (0.76 mm)

- Solder mask opening for solder ball: 0.65 mm

- Single-core, two–metal layer substrate

- Cavity-down design with heat spreader

14.3.2   Electrically and thermally enhanced low-cost package

- Single split-ring segments for power/ground wire bonding

- Power/ground split via connection (SVC) or split wraparound (SWA)

- Flexible I/O to power/ground pad ratio

- Flexible power/ground supporting multiple power supply and noise decoupling

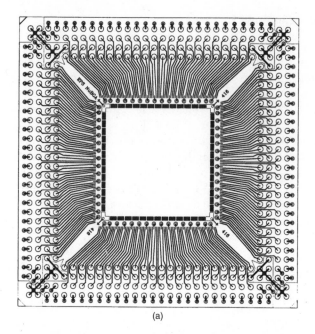

(a)

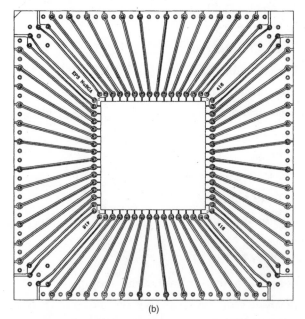

(b)

Figure 14.7 Layout of a 40 × 40-mm, 416-pin SWA NuBGA. (*a*) Top side. (*b*) Bottom side.

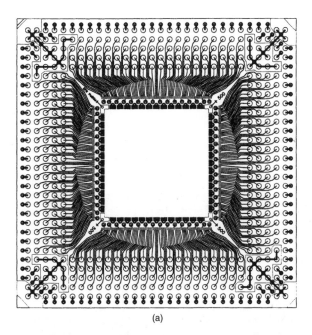

(a)

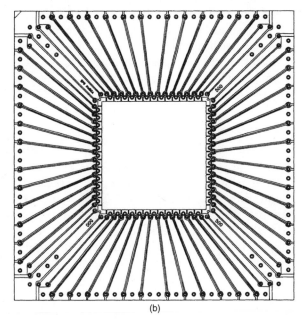

(b)

Figure 14.8   Layout of a 40 × 40-mm, 500-pin SVC NuBGA. (a) Top side. (b) Bottom side.

- Controlled-impedance microstripline and coplanar stripline traces
- Strong coupling between signal to both power and ground to reduce SSO noise
- Low thermal resistance
- Low package warpage

14.4   NuBGA Package Family

As described earlier, with the special design concepts, NuBGA provides outstanding electrical and thermal performance. Table 14.1 shows a NuBGA package family (1.27-mm ball pitch) with different body sizes, four or five rows of solder balls, and different ball counts. Further thermal and electrical enhancement may be achieved by applying additional metal stiffeners or heat sink and thinner core substrates.

14.5   NuBGA Electrical Performance

The package parasitic parameters (R, L, and C) can be extracted from TDR/TDT measurements.[10] Package resistance (R), inductance (L), and capacitance (C) of all the NuBGAs shown in Table 14.1 are calculated by the finite element method. To evaluate the SSO noise performance, system circuits representing a generic point-to-point system net are created. System simulations are performed to compare the system performance of NuBGA with that of the competitive packages: plastic quad flat pack (PQFP), conventional single-core double-sided face-up PBGA (2MPBGA) and four–metal layer enhanced PBGA (4MEPBGA).

14.5.1   NuBGA package parasitic parameters

The finite element method was used to calculate package parasitic parameters such as R (signal resistance), L (signal self-inductance), C (signal self-capacitance), KLm (signal inductive coupling coefficient), KCm (signal capacitive coupling coefficient), Kb (signal backward crosstalk coupling coefficient), $Ksig\_VDD$ (signal-to-VDD inductive coupling coefficient), $Ksig\_VSS$ (signal-to-VSS inductive coupling coefficient), $LVDD$ (effective VDD inductance), $LVSS$ (effective VSS inductance), and Z_0 (signal characteristic impedance) of the NuBGAs shown in Table 14.1. The results of these calculations are shown in Table 14.2. Some important results are summarized as follows:

1. The signal self-inductance (L) depends on the length, thickness, and width of the signal traces. The longer the signal traces, the higher the inductance. Also, the wider the signal traces, the lower the inductance.

TABLE 14.2 NuBGA Package Parasitic Parameters

Body (mm)	Pin Count	L (nH)	KLm	C (pF)	KCm	Kb	Ksig_VDD	Ksig_VSS	LVDD (nH)	LVSS (nH)	R (mΩ)	Z_0 (Ω)
35	352	3.5–7.9	0.02–0.4	0.77–2.7	0.002–0.36	0.006–0.19	0.1–0.4	0.1–0.4	0.05–1.8	0.05–1.8	97–142	54–67
40	416	3.8–8.6	0.02–0.4	0.84–2.9	0.002–0.36	0.006–0.19	0.1–0.4	0.1–0.4	0.06–2.0	0.06–2.0	105–155	54–67
35	420	3.2–8.2	0.03–0.44	0.72–3.0	0.002–0.39	0.006–0.21	0.1–0.4	0.1–0.4	0.04–1.6	0.04–1.6	85–170	54–67
40	500	3.4–9.0	0.03–0.44	0.79–3.3	0.002–0.39	0.006–0.21	0.1–0.4	0.1–0.4	0.05–1.8	0.06–1.8	95–190	54–67

- The inductances of NuBGA 416 and NuBGA 500 (40×40-mm body size) are higher than those of NuBGA 352 and NuBGA 420 (35×35-mm body size) due to the longer signal traces on the larger packages.
- The ranges of L between NuBGA 352 (4-mil line width) and NuBGA 420 (3-mil line width) are very close, because NuBGA 352 has longer and wider signal traces while NuBGA 420 has shorter and narrower signal traces. This is also true between NuBGA 416 and NuBGA 500.

2. The signal self-capacitance (C) depends on the length of the signal traces and the spacing between the signal traces and VDD/VSS. The longer the signal traces, the higher the capacitance. Also, the more space between the signal traces, the lower the capacitance.

- The capacitance of NuBGA 420 and NuBGA 500 (5 rows of solder balls and 3-mil spacing between signal traces) is higher than that of NuBGA 352 and NuBGA 416 (4 rows of solder balls and 4-mil spacing between signal traces) due to smaller spacing between signal traces and VDD/VSS.
- The ranges of C between NuBGA 352 and NuBGA 420 are very close, because NuBGA 352 has longer signal traces and larger spaces between signal traces while NuBGA 420 has shorter signal traces and smaller spacing between signal traces. This is also true between NuBGA 416 and NuBGA 500.

3. The signal resistance (R) depends on the length, thickness, and width of signal traces. The longer the trace, the higher the resistance. Also, the wider the trace, the lower the resistance.

- The R values of NuBGA 416 and NuBGA 500 are higher than those of NuBGA 352 and NuBGA 420 because NuBGA 416 and NuBGA 500 have longer signal traces.
- The ranges of R between NuBGA 352 and NuBGA 420 are close, because NuBGA 352 has longer and wider signal traces while NuBGA 420 has shorter and narrower signal traces. This is also true between NuBGA 416 and NuBGA 500.

4. The signal-to-VDD inductive coupling coefficient ($Ksig\_VDD$) and the signal-to-VSS inductive coupling coefficient ($Ksig\_VSS$) are the same for all NuBGA packages even though their traces have different length and width.

5. Given a fixed configuration of VDD/VSS, the signal inductive coupling coefficient (KLm) and mutual inductance (Lm) depend on the spacing between the signal traces. The more space between the signal traces, the lower the KLm and Lm values.

- The KLm and Lm of NuBGA 352 and NuBGA 416 are lower than those of NuBGA 420 and NuBGA 500 due to larger spacing between signal traces.

6. The signal capacitive coupling coefficient (KCm) and mutual capacitance (Cm) depend on the spacing between the signal traces. The more spacing between the signal traces, the lower the KCm and Cm.

- The KCm and Cm of NuBGA 352 and NuBGA 416 are lower than those of NuBGA 420 and NuBGA 500 due to larger spacing between the signal traces.

7. The signal backward crosstalk coupling coefficient (Kb) is approximately one-quarter of the sum of KLm and KCm.

8. The effective VDD inductance $(LVDD)$ and the effective VSS inductance $(LVSS)$ depend on the length of the traces and the width of the copper plates on the bottom of the substrate. The longer the traces, the higher the inductance. Also, the wider the copper plates, the lower the inductance (assuming the power and ground ring segments have half VDD and half VSS).

- The $LVDD$ and $LVSS$ of NuBGA 416 and NuBGA 500 are higher than those of NuBGA 352 and NuBGA 420 due to the longer traces and narrower copper plates.

- The $LVDD$ and $LVSS$ of NuBGA 352 are higher than those of NuBGA 420. Also, the $LVDD$ and $LVSS$ of NuBGA 416 are higher than those of NuBGA 500.

9. The signal characteristic impedance is the same for all NuBGA packages due to the use of the same design concepts for electrical performance.

The foregoing simulation results for the 35 × 35-mm NuBGA with four rows of solder balls have been verified within 5 percent by TDR and TDT measurements. The TDR and TDT measurement techniques and results can be found in Lau and Chou.[5]

14.5.2   NuBGA package SSO noise

A very simple system-level simulation model using H-SPICE computer software was developed to calculate the signal integrity. It consists of the I/O driver, the 35 × 35-mm NuBGAs with four rows of solder balls, and a system load to emulate a point-to-point situation (Fig. 14.9). There are eight drivers on the NuBGA, with 50 percent of the drivers switching, and eight receivers on the other NuBGA. Between these two NuBGAs, there are eight 1-in.-long PCB traces (50 Ω each). The PCB traces are modeled by eight isolated transmission lines that introduce delay but no

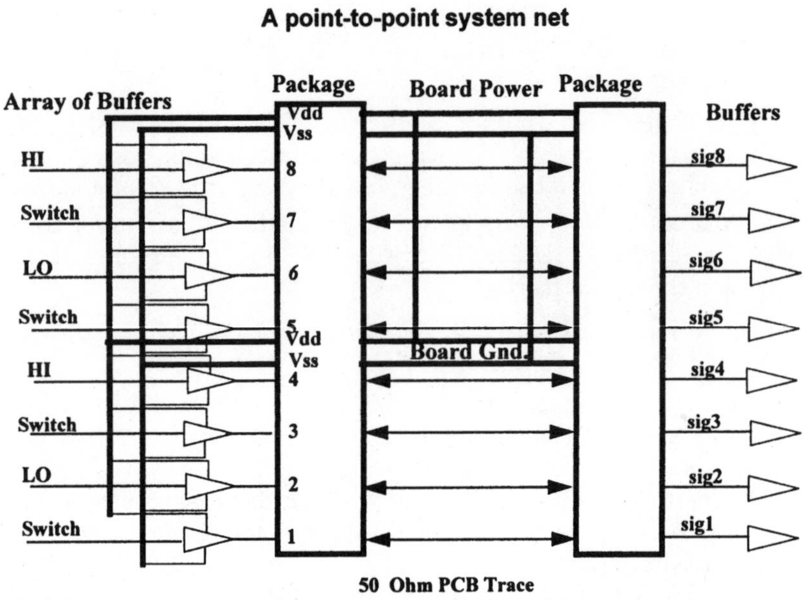

A point-to-point system net

Figure 14.9 A point-to-point system net for performance evaluations.

crosstalk or switching noise on the PCB. Consequently, the resultant noise is primarily from the I/O and package parasitic parameters.

Figure 14.10 shows the SSO noise on quiet low lines (peak noise) of the NuBGA along with other noted packages at a clock frequency of 100 MHz. (Physically, SSO noise on the quiet low lines results from the ground bounce, which is indicative of package signal and ground performance.) It can be seen from Fig. 14.10 that the NuBGA is much better than the PQFP and the conventional PBGA (2MPBGA) and is very close to the four–metal layer enhanced PBGA (4MEPBGA).

Figure 14.11 shows the SSO noise on quiet high lines (dip noise) of the NuBGA along with other packages at a clock frequency of 100 MHz. (Physically, SSO noise on the quiet high lines results from power drop, which is indicative of package signal and power performance.) It can be seen that not only is NuBGA better than PQFP and 2MEPBGA, but it also outperforms 4MEPBGA due to stronger coupling between signal traces and power plane.

14.6   NuBGA Thermal Performance[7–9]

The NuBGA packages shown in Table 14.1 with various sizes of heat spreaders as shown in Fig. 14.12 were studied using the finite element approach. Wind tunnel experiments were performed to verify the modeling results.

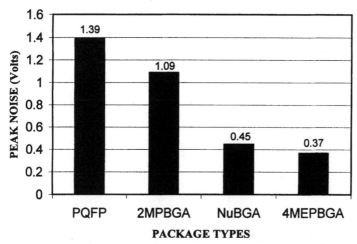

Figure 14.10 Comparisons between NuBGA and other packages of the simulated SSO noise on quiet low lines (peak noise).

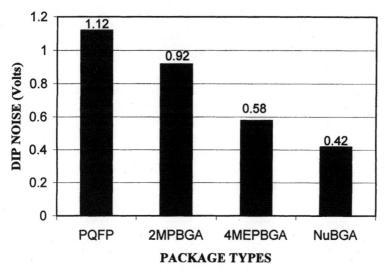

Figure 14.11 Comparisons between NuBGA and other packages of the simulated SSO noise on quiet high lines (dip noise).

14.6.1 Problem definition

Figure 14.13 shows a typical finite element model of the NuBGA package on a multilayer PCB. Due to double symmetry, only one quarter of the system is modeled with the eight-node brick elements. In this model, all the key elements of the packaging system, such as the chip, heat spreader (0.5 mm thick), die attach (0.025 mm thick), BT sub-

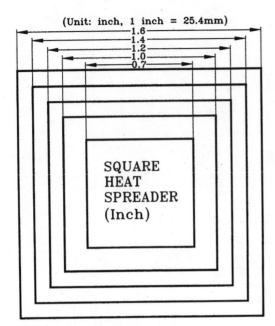

Figure 14.12 Heat spreader sizes.

strate (0.55 mm thick), encapsulant (the gap between the encapsulant and PCB is 0.2032 mm), 63wt%Sn-37wt%Pb (0.76-mm-diameter) solder balls, vias, and PCB are included. The FR-4 PCB dimensions are $105.4 \times 101.6 \times 1.57$ mm including four 0.5-oz (0.018-mm-thick) copper layers. The top and bottom copper layers are for signal and the two embedded layers are for power and ground, respectively. The chip size and cavity size for each case considered are shown in Table 14.1.

The physical and material properties of the packaging system are shown in Table 14.3, in which κ is the thermal conductivity, E is the Young's modulus, α is the coefficient of thermal expansion, and ν is the Poisson's ratio. Table 14.4 lists the coefficients of convective heat transfer (h) for the NuBGA and PCB surfaces calculated using the flat plate correlation.[7–9]

14.6.2 Temperature distribution

Typical temperature distributions for the NuBGA and PCB, chip and copper heat spreader, substrate and solder ball, and PCB only are shown in Fig. 14.14 for an air flow rate of 1 m/s. For the 35×35-mm NuBGA package (four rows) with a full-size heat spreader, the calculation shows that 47 percent of the heat generated by the chip is dissipated through the copper heat spreader; 38.5 percent goes through the

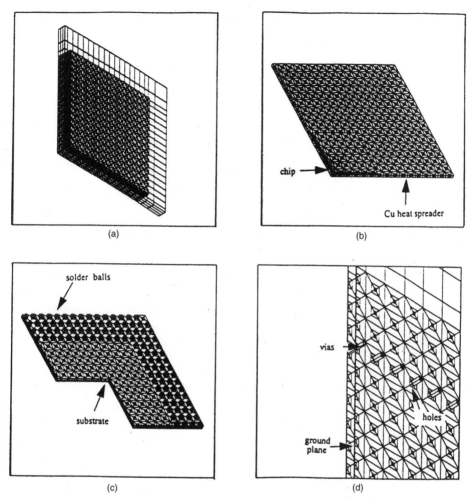

Figure 14.13 Finite element model of the NuBGA. (*a*) One-fourth of the whole packaging system. (*b*) Chip and heat spreader. (*c*) Substrate with solder balls. (*d*) Power and ground planes with via in the PCB.

copper heat spreader, BT substrate, solder ball, and dissipates through the PCB; 1.5 percent goes through the four sides of the copper heat spreader; 9 percent transfers to the PCB through the encapsulant; and 5.5 percent transfers to the PCB through the air gaps between the solder balls as shown in Fig. 14.15*a* (air flow rate = 1 m/s). Figures 14.15*b* through *d* show the temperature distribution, respectively, for the 35 × 35-mm (five rows) and 40 × 40-mm (four and five rows) NuBGA packages with full-size heat spreaders on PCB. It can be seen that the larger the package the greater the heat dissipation through the heat

TABLE 14.3   Physical and Mechanical Properties of the NuBGA Packages and PCB

Material	κ (W/m°C)	E (GPa)	α (ppm/°C)	ν
Silicon	150	128.2	3.6	0.3
Encapsulant	65	11.8	30	0.3
Die attach	1	0.7	40	0.3
Cu heat spreader	260	120	17	0.35
Organic substrate	0.3	18.6(x)	19.1(x)	0.16(xy)
		18.6(y)	19.1(y)	0.42(yz)
		7.6(z)	77.3(z)	0.42(xz)
63Sn-37Pb Solder*	50	10.3	21	0.4
FR-4 PCB	0.3	22(x)	18(x)	0.11(xy)
		22(y)	18(y)	0.28(xz)
		10(z)	70(z)	0.28(yz)
Power/ground planes	390	76	17	0.34
Air	0.024	—	—	—

* Solder is considered temperature dependent.

TABLE 14.4   Convective Heat Transfer Coefficient (*h*) of the NuBGA Packages and PCB

Air flow rate (m/s)	*h* at heat spreader (W/m²-°C)	*h* at four sides of the package (W/m²-°C)	*h* at PCB (W/m²-°C)
0	11.0	7.7	20.5
0.25	24.5	11.0	20.5
0.5	31.1	14.9	20.5
1.0	42.4	21.5	21.5
2.0	52.9	27.6	27.6

spreader, and that the greater the number of solder balls, the greater the heat dissipation into the PCB.

Figures 14.16 and 14.17 show, respectively, the effect of heat spreader size on the temperature distribution from the chip to the back of the heat spreader, from the chip to the four sides of the heat spreader, from the chip to the heat spreader to the BT substrate to the solder balls and then to the PCB, and from the chip to the encapsulant to the air and then to the PCB for the 40 × 40-mm NuBGA packages with four and five rows of solder balls. It can be seen that for all cases under consideration, the heat dissipation from the chip to the four sides of the heat spreader is very small (~1.5 percent).

A major path for heat dissipation is from the chip to the back of the heat spreader. Heat dissipation is at its maximum for a full-size heat spreader and decreases as the heat spreader becomes smaller (Figs. 14.16 and 14.17).

Another major path for heat dissipation is from the chip to the heat spreader to the BT substrate to the solder balls and then to the PCB. Heat dissipation increases as heat spreader size decreases as long as the heat spreader covers most of the solder balls (Figs. 14.16 and

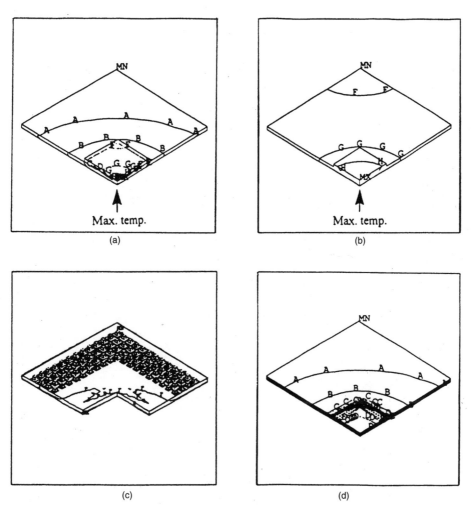

Figure 14.14   Temperature distribution in (*a*) NuBGA and PCB, (*b*) chip and heat spreader, (*c*) substrate and solder balls, and (*d*) PCB only (105.4 × 101.6 × 1.57 mm). For 1 W of heat source, A = 0.9°C, B = 2°C, C = 3°C, D = 4°C, E = 5°C, F = 6.2°C, G = 6.6°C, and H = 7.1°C.

14.17). However, heat dissipation decreases when the heat spreader no longer covers the solder balls.

The heat dissipation from the chip to the encapsulant to the air and then to the PCB (~9 percent) does not increase much as long as the heat spreader is covering most of the solder balls (Figs. 14.16 and 14.17). However, this could become a major heat path (greater than 30 percent) if the heat spreader barely covers the chip and the cavity opening. As a matter of fact, in this case, the heat dissipation from the chip to the encapsulant to the air and then to the PCB is greater than that from the chip to the back of the heat spreader.

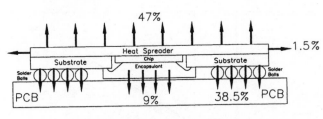

Other heat paths : 4%

(a)

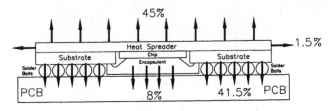

Other heat paths : 4%

(b)

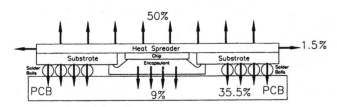

Other heat paths : 4%

(c)

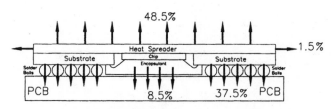

Other heat paths : 4%

(d)

Figure 14.15 Heat dissipation distribution with full-size heat spreader: (a) 35 × 35-mm NuBGA with four rows of solder balls; (b) 35 × 35-mm NuBGA with five rows of solder balls; (c) 40 × 40-mm NuBGA with four rows of solder balls; (d) 40 × 40-mm NuBGA with five rows of solder balls.

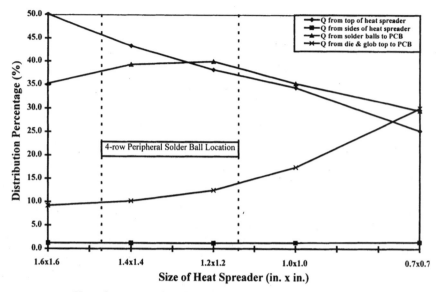

Figure 14.16 Heat dissipation distribution for various sizes of heat spreaders for the 40 × 40-mm NuBGA with four rows of solder balls (air flow rate = 1 m/s).

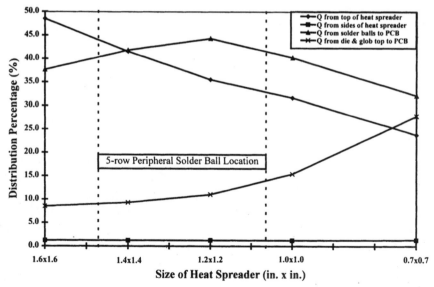

Figure 14.17 Heat dissipation distribution for various sizes of heat spreaders for the 40 × 40-mm NuBGA with five rows of solder balls (air flow rate = 1 m/s).

14.6.3   Thermal resistance

The calculated maximum temperatures from the finite element simulations are used to calculate the thermal resistance of the NuBGA. Usually, assuming a device power of 1 W and a specified ambient temperature of 0, the calculated maximum device junction temperature becomes the thermal resistance of the NuBGA. Figures 14.18 and 14.19 show the thermal resistance of the 35 × 35-mm and 40 × 40-mm NuBGA packages with various sizes of heat spreaders at an airflow rate of 1 m/s. For both package sizes, the thermal resistance of NuBGA with four rows of solder balls is higher than that of NuBGA with five rows of solder balls. Also, for both package sizes, the thermal resistance increases as the size of the heat spreader decreases.

Tables 14.5 and 14.6 summarize, respectively, the thermal resistance of the 35 × 35-mm and 40 × 40-mm NuBGA packages with various sizes of heat spreaders at an airflow rate of 1 m/s. It can be seen that the thermal resistance of the 40 × 40-mm NuBGA packages is smaller than that of the 35 × 35-mm packages, since the latter have less packaging area into which to dissipate the heat out from the chip.

14.6.4   Cooling power

Assuming a device maximum junction temperature of 115°C and an ambient temperature of 55°C, the cooling power of the NuBGA packages for different sizes of heat spreaders at an airflow rate of 1 m/s can be

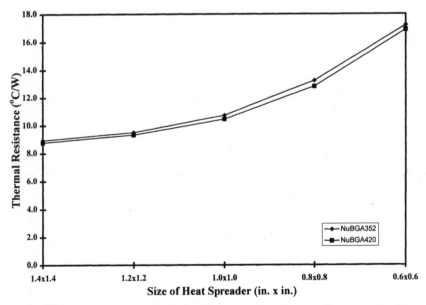

Figure 14.18   Thermal resistance of the 35 × 35-mm NuBGAs (air flow rate = 1 m/s).

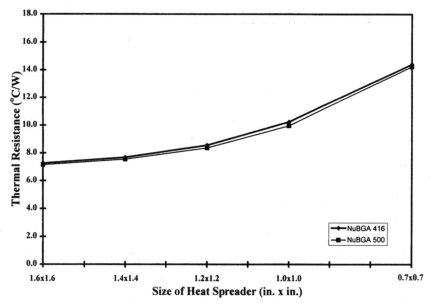

Figure 14.19 Thermal resistance of the 40 × 40-mm NuBGAs (air flow rate = 1 m/s).

obtained and is shown in Figs. 14.20 and 14.21, respectively, for the 35 × 35-mm and 40 × 40-mm packages. Again, for both packages the cooling power is higher for the packages with more solder balls and decreases as the heat spreader size decreases. Also, the cooling power of the 40 × 40-mm NuBGA package is higher than that of the 35 × 35-mm package.

14.6.5 Solder ball temperature

Since it is very difficult to measure the temperature at the solder balls of PBGA PCB assemblies (and thus not much data is available), usually a very large (conservative) temperature acting at the solder balls is assumed for predicting the solder ball reliability under thermal conditions.[11–15] If the predicted thermal fatigue life meets the reliability requirements, then the job is done! However, if the conservatively pre-

TABLE 14.5 Thermal Resistance of the 35 × 35-mm NuBGA Packages

	Thermal Resistance (°C/W)	
Size of heat spreader (in.)	NuBGA 352 (4 rows)	NuBGA 420 (5 rows)
1.4 × 1.4	8.93	8.77
1.2 × 1.2	9.50	9.32
1.0 × 1.0	10.73	10.45
0.8 × 0.8	13.24	12.82
0.6 × 0.6	17.21	16.89

Air flow rate = 1 m/s.

TABLE 14.6   Thermal Resistance of the 40 × 40-mm NuBGA Packages

Size of heat spreader (in.)	Thermal Resistance (°C/W)	
	NuBGA 416 (4 rows)	NuBGA 500 (5 rows)
1.6 × 1.6	7.27	7.16
1.4 × 1.4	7.69	7.56
1.2 × 1.2	8.56	8.37
1.0 × 1.0	10.25	9.97
0.7 × 0.7	14.42	14.25

Air flow rate = 1 m/s.

dicted thermal fatigue life does not meet the reliability requirements, then a more precise method is needed to determine the temperature at the solder balls.

In the present study, since all the solder balls are modeled, the maximum solder ball temperature is calculated. These values are shown in Figs. 14.22 and 14.23, respectively, for the 35 × 35-mm and 40 × 40-mm packages. It can be seen that for both packages, the maximum solder ball temperature is higher for the package with five rows of solder balls and for the smaller heat spreader. Also, the maximum solder ball temperature is higher for the 35 × 35-mm NuBGA packages and is between 6.2 and 8.3°C (this is under 1 W of heat dissipated from the chip and 1-m/s airflow velocity). By adding this temperature to the ambient temperature, we can obtain the maximum solder ball temperature.

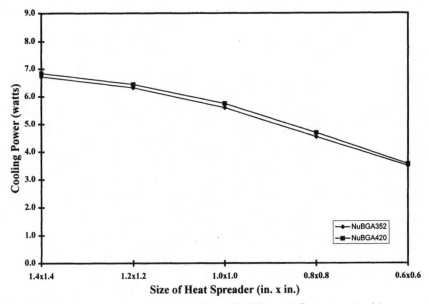

Figure 14.20   Cooling power of the 35 × 35-mm NuBGAs (air flow rate = 1 m/s).

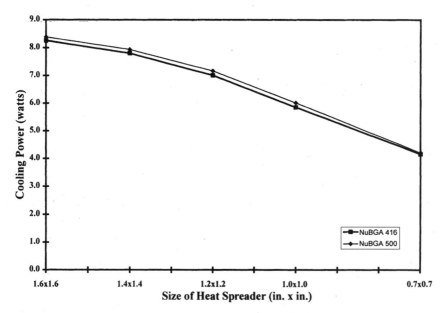

Figure 14.21   Cooling power of the 40 × 40-mm NuBGAs (air flow rate = 1 m/s).

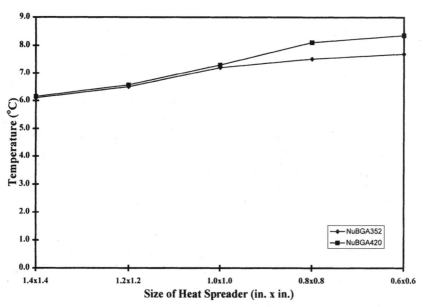

Figure 14.22   Maximum solder ball temperature for the 35 × 35-mm NuBGAs (air flow rate = 1 m/s).

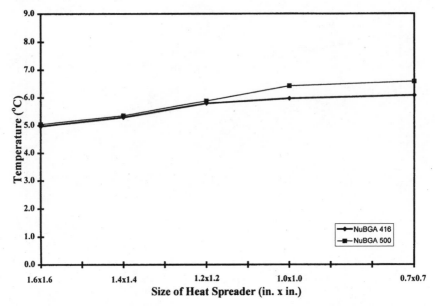

Figure 14.23 Maximum solder ball temperature for the 40 × 40-mm NuBGAs (air flow rate = 1 m/s).

14.7   NuBGA Solder Joint Reliability

The NuBGA packages shown in Table 14.1 are modeled in order to determine the stress and strain in the solder joints on PCB. The material properties of package constituents are given in Table 14.3. All materials are assumed to be linear-elastic, except for the 63wt%Sn-37wt%Pb solder balls, which are elastoplastic and temperature dependent (Figs. 14.24 and 14.25). A uniform temperature change of 85°C is applied to the assembly.

The whole-field deflection of one quarter of the 35 × 35-mm NuBGA package with four rows of solder balls on PCB is shown in Fig. 14.26. It can be seen that the maximum deflection (0.00099 mm) is at the center of the package. This is due to two major sources: (1) the thermal expansion of the solder ball in the vertical direction and the local thermal expansion mismatch between the Si chip and the copper heat spreader, and (2) the global thermal expansion mismatch between the PCB and the remaining parts of the assembly. It should be noted that the deflection of the package center due to source (2) is in the opposite direction of that due to source (1).

Figure 14.27 shows the distribution of the accumulated effective plastic strain in each solder ball for the four-row-perimeter NuBGA

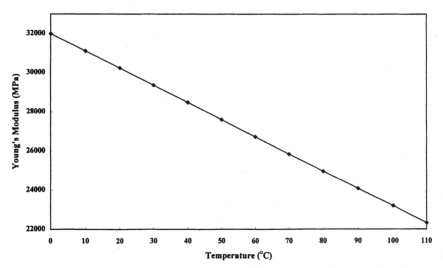

Figure 14.24   Temperature-dependent Young's modulus for 63wt%Sn-37wt%Pb solder.

with a package size of 35 mm. The highest inelastic strain is very small (less than 0.1 percent) and appears in the outermost solder joint. In fact, the strains obtained in all other cases considered in this study are really very small. Thus, it can be concluded that the solder joints of NuBGA packages should be very reliable in applications.

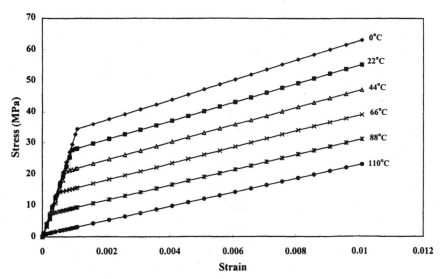

Figure 14.25   Temperature-dependent stress-strain curves for 63wt%Sn-37wt%Pb solder.

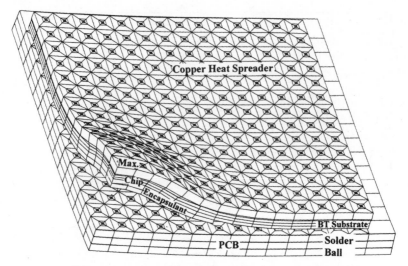

Figure 14.26 Deformation of one quarter of a NuBGA assembly (0 to 85°C).

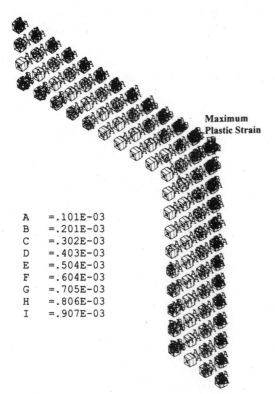

A	=.101E-03
B	=.201E-03
C	=.302E-03
D	=.403E-03
E	=.504E-03
F	=.604E-03
G	=.705E-03
H	=.806E-03
I	=.907E-03

Figure 14.27 Accumulated effective plastic strain in solder balls (0 to 85°C).

14.8   Summary of the Standard NuBGA Packages

A family of face-down, single-core, two–metal layer NuBGA packages for housing IC devices has been designed and analyzed. The back of the IC device is attached to a heat spreader. With the special SVC or SWA design concepts, excellent electrical performance can be achieved through programmable VDD/VSS split-ring segments for wire bonding, shorter wire bond length, and controlled-impedance microstripline and coplanar stripline signal traces on the substrate.

In comparison with the noted packages, NuBGA provides SSO noise performance up to 3 times better than PQFP and up to 2.5 times better than conventional PBGA. Also, NuBGA is comparable with the four-layer PBGA in noise performance on the quiet low line and surpasses the four-layer PBGA in noise performance on the quiet high line.

The accumulated effective plastic strains in the solder joints have been shown to be very small in all the cases considered. Thus, the solder joints of NuBGA are reliable for use in most applications.

The temperature distribution, thermal resistance, and cooling power of the 35 × 35-mm and 40 × 40-mm NuBGA packages with different sizes of heat spreaders and with four and five rows of solder balls (on a 105.4 × 101.6 × 1.57-mm PCB with four 0.5-oz copper layers) have been determined. Some important results (for airflow rates of 1 m/s) are summarized as follows.

- The thermal resistance is 8.93 (four rows) and 8.77 (five rows) for the 35 × 35-mm NuBGA packages with full-size heat spreaders, and 7.27 (four rows) and 7.16 (five rows) for the 40 × 40-mm NuBGA packages with full-size heat spreaders.

- The thermal resistance of the 40 × 40-mm NuBGA packages is smaller than that of the 35 × 35-mm packages with heat spreaders of the same size.

- The thermal resistance of both sizes of packages with four rows of solder balls is higher than that for packages with five rows of solder balls.

- The thermal resistance of both sizes of packages increases as the size of the heat spreader decreases.

- The cooling power of the 40 × 40-mm NuBGA packages is higher than that of the 35 × 35-mm packages with heat spreaders of the same size.

- The cooling power of both sizes of packages decreases as the heat spreader size decreases.

- The cooling power of both sizes of packages is higher for the packages with more solder balls.

- A major path of heat dissipation is from the chip to the back of the heat spreader. Heat dissipation is at its maximum for a full-size heat spreader and decreases as the heat spreader becomes smaller.

- Another major path of heat dissipation is from the chip to the heat spreader to the BT substrate to the solder ball and then to the PCB. Heat dissipation increases with heat spreader size decrease as long as the heat spreader covers most of the solder balls. However, heat dissipation decreases when the heat spreader no longer covers the solder balls.

- The heat dissipation from the chip to the encapsulant to the air and then to the PCB does not increase much as long as the heat spreader is covering most of the solder balls. However, this could become a major heat path (greater than 30 percent) if the heat spreader barely covers the chip and the cavity opening.

- The maximum solder ball temperature for both sizes of packages is higher for the packages with five rows (vs. four rows) of solder balls.

- The maximum solder ball temperature for both sizes of packages is higher for smaller heat spreaders.

- The maximum solder ball temperature for the 35 × 35-mm packages is higher than that of the 40 × 40-mm packages with heat spreaders of the same size.

The thermal and electrical performance of NuBGA can be further improved with thinner substrates and thicker heat spreaders with or without heat sinks. The cost of NuBGA is lower than that of the four-layer PBGA and comparable to that of the conventional PBGA.

14.9 Thinner Substrate and Nonuniform Heat Spreader NuBGA

Figure 14.28 schematically shows a cross section of the new design. Figures 14.29*a* and *b* show, respectively, the layout of the top and bot-

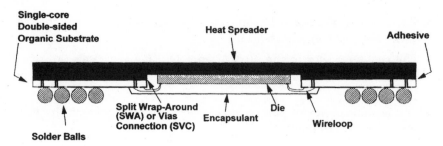

Figure 14.28 Schematic of the cross section of the new NuBGA package.

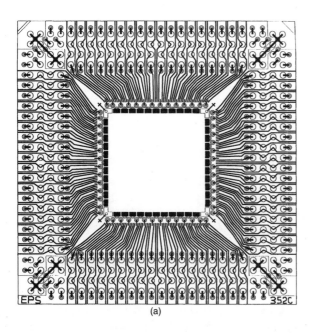

(a)

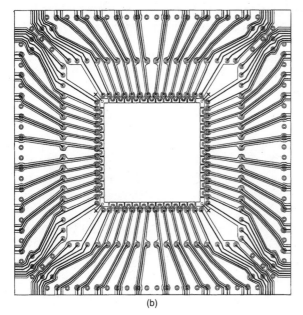

(b)

Figure 14.29 Substrate layout of the 35×35-mm SVC new NuBGA package. (a) Top side. (b) Bottom side.

tom sides of the 35 × 35-mm SVC new package. The key process steps for making the substrate are shown in Fig. 14.30. Cross sections of the new package with a thermal die soldered on the PCB are shown in Fig. 14.31. It can be seen that, unlike the NuBGA reported in Refs. 4 through 9 and shown in Fig. 14.2, the new package's substrate is very thin (<0.15 mm) and the heat spreader is thicker in the outer ring to compensate for the thinner substrate.

14.10   Thermal Performance of the New NuBGA Package

The thermal performance of the new package will be determined by finite element analysis and wind tunnel measurement. The finite element model of the new package on a multilayer PCB (not shown) is very similar to that in Fig. 14.13. Due to the double symmetry, only one quarter of the system is modeled with the eight-node brick elements. This model includes all the key elements of the packaging system, such as the chip, heat spreader, die attach (0.025 mm thick), BT substrate (<0.15 mm thick), encapsulant (the gap between the encapsulant and PCB is 0.2032 mm), 63wt%Sn-37wt%Pb (0.76-mm) solder balls, via, and PCB. The FR-4 PCB dimensions are 105.4 × 101.6 × 1.57 mm including four 0.5-oz (0.018-mm-thick) copper layers. The top and bottom copper layers are for signal and the two embedded layers are for power and ground, respectively.

The physical and material properties of the new packaging system and the coefficients of convective heat transfer (h) for the heat spreader and PCB surfaces are assumed to be the same as those reported earlier in this section.

14.10.1   Temperature distribution

Typical temperature distributions for the new NuBGA and PCB, chip and copper heat spreader, substrate and solder ball, and PCB only are shown in Fig. 14.32 for an airflow rate of 1 m/s. The calculation shows that 45.5 percent of the heat generated by the chip is dissipated through the copper heat spreader; 41 percent goes through the copper heat spreader, the BT substrate, and the solder balls, and dissipates through the PCB; 1.5 percent goes through the four sides of the copper heat spreader; 8 percent transfers to the PCB through the encapsulant, and 4 percent transfers to the PCB through the air gap between the solder balls, as shown in Fig. 14.33 (airflow rate = 1 m/s).

14.10.2   Thermal resistance

Figure 14.34 shows the thermal resistance of the 35 × 35-mm new NuBGA at different airflow rates. It can be seen that the thermal resis-

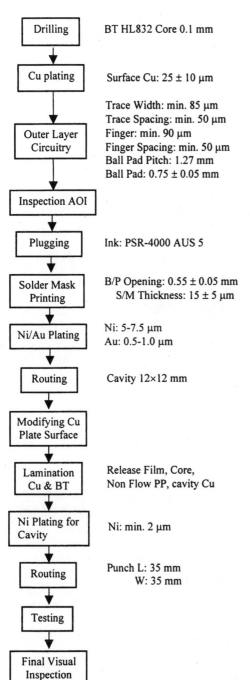

Process	Specifications
Drilling	BT HL832 Core 0.1 mm
Cu plating	Surface Cu: 25 ± 10 μm
Outer Layer Circuitry	Trace Width: min. 85 μm Trace Spacing: min. 50 μm Finger: min. 90 μm Finger Spacing: min. 50 μm Ball Pad Pitch: 1.27 mm Ball Pad: 0.75 ± 0.05 mm
Inspection AOI	
Plugging	Ink: PSR-4000 AUS 5
Solder Mask Printing	B/P Opening: 0.55 ± 0.05 mm S/M Thickness: 15 ± 5 μm
Ni/Au Plating	Ni: 5-7.5 μm Au: 0.5-1.0 μm
Routing	Cavity 12×12 mm
Modifying Cu Plate Surface	
Lamination Cu & BT	Release Film, Core, Non Flow PP, cavity Cu
Ni Plating for Cavity	Ni: min. 2 μm
Routing	Punch L: 35 mm W: 35 mm
Testing	
Final Visual Inspection	

Figure 14.30 Key process flow of the substrate of the new NuBGA package.

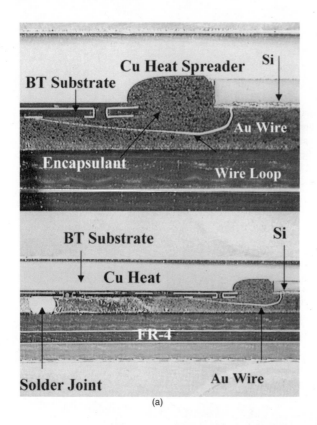

(a)

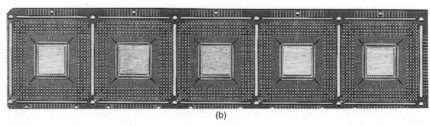

(b)

Figure 14.31 (*a*) Cross-sectional view of PCB assembly of the new NuBGA package. (*b*) Top view of the new NuBGA package substrate.

tance decreases as the airflow rate increases. At zero airflow, the thermal resistance is 10.66°C/W.

14.10.3 Cooling power

Assuming a maximum device junction temperature of 115°C and an ambient temperature of 55°C, the cooling power of the package at an airflow rate of 1 m/s can be obtained, as shown in Fig. 14.35 for the 35 ×

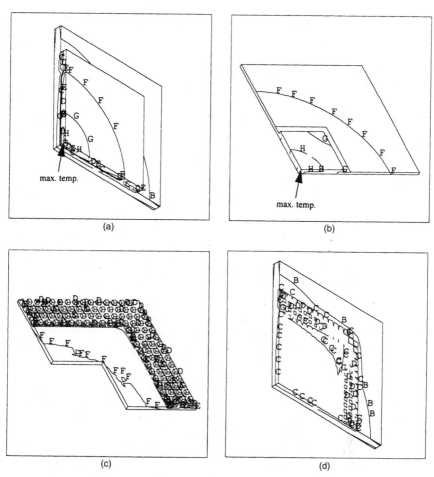

Figure 14.32 Temperature distribution in (*a*) the new NuBGA package and PCB; (*b*) the chip and heat spreader; (*c*) the substrate and solder balls; (*d*) the PCB only (105.4 × 101.6 × 1.57 mm). It is noted that for 1 W of heat source, A = 0.91°C, B = 2.2°C, C = 3.1°C, D = 3.9°C, E = 4.8°C, F = 6°C, G = 6.5°C, and H = 6.9°C.

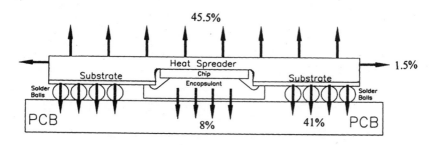

Figure 14.33 Heat dissipation distribution of the new NuBGA package on PCB.

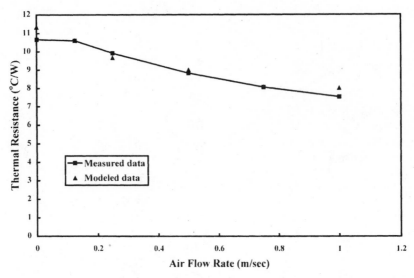

Figure 14.34   Thermal resistance of the 35 × 35-mm new NuBGA package.

35-mm package. It can be seen that the cooling power increases as the airflow rate increases and that the cooling power at still air is 5.7 W.

14.10.4   Wind tunnel experimental analysis

The thermal resistance and cooling power of the 35 × 35-mm new NuBGA, determined in the previous sections by the finite element method, will be verified here by experimental measurements.

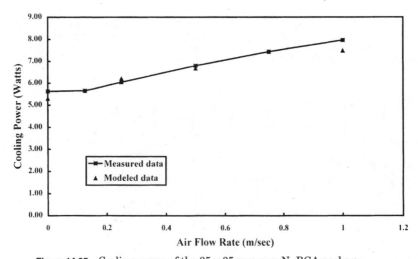

Figure 14.35   Cooling power of the 35 × 35-mm new NuBGA package.

Test board. The PCB dimensions are 105.4 × 101.6 × 1.57 mm with four 0.5-oz copper layers. There are six pairs of solder ball pads (Fig. 14.36) for thermal test (accessibility) purposes, e.g., the resistor pads on the chip can be connected to an external power supply, and the diode pads on the chip can be connected to an external sense current circuit.

Test die. The test die (PST6) (Fig. 14.37) used in this study is designed and manufactured by Delco and measures 10 × 10 mm. It provides one pair of bridge diodes and one pair of heating resistors. The test die is wire-bonded (6 pairs) on the bonding fingers of the 325-pin new NuBGA substrate. These six pairs of bonding wires correspond to the six pairs of solder-ball pads as shown in Fig. 14.36 on the test board, which are routed out for accessibility purposes.

Device calibration. The junction temperature in the package measured is based on the temperature and voltage dependency exhibited by semiconductor diode junctions. The voltage-temperature relationship is an intrinsic electrothermal property of semiconductor junctions. Figure 14.38 shows the voltage-temperature relationship of the present test die under constant sense current (T_j is the junction temperature and V_f is the forward-biased voltage drop). It can be seen that the slope is equal to 528°C/V and the temperature ordinate intercept is equal to 411°C.

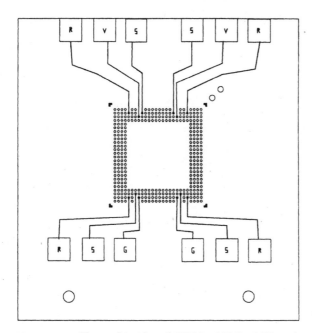

Figure 14.36 Thermal test board (105.4 × 101.6 × 1.57 mm).

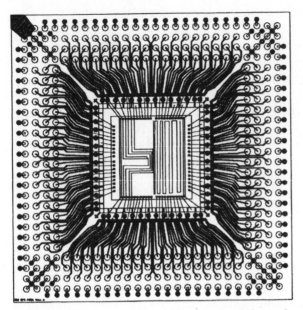

Figure 14.37 Thermal test die (PST6) wire-bonded on substrate.

Thermal resistance measurement results. Steady-state thermal resistance is measured for the 35×35-mm new NuBGA package at several power levels (1, 2, 5, and 7 W) and different airflow rates (0, 0.12, 0.25, 0.5, 0.75, and 1 m/s). The results are shown in Fig. 14.34. It can be seen that the discrepancy between the measurement and simulation results

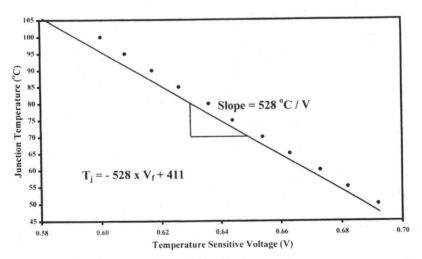

Figure 14.38 Temperature-voltage relationship.

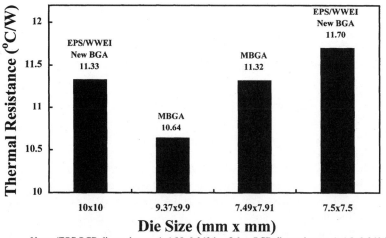

Note: *(EPS PCB dimensions = 4x4.15x0.062 in; Other PCB dimensions = 4x4.5x0.062 in)*

Figure 14.39 Thermal resistance of various types of packages (finite element results).

is within 4 percent. Due to the thinner substrate and the thicker heat spreader of the new NuBGA, the present results are at least 20 percent better than those reported earlier for the NuBGA.

It is interesting to compare the present measurement and modeling results with those reported[2] for other types of packages (Figs. 14.39 and 14.40). It can be seen that, based on the experimental results, the metal BGA (MBGA) package performs the best and the new NuBGA package performs better than the 3M tape BGA (TBGA) and the Viper

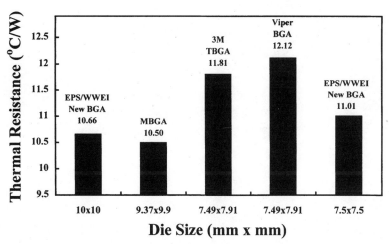

Note: *(EPS PCB dimensions = 4x4.15x0.062 in; Other PCB dimensions = 4x4.5x0.062 in)*

Figure 14.40 Thermal resistance of various types of packages (wind tunnel results).

BGA. Also, based on the modeling results, the MBGA is slightly better than the new package.

14.11 Solder Joint Reliability of the New NuBGA Package

The package shown in Figs. 14.28 through 14.33 is modeled for determining the stress and strain in the solder joints on PCB. The material properties of the package constituents are the same as those for the NuBGA. Five watts of heat is applied to the chip. The whole-field deflection (not shown) of the 35 × 35-mm new package is very similar to that shown in Fig. 14.26. Figure 14.41 shows the distribution of the accumulated effective plastic strain in each solder ball of the package. The highest inelastic strain is very small (less than 0.1 percent) and appears in the outermost solder joint. Again, it can be concluded that the solder joints of the new NuBGA package should be very reliable in most applications.

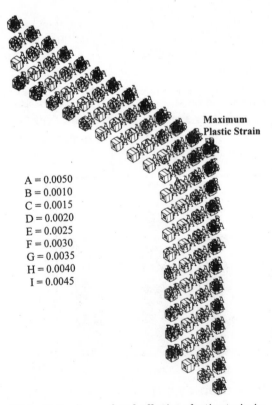

Maximum
Plastic Strain

A = 0.0050
B = 0.0010
C = 0.0015
D = 0.0020
E = 0.0025
F = 0.0030
G = 0.0035
H = 0.0040
I = 0.0045

Figure 14.41 Accumulated effective plastic strain in solder joints (5 W of heat in the chip).

14.12   Electrical Performance of the New NuBGA Package

Since the new NuBGA package has a thinner substrate, a smaller inductance is expected. The parasitic inductance of the package is one of the most significant reasons for the ground bounce; thus reduction in the inductance will yield better electrical performance. The electrical capacitance and inductance of the new package, shown in Figs. 14.28 through 14.33, are extracted from TDR measurement.[10] The lines measured are marked in Fig. 14.42.

14.12.1   Capacitance

$$C = \frac{1}{2Z_0 V_{step}} \int_0^\infty (W_{open} - W_{pkg}) dt$$

where Z_0 is the characteristic impedance of the probe (50 Ω) and V_{step} is 250 mV. W_{open} is the waveform of the probe with open end and W_{pkg} is the impedance of the package trace. The waveforms of a microstripline for capacitance measurement and the capacitance are shown in Fig. 14.43a and b, respectively.

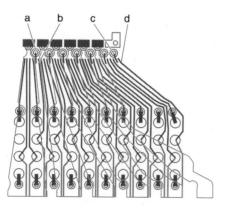

	Trace measured	C (pF)	L (nH)
Corner co-planar line	c	2.54	5.86
Corner μ-stripline	d	1.40	4.92
Center co-planar line	a	1.01	4.24
Center μ-stripline	b	1.64	3.96

Figure 14.42   Locations for electrical measurements and results.

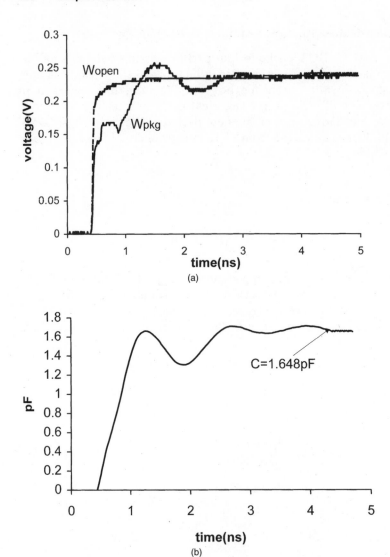

Figure 14.43 (*a*) TDR waveform of a microstripline for capacitance measurement. (*b*) TDR integration value of capacitance.

14.12.2 Inductance

$$L = \frac{Z_0}{2V_{\text{step}}} \int_0^\infty (W_{\text{pkg}} - W_{\text{short}}) dt$$

where W_{short} is the waveform when measuring the ground plane. The waveforms of a microstripline for inductance measurement and the integration value of inductance are shown in Figs. 14.44a and b, respectively. The capacitance and inductance of traces at the corner and in the center are listed in Table 14.1. It can be seen that they are more than 25 percent better than those for the NuBGA with a thicker substrate. The characteristic impedance Z_0 of the center coplanar stripline (trace a in Fig. 14.42) is 62.43 Ω, and that of the center microstripline (trace b in Fig. 14.42) is 49.08 Ω.

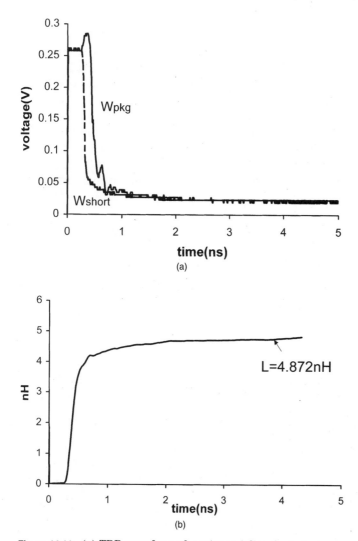

Figure 14.44 (a) TDR waveform of a microstripline for inductance measurement. (b) TDR integration value of inductance.

14.13   Summary of the New NuBGA Package

A face-down, very thin single-core, two–metal layer package for housing an IC device has been designed and analyzed. The back of the IC device is attached to a nonuniform sawtooth heat spreader. With the special design concept, excellent electrical performance can be achieved through programmable VDD/VSS split-ring segments for wire bonding, shorter wire bond length, and controlled-impedance microstripline and coplanar stripline signal traces on the substrate. The electrical performance of the thinner substrate is much better than that of the thicker substrate.

The temperature distribution, thermal resistance, and cooling power of the 35 × 35-mm new NuBGA package have been determined. The thermal performance is not as good as that of the MBGA, but is better than those of some other well-known packages. The thermal performance can be further improved by using thicker heat spreaders with or without heat sinks. The cost of the new package is lower than that of the four-layer PBGA and comparable to that of the conventional PBGA.

The accumulated effective plastic strains in the solder joints on PCB have been shown to be very small in all the cases considered. Thus, the solder joints of the new package are reliable for use in most applications.

Acknowledgments

The author would like to thank T. Chen, T. Tseng, D. Cheng, Eric Lao, and Dyna Lin of WWEI and UMTC; T. Chen and T. Chou of Sun-Microsystems; and C. Chen, H. Chen, and C. Chang of EPS for their useful help and constructive comments.

References

1. Lau, J. H., *Ball Grid Array Technology*, McGraw-Hill, New York, 1995.
2. Solomon, D., E. Rosario, E. Opiniano, B. Bright, and A. Kenyon, "Thermal Characterization of Metal Enhanced BGA Packages," *SEMICON West*, G4 Session 2, pp. 1–12, 1998.
3. Mattei, C., and R. Marrs, "SuperBGA: Design for Enhanced Performance," *Proceedings of NEPCON West*, pp. 687–698, February 1996.
4. Lau, J. H., et al., "Single-Core Two-Side Substrate with μ-Strip and Co-Plane Signal Trace, and Power and Ground Planes Through Split-Wrap-Around (SWA) or Split-Via-Connections (SVC) for Packaging IC Devices," U.S. Patent No. 5,825,084, 1998.
5. Lau, J. H., and T. Chou, "Electrical Design of a Cost-Effective Thermal Enhanced Plastic Ball Grid Array Package—NuBGA," *IEEE Trans. Component, Packaging, and Manufacturing Technology*, part B, *Adv. Packaging*, 21 (1): 35–42, February 1998.
6. Lau, J. H., T. Chen, and T. Chou, "Design, Analysis and Measurement of the Cost-Effective Substrate of a Plastic Ball Grid Array Package—NuBGA," *Circuit World*, 25 (2): 41–48, 1999.
7. Lau, J. H., T. Chen, and R. Lee, "Effect of Heat-Spreader Sizes on the Thermal Performance of Large Cavity-Down Plastic Ball Grid Array Packages," *ASME Trans., J. Electronic Packaging*, 242–248, December 1999.

8. Lau, J. H., and T. Chen, "Cooling Assessment and Distribution of Heat Dissipation of a Cavity Down Plastic Ball Grid Array Package—NuBGA," *IMAPS Trans., Int. J. Microelectronics & Electronic Packaging,* **21** (1): 109–118, 1998.

9. Lau, J. H., and K. L. Chen, "Thermal and Mechanical Evaluations of a Cost-Effective Plastic Ball Grid Array Package," *J. Electronic Packaging, Trans. ASME,* **119:** 208–212, September 1997.

10. Tektronix, "TDR Tools in Modeling Interconnects and Packages," Application Note, Tektronix, Beaverton, OR, 1993.

11. Lau, J., and Y. Pao, *Solder Joint Reliability of BGA, CSP, Flip Chip, and Fine Pitch SMT Assemblies,* McGraw-Hill, New York, 1997.

12. Lau, J. H., C. P. Wong, J. L. Prince, and W. Nakayama, *Electronic Packaging: Design, Materials, Process, and Reliability,* McGraw-Hill, New York, 1998.

13. Lee, S.-W. R., and J. H. Lau, "Effect of Chip Dimension and Substrate Thickness on the Solder Joint Reliability of Plastic Ball Grid Array Packages," *Circuit World,* **23** (1): 16–19, 1996.

14. Lee, S.-W. R., and J. H. Lau, "Design for Plastic Ball Grid Array Solder Joint Reliability," *Circuit World,* **23** (2): 11–14, 1997.

15. Lee, S.-W. R., and J. H. Lau, "Solder Joint Reliability of Plastic Ball Grid Array with Solder Bumped Flip Chip," ASME Paper No. 97WA/EEP-6, *ASME Winter Annual Meeting,* Dallas, TX, November 1997.

Solder-Bumped Flip Chip in PBGA Packages

15.1   Introduction

The solder-bumped flip chip on board technology[1–20] has been used to package the memory devices and low-power, low-pin-count ASICs in the CSP and WLCSP configurations.[15,16] Recently, it has been used to house the microprocessors on organic instead of the ceramic substrates and in a PBGA format. There are at least three major reasons: (1) lower cost; (2) better electrical performance (relative dielectric constant ~ 4); and (3) great similarity to the second-level SMT. In this chapter, the solder-bumped flip chip in PBGA packages developed and manufactured by Intel, Mitsubishi, IBM, and Motorola are presented. Emphasis is placed on the design, materials, process, thermal management, and reliability of these packages.

15.2   Intel's OLGA Package Technology[1,2]

The organic land grid array (OLGA) package technology developed by Intel for housing its top-of-the-line area-array solder-bumped flip chip microprocessors is presented in this section. Figure 15.1 shows the application of OLGA packages in the Xeon processors. Over 15 million OLGA units have been shipped in the first year of introduction of this technology.[1,2]

The area-array solder-bumped flip chip OLGA package technology is schematically shown in Figs. 15.2 and 15.3. The key attributes of this technology are:[1,2]

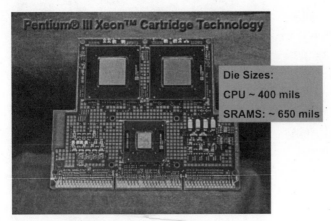

Figure 15.1   OLGA packages in Xeon processors.

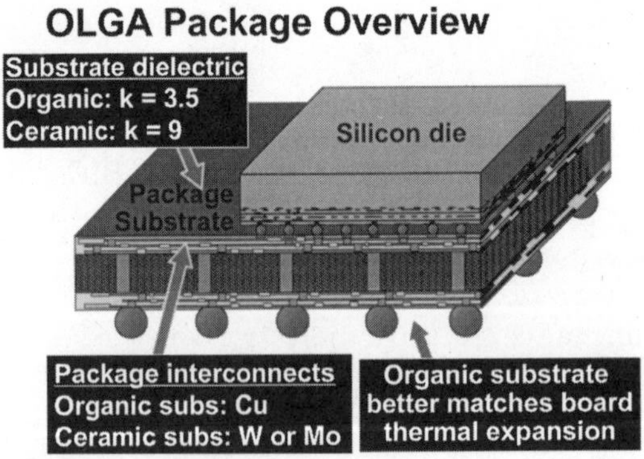

OLGA Package Overview

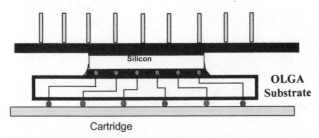

Figure 15.2   OLGA package overview.

Figure 15.3   Cross-sectional schematic of the silicon die flip chip attached to a package, which in turn is surface-mounted onto a cartridge. The thermal solution is integrated in the cartridge design.

- Eutectic tin-lead solder bonding between the high-lead bumped chip (225-μm minimum pitch) and organic substrate
- Full area-array chip design and multichip capability
- Low-alpha solder/materials to allow the use of bumps on the entire chip surface
- Underfill encapsulant to allow chip sizes as large as 20×20 mm^2
- OLGA substrate made of a BT core with SBU technology (up to 10 interconnect layers)
- Compliance with Intel's standards for CPU component package reliability
- OLGA package with SMT-compatible solder balls
- Back side of chip for thermal management

15.2.1   OLGA package design

Figure 15.4 shows a small portion of the chip. It can be seen that the chip has peripheral pads and that the solder bumps are several rows deep. The first 3 to 4 rows of peripheral bumps are for I/Os, and the bumps at the center portion of the chip (not shown) are for power/ground. Such topology helps to minimize the number of routing layers and allow the use of "array" probe cards to electrically probe the wafers successfully.

15.2.2   OLGA wafer bumping

The wafer-bumping process is shown in Fig. 15.5. It can be seen that the 97wt%Pb-3wt%Sn solder bumps are made by the electroplating method with a proprietary etch process. The UBM on the 200-mm wafer is Ti-Ni. Some key attributes of the process are shown in Table 15.1. Tight bump height control is a must for high assembly yields.

Figure 15.4 I/O bump layout on a silicon product showing dense packing of solder bumps. This die was a die originally laid out for wire bonding (the bond pads are visible as the peripheral row).

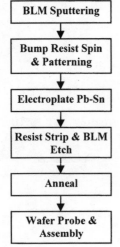

Figure 15.5 Schematic process flow for bumping wafers after passivation via opening.

15.2.3   OLGA substrate technology

In order to meet the routing and electrical performance requirements, in addition to the 225-μm pitch of the solder bumps, a special substrate shown in Figs. 15.6 and 15.7 is designed and made. The key attributes are shown in Table 15.2. It can be seen that the OLGA substrate has a BT epoxy resin core and up to three buildup layers per side. The microvias are made by the photodefining method (Chap. 4) and the hole size is as small as 86 μm. OLGA also allows VIP. The copper pads are coated with eutectic 63Sn-37Pb solder with stringent coplanarity requirements.

15.2.4   OLGA package assembly

The solder-bumped flip chip on OLGA substrate assembly process is shown in Figs. 15.8 and 15.9. Basically, it consists of four major tasks: die preparation, SCAM, deflux, and underfill.

TABLE 15.1   Key Attributes of Solder Bump Process

Attribute	Value
Bump diameter	118 ± 5 μm
Bump height	100 ± 12 μm
Composition	97% Pb-3% Sn
Contact resistance	$\leq 1 \times 10^{-6}$ Ω/cm^2
Bump adhesion	>3 kg/mm^2
Surface leakage	$<1 \times 10^{-10}$ amps at 10 V
Die yield loss due to bumping	0%
Wafer losses due to bumping	$<0.15\%$

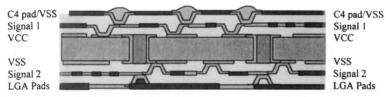

C4 pad/VSS
Signal 1
VCC

VSS
Signal 2
LGA Pads

C4 pad/VSS
Signal 1
VCC

VSS
Signal 2
LGA Pads

Figure 15.6   Cross-sectional view of the OLGA package. This is a multilayer buildup package with a BT epoxy resin core and photodefined blind/buried vias.

Figure 15.7   Cross-sectional micrograph showing VLSI silicon OLGA package interconnect layers. Two buildup layers per side of OLGA are visible.

Die preparation.   The die preparation process flow is shown in Fig. 15.10. It consists of three steps: mounting the wafers, sawing the wafers, and placing the dies onto tape and reel (or waffle pack). Typically, the die is in a face-down configuration.

Chip joining process.   The chip joining process, SMEMA chip attach module (SCAM), consists of three steps, flux application, chip placement, and chip joining, as shown in Fig. 15.11. SMEMA is the protocol

TABLE 15.2   OLGA Substrate Key Attributes

Attribute	Value
Minimum bump pitch	225 μm
Core thickness	0.8 mm
Buildup dielectric thickness	33 μm
Copper thickness	15 μm (buildup)
Photovia size	86 μm
Core via pitch	600 μm
Via-in-pad	Allowed
Buildup Cu pitch	68 μm

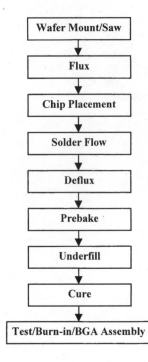

Figure 15.8 Flip chip OLGA assembly process flow.

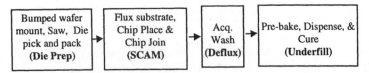

Figure 15.9 Flip chip assembly process flow.

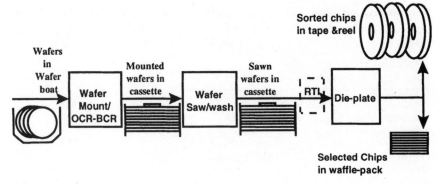

Figure 15.10 Die preparation process flow.

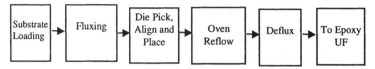

Figure 15.11   Chip joining (SCAM) process flow.

linking all the tools. The flux used is water soluble and is applied in a controlled quantity onto the OLGA copper pads. The chip placement machine accuracy is better than ±25 μm (at 3σ limit). The chip joining is performed in an O_2-controlled convection oven with high-purity N_2 gas. The peak temperature during reflow is 230°C. Figure 15.12 shows a typical cross section of the assembled solder-bumped flip chip on OLGA substrate. The yield improvements for the SCAM process line are shown in Fig. 15.13. It can be seen that Intel brought the line from ~40 percent to greater than 99 percent yield in less than a year.

Deflux process.   Since water-soluble flux is used for the OLGA package technology, defluxing (dissolving and removing the flux residues) is necessary. It is essential to avoid the potential for corrosion that may lead to electrical issues or create voids in the underfill encapsulant. Intel found that the major factors for undissolved flux are: (1) wash/rinse water temperature and (2) dwell time (i.e., conveyor belt speed). Since there is a limitation on the water temperature, in order to increase the throughput, radical change in the nozzle design is needed; modifications in the design resulted in a fourfold improvement in the units per hour and run rate.

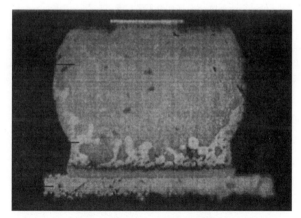

Figure 15.12   Cross section showing the C4 interconnect.

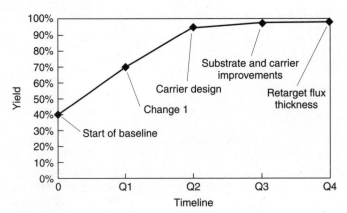

Figure 15.13 Assembly line yield for SCAM as a function of time.

Underfill process.   The underfill process consists of three steps: (1) prebaking; (2) underfill dispensing; and (3) underfill curing. In order to reduce voids in the underfill, moisture in the OLGA package substrate is baked out in a vertical oven. After prebaking, the package is brought to a temperature of 85 to 95°C and the underfill material is dispensed using a multipass dispensing method. After underfill dispensing, the package is put into a vertical oven for curing according to the underfill's curing requirements.

15.2.5   OLGA package reliability

All of Intel's CPU component packages have to pass the qualification tests shown in Tables 15.3 and 15.4. According to Refs. 1 and 2, OLGA packages meet all of these stringent requirements. However, this result did not come easily. During the development of OLGA package technology, Intel encountered and resolved the following key failure modes:

- Die cracking
- Solder mask cracking
- Via delamination in the OLGA substrate
- Solder bump extrusion
- Package insulator cracking
- Interlayer shorts
- Lifted/corroded metal lines
- Underfill cracking
- Delamination at the interface between the underfill and the polyimide

TABLE 15.3   OLGA Package Reliability Requirements

Reliability stress	Description	Pass requirements
Temperature cycle, condition B, 1000 cycles	Mil std. 883D method 1010.7; –55°C to +125°C, packages preconditioned with humidity exposure and 5 SMT reflow cycles	<1% failure at 90% Upper Control Limit (UCL)
Biased Highly Accelerated Stress Test (HAST), 100 h	130°C/85% RH, 1.8 V	<1% failure at 90% UCL
Steam, 168 h	121°C, 100% RH @ 2 atm	<3% area delamination at underfill/die or underfill/package interfaces (C-SAM based)
High-Voltage Extended Life Test (HVELT), 1008 h	100°C, 1.8 V	<1% failure at 90% UCL
Moisture sensitivity	Out-of-bag requirements during SMT assembly	Level 3 precon capable

- Delamination at the interface between the underfill and the solder mask

Interested readers are encouraged to read Refs. 1 and 2 for more useful information about Intel's solder-bumped flip chip on OLGA substrate for its microprocessors.

15.3   Mitsubishi's FC-BGA Package[3–5]

Unlike Intel's solder bumps, Mitsubishi's are made of the low-alpha (<1 cph/cm^2) eutectic 63wt%Sn-37wt%Pb solder. Figure 15.14 shows a

TABLE 15.4   Some Key Reliability Stresses for OLGA Package

Stress	Condition
Preconditioning + temperature cycle B	*Preconditioning:* 1. 24 h bake at 125°C 2. 5 cycles 3. 216 h of 30°C/60% RH 4. 5 passes of convection reflow *Temperature cycle B (–55°C to +125°C):* 1000 cycles at 2.5 cycles/h
Biased HAST	130°C/85% RH for 100 h. Voltage is 1.8 V
Steam	121°C at 2 atm for 168 h

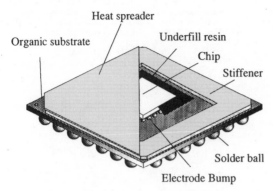

Figure 15.14   Perspective view of Mitsubishi's flip chip BGA.

perspective view of a solder-bumped flip chip on organic substrate in a PBGA package called PC-BGA. Figure 15.15 shows the package family available today. It can be seen that the packages can support pin counts of up to 1681. Dimensions of the packages are shown in Fig. 15.16, and a typical cross section is schematically shown in Fig. 15.17. It can be seen that the package has a BT core with a few buildup layers on both sides. In this section, the wafer bumping, SBU substrate, assembly process, thermal management, electrical performance, and qualification tests and results of Mitsubishi's PC-BGA are presented.

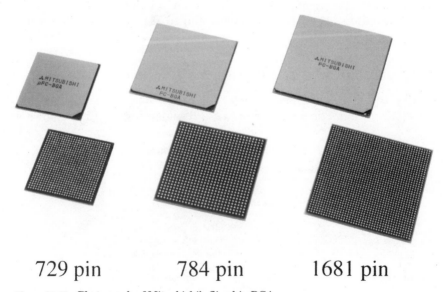

729 pin   784 pin   1681 pin

Figure 15.15   Photograph of Mitsubishi's flip chip BGA.

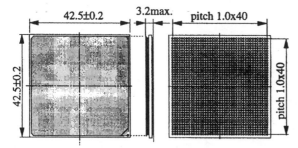

Figure 15.16   Outline of 1681-pin flip chip BGA.

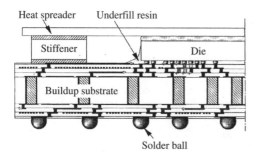

Figure 15.17   Cross section of flip chip BGA.

15.3.1   Wafer bumping

Mitsubishi's 63Sn-37Pb solder bumps (220-μm staggered pitch), as shown in Fig. 15.18, are made by the electroplating method. The UBM is made of Au-Ni-Cu-Ti. The function of Ti is to encourage good adhesion with the aluminum pads on the chip. The Cu film can reduce the internal stress in the Ni layer when the substrate is silicon. Ni is the wetting layer because it has better dissolution life compared with Cu for eutec-

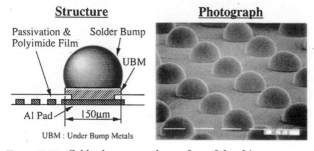

Figure 15.18   Solder bumps on the surface of the chip.

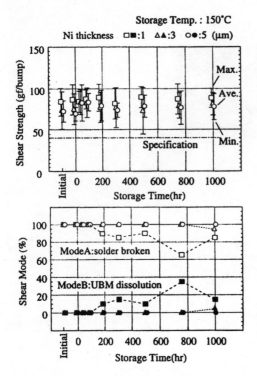

Figure 15.19 Solder bump shear strength after high-temperature storage.

tic solder. Au film is the oxidation barrier layer and protects the Ni from oxidation. After a few experiments, Mitsubishi concluded that 0.1 μm Au-0.6 μm Cu-0.1 μm Ti-5 μm Ni is the best UBM combination, as shown by the shear test results in Figs. 15.19 to 15.21. Figures 15.19 through 15.21 show the shear test results for temperature storage multiple reflows and thermal cycling, respectively. It can be seen that the measured strengths are all greater than 50 g.

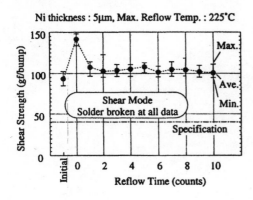

Figure 15.20 Solder bump shear strength after reflow.

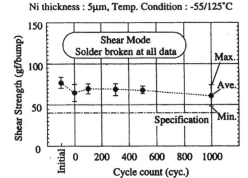

Ni thickness : 5μm, Temp. Condition : -55/125°C

Figure 15.21 Solder bump shear strength after temperature cycling.

15.3.2 Mitsubishi's SBU substrate

Figure 15.22 shows the design rule and a cross section of Mitsubishi's SBU substrate with microvias. It can be seen that it consists of three metal buildup layers on each side of the single-core, two-metal layer. In Fig. 15.22, Step 1 means current production phase and Step 2 means that units have been qualified on the 1681-pin and other PC-BGA as shown in Fig. 15.15. In general, high-density buildup layers are designed as the signal routing layers and the core layers are designed as power/ground planes. Mitsubishi's SBU substrates passed all the qualification tests, as shown in Fig. 15.23. It can be seen that conductive resistance change is less than 10 percent and insulation resistance is more than 100 MΩ. The moisture absorption of the SBU in room temperature is from 0.7 to 0.9 wt%.

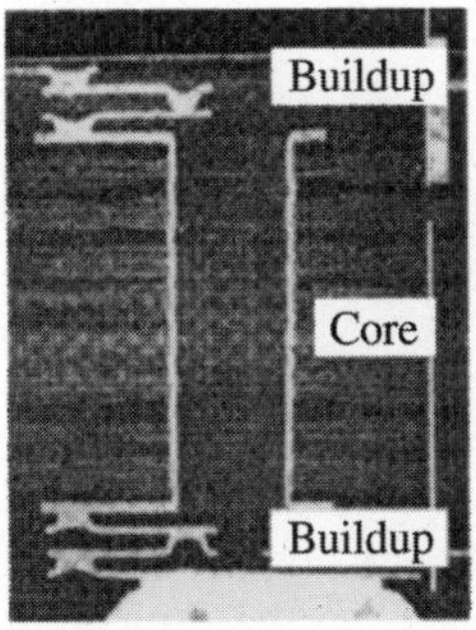

unit : μm

			Step 1	Step 2
Buildup Layer		Line width	50	30
		Spacing width	50	30
		Via diameter	100	75
		Land diameter	150	125
Core Layer		Line width	100	
		Spacing width	100	
		Via diameter	300	
		Land diameter	500	

Cross-Sectional View

Figure 15.22 Design rule of buildup substrate.

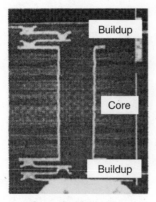

Test Item (Condition)	Conductive Resistance (Spec. ΔR < 10%)	Insulation Resistance (Spec. R > $10^8\Omega$)
Temp. Cycling (−55/125°C)	1000 cycle ΔR < 10%	1000 cycle R > $10^{13}\Omega$
Thermal Shock (−55/125°C)	500 cycle ΔR < 10%	500 cycle R > $10^{13}\Omega$
Pressure Cooker Storage (130°C/85%)	240 hrs. ΔR < 10%	240 hrs. R > $10^9\Omega$

*Precondition: JEDEC level 3

Cross-Sectional View

Figure 15.23 Reliability testing of substrate.

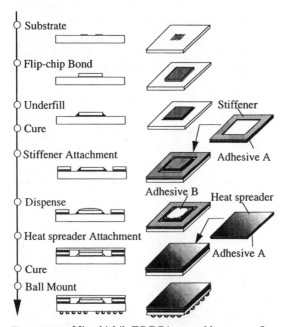

Figure 15.24 Mitsubishi's FC-BGA assembly process flow.

15.3.3 PC-BGA assembly process

The assembly process for solder-bumped flip chip on the SBU substrate in a PC-BGA package is shown in Fig. 15.24. In order to perform fluxless assembly, the low-alpha eutectic 63Sn-37Pb solder paste is printed onto the SBU substrate (copper pads 150 μm in diameter),

(a) Eutectic Solder Bump by Flux-less Flip-chip Bonding	(b) High M.P. Solder Bump by Flip-chip Bonding with Flux

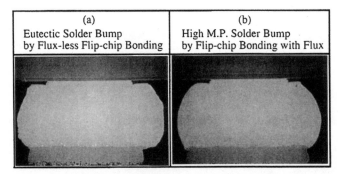

Figure 15.25   Cross-sectional photograph of solder joint.

forming 75-μm-high solder bumps. The eutectic solder bumps on both the chip and the substrate are bonded using the fluxless flip chip solder joining technology such as that shown in Chap. 2 of Lau[18]. Figure 15.25a shows a typical cross section of the fluxless solder joint. In order to avoid delaminations between the insulation layer and the metal layer, Mitsubishi recommends that the substrate be controlled in the process to less than 240°C heating or less than 0.4wt% water absorption.

Underfill is then dispensed on either one side or two adjacent sides of the chip. Two underfills are considered: resin A and resin B. Figure 15.26 shows a typical cross section of the fluxless eutectic solder joint after 510 cycles (thermal shock, −55 to +125°C). It can be seen that there is no crack in the solder joint. On the other hand, the flux-reflowed high-lead solder joint has cracks.

A stiffener (0.6 mm thick) is then attached on the substrate, and the heat spreader (0.25 mm) is attached on the back of the chip with a 125-μm-thick insulating and thermal conductivity (1.3 W/m·K) resin. Both the stiffener and heat spreader are made from the same oxygen-free copper (391 W/m·K) and are attached to the SBU substrate by a heat-resistant tape to maintain the adhesion after reflow. Finally, the solder balls are mounted onto the bottom side of the substrate. Figure 15.27 shows a cross section of the PCB assembly of the solder-bumped flip chip on SBU substrate in a PC-BGA package. The tiny solder joints between the chip and the substrate and the larger solder joints between the substrate and the PCB are clearly shown.

15.3.4   Thermal management

The thermal resistance of Mitsubishi's 784-pin FC-BGA assemblies with and without heat spreaders has been determined by the finite ele-

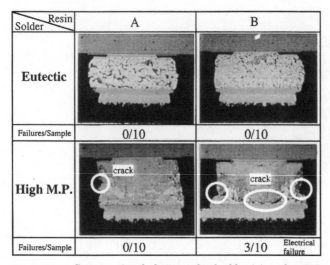

Solder \ Resin	A	B
Eutectic		
Failures/Sample	0/10	0/10
High M.P.		
Failures/Sample	0/10	3/10 Electrical failure

Figure 15.26 Cross-sectional photograph of solder joint after 510 cycles in thermal shock test, −55 to +125°C.

ment model shown in Fig. 15.28. The material properties of the assembly are shown in Table 15.5. The dimensions of the stiffener are 35.5 mm² with a width of 7.5 mm. The results for the 784-pin FC-BGA package are shown in Fig. 15.29. It can be seen that the heat spreader lowers the thermal resistance by more than 40 percent. Accordingly, an IC device with 6 W of heat can be packaged in the 784-pin FC-BGA without a fin under a condition of $\Delta T = 55°C$ and at 1 m/s air velocity. Results for packages with other pin counts are shown in Table 15.6.

Figure 15.30 shows the parameter studies to make θ_{jc} less than 1°C/W. In order to achieve this, the thermal conductivity of the conductive adhesive should be more than 1 W/m·K and the thickness should be no less than 125 µm.

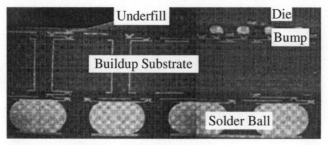

Figure 15.27 Photograph of FC-BGA on PCB.

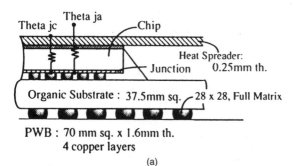

PWB : 70 mm sq. x 1.6mm th.
4 copper layers

(a)

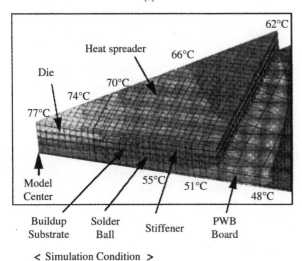

< Simulation Condition >

Package : 37.5 x 37.5 mm, 784 pins   Power : 6 W
PWB Board : 70 x 70 mm, 4 metal   Air velocity : 1 m/s
                                          T_a : 25°C

(b)

Figure 15.28   Thermal simulation model.

15.3.5   Electrical performance

Most of the solder-bumped flip chip in PBGA packages have outstanding electrical performance because of lower inductance and relative dielectric constant. The Mitsubishi PC-BGA packages perform even better since the power and ground planes in the SBU layers are supported to reduce the crosstalk noise, as shown in Fig. 15.31. It can be seen from the chip side that the metal layers are designated D1 through D6. The core layers are called L1 and L2. The signal lines are mainly routed in D1 and D3, the ground planes are inserted in D2 and L2, and the power planes are inserted in L1 and D5.

TABLE 15.5   Material Properties Used in Simulation

Material		E (GPa)	v	CTE (ppm/K)	Yield stress (MPa)
Chip		166.7	0.28	3.5	—
HS	Copper	127.5	0.3	17	—
	Alloy 42	145.1	0.3	5	—
Underfill resin		7.85	0.35	20	—
Adhesive A		7.55	0.35	36	—
Adhesive B		7.85	0.35	40	—
Substrate		19.6	0.18	15	—
Bump	RT	30.9	0.35	24.5	24.5
	125°C	18.1	0.36	24.5	7.85

HS: Heat Spreader. *E:* Young's modulus. *v:* Poisson's ratio

Table 15.7 shows the comparison of electrical performance for various PBGA packages such as the 1681-pin FC-BGA, the 520-pin cavity (face-down) PBGA, and the 456-pin overmolded PBGA. It can be seen that, compared with the other PBGA packages, the FC-BGA (column 2 of Table 15.7) has excellent performance on the crosstalk noise.

15.3.6   Qualification tests and results

The results of qualification tests performed on the 1681-pin FC-BGA package are shown in Table 15.8. It can be seen that this package meets all the requirements. This could be attributed to the optimized design of the package structural materials and geometry. The PCB solder joint reliability of the 784-pin FC-BGA has been demonstrated by thermal cycling testing.[21]

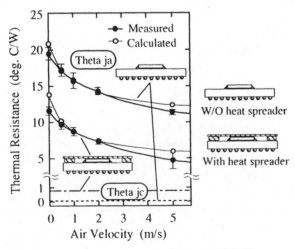

Figure 15.29   Thermal resistance of 784-pin FC-BGA.

TABLE 15.6 Thermal Performance

Pin count	Body size (mm^2)	Die size (mm^2)	R_{jc} (°C/W)	R_{ja} (°C/W)
729	29.0	10.2	0.97	10.5
961	33.0	13.4	0.55	9.7
1296	37.5	15.0	0.38	7.4
1681	42.5	14.6	0.47	7.4

R_{ja}: Air velocity 1.0 m/s

15.4 IBM's FC-PBGA Package[6-11]

The structure of solder-bumped flip chip in PBGA packages by IBM and Intel is very similar in the sense that both use the C4 solder bumps and the Sn-Pb solder-coated built-up substrates. In this section, some useful thermal management and second-level PCB solder joint reliability data for a FC-PBGA assembly obtained by IBM are presented.

15.4.1 Problem definition

Figure 15.32a schematically shows a cross section of an FC-PBGA assembly. The FSRAM silicon chip dimensions are $14 \times 10 \times 0.75$ mm. The BT substrate ($22 \times 14 \times 0.7$ mm) has 7×17 solder balls arrayed at

TABLE 15.7 Comparison of Package Electrical Characteristics

	Flip chip BGA	Cavity BGA	Mold BGA
Package type			
Pin count	1681	520	456
Body size	42.5 mm^2	41 mm^2	35 mm^2
L_s (nH)	1.4–6.6	4.4–9.7	5.2–12.0
C_t (pF)	1.2–3.9	1.3–3.0	2.6–3.3
R (Ω)	0.15–1.5	0.19–0.47	0.27–0.59
L_m (nH)	0.1–0.6	0.8–2.4	2.1–2.7
C_m (pF)	0.01–0.2	0.2–0.6	0.3–0.4
Circuit simulation			

TABLE 15.8   Reliability Test Results

Item	Condition	Results
Temperature cycling	–55/+125°C	1000 cycles, pass
Thermal shock	–55/+125°C	300 cycles, pass
High-temperature storage	150°C	1000 h, pass
Pressure cooker storage	130°C/85%	168 h, pass
High-temperature humidity bias	85°C/85%/3.6 V	1000 h, pass

Precondition: JEDEC level 3.

1.27-mm pitch. The substrate is assumed to have metal layers and a thermal conductivity of 3 W/m·K. The FR-4 PCB is 76 × 76 × 1.57 mm in size and has solder mask defined (SMD) copper pads as shown in Fig. 15.32a. Detailed material properties and geometry are given in Table 15.9.

The finite element model for the analysis is shown in Fig. 15.32b. It can be seen that only a single array of FC-PBGA solder joints and hybrids in the center is modeled. The modeled package is power-cycled under a convective cooling condition with airflows of 0.1, 0.2, and 0.5 m/s. A cyclic power load of 0/3 W with 100-s ramps and 800-s dwells is assumed in the modeled package. Both power and temperature cycling conditions have a frequency of 2 cycles per h.

In order to study the effect of airflow on the solder joint reliability, the use of integrated flow-thermal-mechanical analysis method is necessary. This method makes possible the integration of computational fluid dynamics (CFD), transient heat transfer, and thermal strain solutions.

15.4.2   CFD analysis for thermal boundary conditions

Assume that the FC-PBGA is inside a computer cabinet (156 mm in the 1-direction, 107 mm in the 3-direction, and 36 mm in the 2-direction,

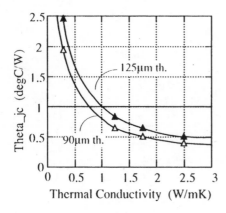

Figure 15.30 Relation between θ_{jc} and adhesive B.

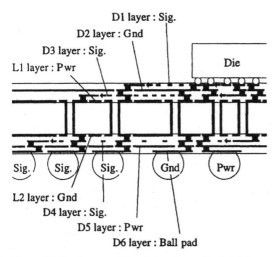

D1 layer : Sig.

D2 layer : Gnd

D3 layer : Sig.

L1 layer : Pwr

Die

Sig.   Sig.   Sig.   Gnd   Pwr

L2 layer : Gnd

D4 layer : Sig.

D5 layer : Pwr

D6 layer : Ball pad

Figure 15.31   Cross-sectional structure of substrate.

Fig. 15.32*b*), which is the CFD domain. CFD calculates the profiles of local velocity, temperature, and heat transfer coefficients along the package. The local heat transfer coefficient distributions at both the top and bottom surfaces of the FC-PBGA are shown, respectively, in Figs. 15.33*a* and *b*. It can be seen that the heat transfer coefficients increase as the air flow velocity (*U*) increases, and they will be imposed as the thermal boundary conditions in the nonlinear finite element analysis. The locations A through J are shown in Fig. 15.32*a*.

15.4.3 Nonlinear finite element stress analysis

The nonlinear strains of the solder joints can be determined by a sequential transient heat transfer and nonlinear stress analysis using a commercial finite element code such as ABAQUS, MARC, or ANSYS. In the sequence of analysis, CFD-determined thermal boundary conditions of local heat transfer coefficients and cyclic power loads of 0/3 W and 2 cycles per hour, respectively, are imposed on the transient heat transfer model to determine the local temperatures of the modeled package. The resulting temperature time history is stored in a file. This file is then treated as the input file (cyclic temperature loads) for the subsequent thermal stress model to determine the nonlinear responses induced by power cycling. Usually, even with the same finite element code, two different runs with two different types of elements are needed, respectively, for the transient heat transfer and thermal stress analyses.

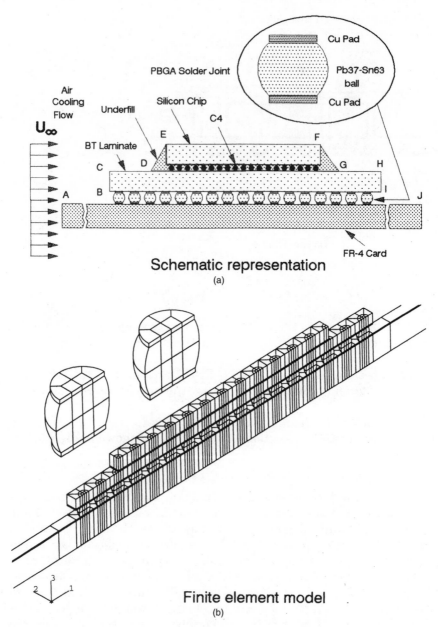

Schematic representation

(a)

Finite element model

(b)

Figure 15.32 The 119-ball flip chip PBGA package.

TABLE 15.9 Material Properties of FC-PBGA Package

Material	Dimension (mm)	Temperature (K)	Elastic modulus (MPa)	Poisson ratio	Yield strength (MPa)	CTE (ppm/K)	Thermal conductivity (W/m-K)	Specific heat (J/kg-K)	Density (kg/m³)
BT laminate	22 × 14 × 0.7	298	19,000	0.22	Elastic	15.7	3	1,190	1,995
C4/underfill	14 × 10 × 0.1	298	14,470	0.28	Elastic	20	1.6	674	6,080
Silicon chip	14 × 10 × 0.75	298	162,000	0.28	Elastic	2.3	110	712	2,330
Solder mask	0.05 thick	298	3,448	0.35	Elastic	30	0.2	1,190	1,995
FR-4 card	76 × 76 × 1.57	298	18,200	0.25	Elastic	19.0	13	879	1,938
Copper pad	0.68 diameter 0.025 thickness	298	68,900	0.34	69.0	16.7	389	385	8,942
Pb37-Sn63 solder ball	0.76 diameter 0.60 height	273 323 373	26,447 12,521 6,909	0.360 0.365 0.378	36.4 15.2 9.6	25.2 26.1 27.3	51	150	8,470

$$\dot{\varepsilon}^{cr} = 12,423.2 \, (\sinh 0.125938 \, \sigma)^{1.88882} \exp(-61,417/RT)$$

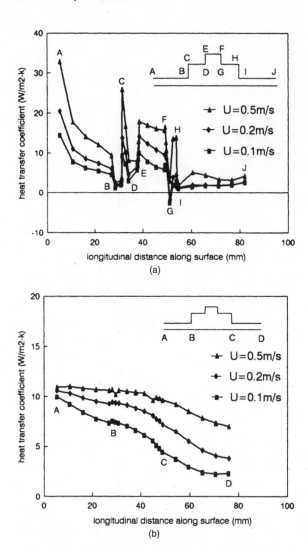

Figure 15.33 Local heat transfer coefficient distribution at the center line of the PBGA package.

15.4.4 Simulation results

The temperature time histories of the FC-PBGA, chip, and solder joints are presented in this section. Also, the creep strain range and mean thermal fatigue life of the FC-PBGA solder joints determined by IBM are provided.

Temperature time-history responses. Figure 15.34 shows the time history of peak local temperature of the FC-PBGA component subjected to an

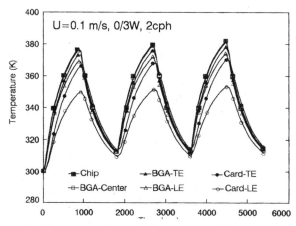

Figure 15.34 Temperature history of package components for the case of $U = 0.1$ m/s. LE: leading edge; TE: trailing edge.

airflow of $U = 0.1$ m/s (ambient temperature is 300 K). It can be seen that (1) the temperature time history at various locations of the assembly fluctuates as a wave shape and tends to reach a saturated state after three power cycles; (2) the chip has the largest response and saturates in a range of 312 to 380K; (3) the center solder joints have almost the same temperature as the chip; and (4) the maximum difference in local temperature of the PCB can be as high as 17K at an airflow of $U = 0.1$ m/s.

Figure 15.35 shows the effect of airflow velocity on the temperatures of the chip. As expected, the chip temperatures decrease as the airflow velocities increase.

Figure 15.36 shows the temperature time histories of the FC-PBGA solder joints at three different locations with respect to the air flow direction. As expected, the higher the airflow velocities, the lower the temperatures in the solder joints. Denote the center solder joint #0, the leading edge (LE) solder joint #–8, and the trailing edge (TE) solder joint #8. Figure 15.37 shows that the center solder joint has the highest temperature.

Thermal resistance. The junction-to-ambient thermal resistance (R_{ja}) of the FC-PBGA package has been determined by CFD and finite element methods, including steady-state and transient-state analyses. Assuming the chip power is 3 W and the ambient temperature of approaching air is 300K, Table 15.10 lists the results for various air flow velocities. It can be seen that (1) all the analysis tools yield almost the same results, and (2) the higher the air flow velocity, the lower the thermal resistance.

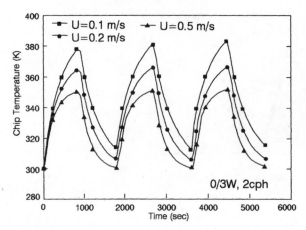

Figure 15.35   Temperature history of chip in a 119-ball PBGA package. Effect of air flow velocity on the chip temperature is significant.

15.4.5   Solder joint thermal fatigue life prediction

The peak values of equivalent creep strain range $\Delta\varepsilon^{cr}_{eq}$ for the FC-PBGA solder joints as a function of ball site at various cycle conditions are shown in Fig. 15.38a. It can be seen that the maximum creep strain range does not occur at the end solder joints but near the center solder joints. This is due to the large thermal expansion mismatch between the silicon chip (2.3 ppm/°C), the BT substrate (15.7 ppm/°C), the solder ball (26 ppm/°C), and the FR-4 PCB (19 ppm/°C), as well as the deflection (or warpage) of the whole FC-PBGA assembly.

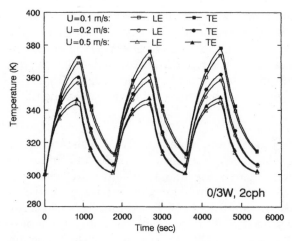

Figure 15.36   Temperature histories of solder joints in a 119-ball PBGA under a cyclic power load of 0 to 3 W at 2 cycles per hour. Three cycles are simulated.

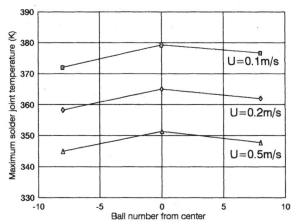

Figure 15.37   Maximum solder joint temperature as a function of ball site and the air flow velocity as shown in Fig. 15.36.

Based on a simple power law equation for the 63Sn-37Pb solder (see the last row of Table 15.9), the mean thermal fatigue life (N50) of FC-PBGA solder joints as a function of ball site is determined and is given in Fig. 15.38b. It can be seen that (1) the higher the airflow velocity, the longer the solder joint thermal fatigue life; (2) the fatigue life is larger for those solder joints near the ends of the package; and (3) the center solder joint has the shortest fatigue life.

15.5   Motorola's FC-PBGA Packages[12, 13]

Motorola has studied at least two different FC-PBGA packages. One uses C4 solder bumps (similar to the one by Intel and IBM) and the other uses E3 (see Sec. 6.2) solder bumps.[15] In this section, (1) the thermal management of the FC-PBGA with E3 solder bumps and (2) the reliability of controlled-collapse chip carrier connection (C5) solder joints on the FC-PBGA with C4 solder bumps are presented. Motorola has designated the joints between the chip and the substrate C4 joints and the joints between the substrate and the PCB C5 joints.

TABLE 15.10   Verification of Analysis Tools and Results

		R_{ja} (C/W)*	
		FEA (ABAQUS)	
U (m/s)	CFD (FLUENT)	Steady state	Transient state
0.1	26.4	26.83	26.72
0.2	22.27	22.85	21.96
0.5	18.9	19.36	18.24

* Comparison for the calculated thermal resistance of package.

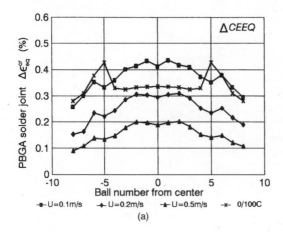

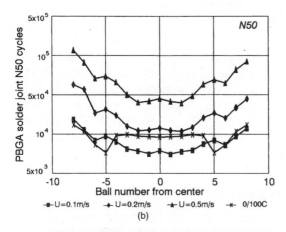

Figure 15.38 (*a*) Creep strain range of PBGA solder joints as a function of ball site. (*b*) Mean fatigue life of PBGA solder joints as a function of ball site. Comparison is made between three cases of power cycling (0/3 W, 2 cycles per hour) and that of a temperature cycling (0 to 100°C, 2 cycles per hour) for 119-ball PBGA.

15.5.1 Thermal management of FC-PBGA assemblies with E3 bumps[12]

Figure 15.39 schematically shows a cross section and a simplified geometric representation for finite element modeling of the FC-PBGA with E3 solder bumps on a $10.5 \times 7 \times 0.75$-mm FSRAM. Three different thicknesses of substrate are considered: 0.1, 0.5, and 1 mm. The overall dimensions of the 164-pin FC-PBGA assembly are shown in Table 15.11.

Junction to case thermal resistance. Figure 15.40 shows a simplified schematic of a ring-style cold-plate apparatus for measuring thermal

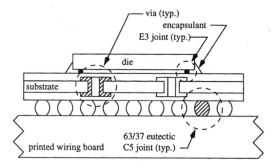

Typical FC-PBGA package

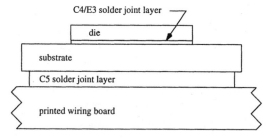

FC-PBGA package representation

Figure 15.39 Physical and simplified geometric representation of the FC-PBGA packages.

resistance. It is found that the junction-to-case thermal resistance is very small (less than 0.3°C/W). This is because the FC-PBGA package has the back of the chip exposed on the top of the package. The measurement results for the junction-to-ambient thermal resistance are shown in Table 15.12. The dimensions and thermal conductivity of the single-layer and four-layer test board are shown in Table 15.13.

Finite element results.   The finite element model of the key interests of the FC-PBGA assembly (with 0.5-mm-thick substrate) is shown in Fig. 15.41, where typical temperature contours resulting from 1 W of power in a 0°C ambient is shown. The input parameters are shown in Table

TABLE 15.11   Overall Dimensions of the 164-Lead FC-PBGA Test Vehicle Assembly

Dimension (mm)	Die	Substrate	PCB
Length	10.5	13.5	100
Width	7.0	10.5	100
Height	0.75	0.1, 0.5, 1.0	1.57

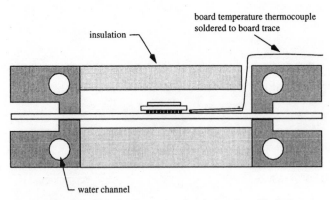

Figure 15.40 Simplified schematic of a ring-style cold-plate measurement apparatus.

15.13, and the results for different substrate thicknesses are shown in the upper portion of Table 15.12 for the package mounted on a single-layer test PCB and in the lower portion of Table 15.12 for the package mounted on a four-layer thermally enhanced test PCB. It can be seen that (1) for all the substrate thicknesses, the thermal resistance declines with the airflow velocity increases, and (2) the four-layer thermally enhanced test PCB reduces the thermal resistance by ~50 percent compared to the single-layer test board.

CDF results. An oblique view of the FC-PBGA CFD model with an 0.5-mm-thick substrate (PCB not shown) is schematically shown in Fig. 15.42. Table 15.14 shows the CFD model parameters for all three substrates. Figure 15.43 and Table 15.12 show the CFD thermal resistance results along with the finite element and measurement results. It can be seen that, in general, the finite element results are closer to the experimental results, and the maximum error of the simulations is less than 25 percent. For the ring-style cold plate shown in Fig. 15.40, the thermal resistance determined by the CFD method yields an error of 26 percent, as shown in Table 15.15.

CFD of CPU FC-PBGA with heat sinks. Figure 15.44 shows a schematic of one of the six FSRAMs packaged in a 164-pin FC-PBGA in a generic central processor unit (CPU) with the airflow heading into the page. The dimensions and parameters of this assembly are shown in Table 15.16. All simulations are performed with a 35°C ambient temperature and 0.5- to 2-m/s airflow applied 13 mm upstream of the module board edge.

The critical temperatures and heat flows for the 164-pin FC-PBGA FSRAMs powered at 3 W each, 30 W in the CPU, with 1-m/s airflow and

TABLE 15.12 Comparison of Numerical and Measurement Results

Substrate thickness (mm)	Air flow (m/s)	θ_{JA} Experiment (°C/W)	Finite element (°C/W)	Difference (%)	CFD (°C/W)	Difference (%)	ψ_{JB} Experiment (°C/W)	Finite element (°C/W)	Difference (%)	CFD (°C/W)	Difference (%)
					Package Mounted on a Single-Layer Test Board						
0.1	0	57.5	56.3	2.1	54.6	5.0	20.8	22.1	6.3	23.2	12
	0.5	48.9			49.6	1.4	20.3			22.6	11
	1	44.7			44.4	0.7	19.8			22.1	12
	2	39.8			39.5	0.8	18.9			21.2	12
	4	32.1			34.2	6.5	17.5			20.1	15
0.5	0	57.6	54.4	5.6	56.3	2.3	22.1	20.7	6.3	24.5	11
	0.5	48.7			50.3	3.3	21.1			23.0	9.3
	1	44.3			43.9	0.9	20.4			21.6	5.7
	2	39.2			37.6	4.1	19.5			19.9	1.8
	4	32.1			30.8	4.0	17.8			17.3	2.8
1.0	0	na	55.8		56.7		NA	22.6		24.8	
	0.5				50.2					23.0	
	1				43.5					21.5	
	2				39.4					20.2	
	4				31.9					18.1	
					Package Mounted on a Four-Layer Thermally Enhanced Test Board						
0.1	0	29.8	28.4	4.7	25.1	16	15.6	16.0	2.9	12.0	23
	0.5	27.0			22.7	16	15.2			12.0	21
	1	25.7			21.1	18	14.9			12.1	19
	2	23.9			19.6	18	14.5			11.9	18
	4	21.5			18.1	16	13.9			11.7	16
0.5	0	30.3	25.5	16	29.9	1.3	16.4	13.1	20	16.2	1.2
	0.5	27.3			26.9	1.5	16.0			15.6	2.3
	1	25.9			24.7	4.6	15.6			15.1	3.3
	2	24.3			22.5	7.4	15.2			14.5	4.5
	4	21.9			19.9	9.1	14.4			13.1	9.3
1.0	0	33.8	27.4	19	31.1	8.0	19.8	15.1	24	17.2	13
	0.5	30.3			27.8	8.3	19.1			16.4	14
	1	28.7			25.5	11	18.6			15.9	15
	2	26.7			23.3	13	17.8			15.3	14
	4	24.3			21.1	13	17.0			14.3	16

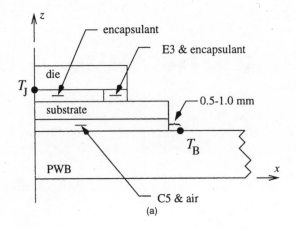

(a)

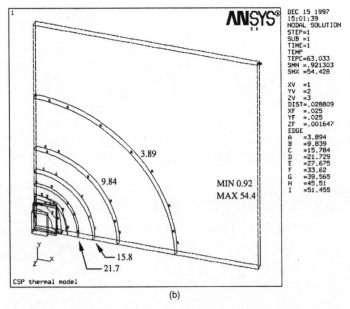

(b)

Figure 15.41 (a) 164 FC-PBGA FE model temperatures of interest. (b) Typical temperature contours resulting from 1 W of power in a 0°C ambient.

35°C ambient temperature are shown in Fig. 15.45. It can be seen that the maximum temperature (103.2°C) occurs at the FSRAM furthest downstream on the bottom side of the module board. This is because of the heating of the air as it flows around the two upstream FC-PBGAs and the low effectiveness of the main PCB as a heat sink. Figure 15.46 shows the FSRAM junction temperatures for various airflows with a heat sink only on top FSRAMs and CPU. The FSRAMs are labeled 1, 2, and 3, starting from the upwind side of the module. It can be seen that

TABLE 15.13   Finite Element Model Parameters

Component	Length (mm)	Width (mm)	Height (mm)	Thermal conductivity W/m·K k_{xy}	k_z
Die	10.5	7.0	0.75	119	119
E3 layer (perimeter)	10.5	7.0	0.1	0.45	15.0
E3 encapsulant	10.1	6.6	0.1	0.45	0.45
Substrate (0.1 mm)	13.5	10.5	0.1	14.3	0.61
Substrate (0.5 mm)	13.5	10.5	0.5	50.7	2.97
Substrate (1.0 mm)	13.5	10.5	1.0	14.4	2.57
C5 layer (BT substrates)	13.5	10.5	0.4	0.027	0.37
C5 layer (polyimide substrate)	13.5	10.5	0.4	0.027	0.74
Single-layer PCB	100	100	1.6	3.4	0.3
Four-layer PCB	100	100	1.6	17.7	0.3

(1) as airflow increases, the FSRAM junction temperature decreases, (2) for all airflows being considered, the upper FSRAM 1 has the smallest junction temperature; and (3) for all airflows being considered, the lower FSRAM 3 has the largest junction temperature. The CPU temperatures are 89.4, 77.1, and 69.9°C for 0.5, 1.0, and 2.0 m/s airflow, respectively. The FSRAM junction temperatures for an airflow of 1 m/s and a pad conductivity of 0.9 W/m·K with three different heat sink configurations are shown in Table 15.17. Two different heat sinks attached to the bottom of the main board are modeled. The only differences between the two heat sinks are the fin thickness and the fin pitch. The first heat sink has a fin thickness of 1 mm and a pitch of 5 mm, as with the upper heat sink. The second heat sink has a fin thickness of 0.75 mm and a pitch of 2.75 mm. It can be seen that the case with the heat sink on top and 2.75-mm pitch yields the best results.

The effects of the thermal conductivity of the interface pads are presented in Fig. 15.47. It can be seen that the pad thermal conductivity is significant only from 0.9 to 1.8 W/m·K. Figure 15.48 shows the critical temperatures and heat flows for the FSRAMs on the module with 15 W of power in the CPU and 3 W of power in each FSRAM at 1-m/s airflow

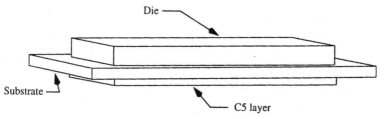

Figure 15.42 Oblique view of the FC-PBGA CFD model with 0.5-mm substrate. PWB not shown.

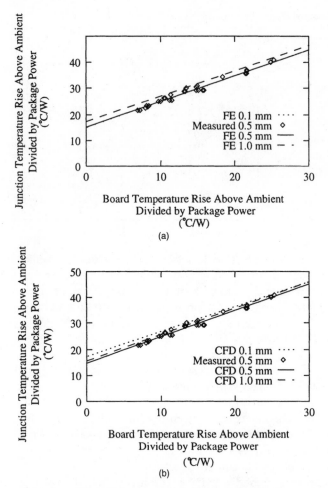

Figure 15.43 Measured and simulated thermal performance of the 164-lead FC-PBGA package. (*a*) FE simulations. (*b*) CFD simulations.

using two heat sinks and the higher-conductivity interface pads (1.8 W/m·K).

15.5.2 Solder joint reliability of FC-PBGA assemblies with C4 bumps[13]

Figure 15.49 schematically shows a cross section of a 164-pin C4 FC-PBGA assembly. The objective is to determine the C5 solder joint reliability on various substrate thickness and PCB pad sizes. Again, the chip is the FSRAM and the substrate thicknesses are 0.1, 0.5, and 1 mm. The pitch of the C5 solder joints is 0.75 mm.

TABLE 15.14   Computational Fluid Dynamics Model Parameters for All
Three Substrates

Component	Model primitive	Length (mm)	Width (mm)	Height (mm)	Thermal conductivity (W/m·K)
Die	Block	10.5	7.0	0.75	119
E3 layer	Plate	10.5	7.0	0.1	1.94
Substrate (0.1 mm)	Block	13.5	10.5	0.1	1.17
Substrate (0.5 mm)	Block	13.5	10.5	0.5	5.61
Substrate (1.0 mm)	Block	13.5	10.5	1.0	4.36
C5 layer (BT substrates)	Block	23.5	13.3	0.6	0.37
C5 layer (polyimide substrate)	Block	23.5	13.3	0.6	0.74
Single-layer PCB	Block	100	100	1.6	3.4
Four-layer PCB	Block	100	100	1.6	17.7

TABLE 15.15   Junction-to-Board Thermal Performance of the 164-Lead FC-PBGA
Using the Ring-Style Cold-Plate Method

Substrate	θ_{JB} (°C/W)				
	Exp	FE	Diff (%)	CFD	Diff (%)
0.1 mm	16.6	14.3	14	12.2	26
0.5 mm	17.5	15.6	11	16.2	7
1.0 mm	21	17.9	15	17.2	18

Effect of substrate thickness on C5 solder joint reliability.   The effect of
substrate thickness on C5 solder joint reliability is shown in Fig. 15.50.
It can be seen that the thicker the substrate, the better the solder joint
reliability. This fact has been demonstrated by thermal cycling tests of
other FC-PBGA packages, as shown in Fig. 15.51. It can be seen that
packages with 1-mm substrate have much longer life than those with
0.7-mm substrate.

Effect of PCB pad sizes on C5 solder joint reliability.   The effect of PCB
pad sizes (0.25, 0.3, 0.35, and 0.4 mm) on the solder joint reliability is
shown in Fig. 15.52. It is assumed that the pad sizes on the substrate
are the same as those on the PCB. It can be seen that for all the PCB
thicknesses, the larger the pad size, the longer the C5 solder joint life.
Also, the pad size is more sensitive for thicker substrates. Fig. 15.52
also demonstrates that the thicker the substrate, the longer the C5 solder joint life.

Acknowledgments

The author would like to thank R. Shukla, V. Murali, and A. Bhansali,
V. K. Nagesh, R. Peddada, S. Ramalingam, B. Sur, and A. Tai of Intel;

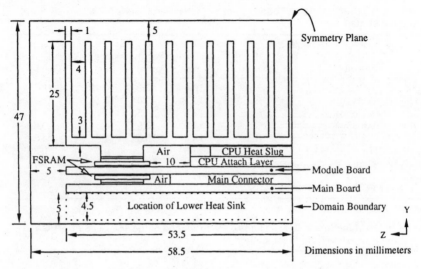

Figure 15.44 Schematic of the CPU module.

H. Matsushima, S. Baba, Y. Tomita, M. Watanabe, E. Hayashi, Y. Take-moto, and Q. Wu of Mitsubishi; B. Hong, and T. Yuan of IBM; and M. Eyman, Z. Johnson, B. Joiner, L. Mercado, V. Sarihan, Y. Guo, and A. Mawer of Motorola for sharing their useful knowledge with the industry.

TABLE 15.16 **Model Dimensions and Parameters**

Component	Length (mm)	Width (mm)	Height (mm)	K (W/m·K)
164-Lead FC-PBGA				
C5 layer	10.5	7.5	0.4	0.37
Substrate	13.5	10.5	0.5	16
E3 layer	10.5	7	0.1	1.94
Die	10.5	7	0.75	119
Interface pad	—	—	0.6	0.9–7.2
CPU (Modeled Length is ½ Actual)				
Attach layer	25	46	2.5	5
Substrate	25	46	2.5	18
Heat slug	15	30	2.5	200
Die	7.5	15	0	Planar source
Interface pad	—	—	0.2	0.9
Miscellaneous				
Main board	90	60	2.0	30
Module board	70	60	1.75	30
Upper heat sink	70	60	25.0	156
Lower heat sink	70	60	5.0	156
Main connector	30	60	2.7	2.36

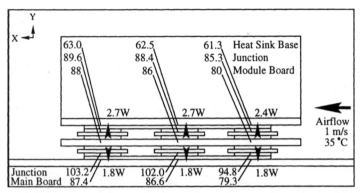

Figure 15.45   Critical temperatures and heat flows for the 164 FC-PBGA FSRAMs powered at 3 W each with 1-m/s airflow and 35°C ambient temperature.

References

1. Shukla, R., V. Murali, and A. Bhansali, "Flip ChipCPU Package Technology at Intel: A Technology and Manufacturing Overview," *IEEE Proceedings of Electronic Components and Technology Conference,* pp. 945–949, June, 1999.
2. Nagesh, V. K., R. Peddada, S. Ramalingam, B. Sur, and A. Tai, "Challenges of Flip Chip on Organic Substrate Assembly Technology," *Proceedings of Electronic Components and Technology Conference,* pp. 975–978, June, 1999.

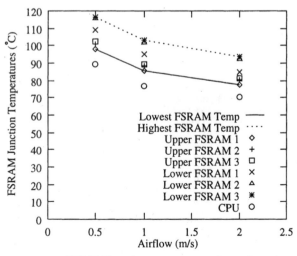

Figure 15.46   FSRAM junction temperatures for various airflows with a heat sink only on top FSRAMs and CPU. K_{pad} = 0.9 W/m·K. FSRAMs were powered at 3 W each and ambient temperature was set to 35°C.

TABLE 15.17   Temperatures of FSRAM Junctions with Three Different Heat Sink Configurations

Heat sink	Top 1	Top 2	Top 3	Bottom 1	Bottom 2	Bottom 3	CPU
Top only	85.3	88.4	89.6	94.8	102.0	103.2	77.1
Top and 5-mm pitch	81.7	84.2	85.1	86.5	89.8	91.0	73.7
Top and 2.75-mm pitch	80.1	82.7	83.7	84.2	87.7	89.0	72.2

Air flow = 1 m/s; pad conductivity = 0.9 W/m·K.

3. Matsushima, H., S. Baba, Y. Tomita, M. Watanabe, E. Hayashi, and Y. Takemoto, "Thermally Enhanced Flip-Chip BGA with Organic Substrate," *Proceedings of Electronic Components and Technology Conference,* pp. 685–691, June, 1998.

4. Baba, S., Q. Wu, E. Hayashi, M. Watanabe, H. Matsushima, Y. Tomita, and Y. Takemoto, "Flip-Chip BGA Applied High-Density Organic Substrate," *Proceedings of Electronic Components and Technology Conference,* pp. 243–249, June, 1999.

5. Baba, S., Y. Tomita, M. Matsuo, H. Matsushima, N. Ueda, and O. Nakagawa, "Molded Chip Scale Package for High Pin Count," *IEEE Transactions on Components, Packaging, and Manufacturing Technology,* part B, **21** (1): 28–34, 1998.

6. Hong, B., and T. Yuan, "Integrated Flow-Thermomechanical and Reliability Analysis of a Low Air Cooled Flip Chip-PBGA Package," *Proceedings of Electronic Components and Technology Conference,* pp. 1354–1360, June, 1998.

7. Hong, B., and T. Yuan, "Integrated Flow-Thermo-Mechanical Analysis of Solder Joints Fatigue in a Low Air Flow C4/CBGA Package," *Proceedings of ISPS Symposium,* pp. 209–214, 1997.

8. Hong, B., and T. Yuan, "Integrated Flow-Thermo-Mechanical Analysis of Solder Joints Fatigue in a Densely Packed C4/CBGA Package," *Proceedings of I-Therm Conference,* pp. 220–228, May 1998.

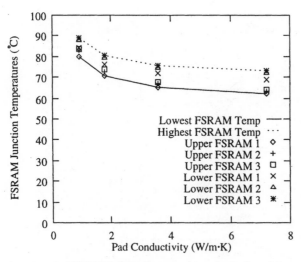

Figure 15.47   FSRAM junction temperatures for various interface pad conductivities at 1-m/s airflow with a 2.75-mm pitch heat sink at the bottom of the main board. FSRAMs were powered at 3 W each and ambient temperature was set to 35°C.

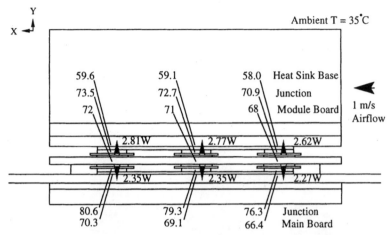

Figure 15.48 Critical temperatures and heat flows for the 164 FC-PBGA FSRAMs powered at 3 W each with 1-m/s airflow and 35°C ambient temperature. Two heat sinks and thermal interface pads of 1.8 W/m·K were used.

9. Hong, B., and T. Yuan, "Heat Transfer and Nonlinear Thermal Stress Analysis of a Convective Surface Mount Package," *IEEE Trans. Components, Packaging, and Manufacturing Technology*, part A, **20** (2): 213–219, June 1997.
10. Hong, B., and L. Burrell, "Nonlinear Finite Element Simulation of Thermoviscoplastic Deformation of C4 Solder Joints in High Density Packaging Under Thermal Cycling," *IEEE Trans. Components, Packaging, and Manufacturing Technology*, part A, **18** (3): 585–591, September 1995.
11. Hong, B., and L. Su, "On Thermal Stresses and Reliability of a PBGA Chip Scale Package," *Proceedings of Electronic Components and Technology Conference*, pp. 503–510, June 1998.
12. Eyman, M., Z. Johnson, and B. Joiner, "Thermal Simulation and Validation of the Fast Static RAM 164-Lead FC-PBGA Package with Investigation of Package Thermal Performance in a Generic CPU Module," *Proceedings of Electronic Components and Technology Conference*, pp. 62–69, June 1998.
13. Mercado, L., V. Sarihan, V., Y. Guo, and A. Mawer, "Impact of Solder Pad Size on Solder Joint Reliability in Flip Chip PBGA Packages," *Proceedings of Electronic Components and Technology Conference*, pp. 255–259, June 1999.
14. Lee, S. W., and J. H. Lau, "Solder Joint Reliability of Plastic Ball Grid Array with Solder Bumped Flip Chip," ASME Paper No. 97WA/EEP-6, November 1997.
15. Lau, J. H., and Y. H. Pao, *Solder Joint Reliability of BGA, CSP, Flip Chip, and Fine Pitch SMT Assemblies*, McGraw-Hill, New York, 1997.

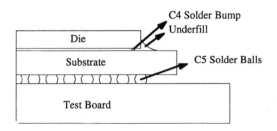

Figure 15.49 Package geometry.

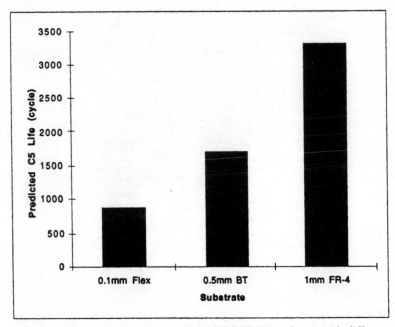

Figure 15.50 C5 solder joint reliability for FC-PBGA packages with different substrate thicknesses.

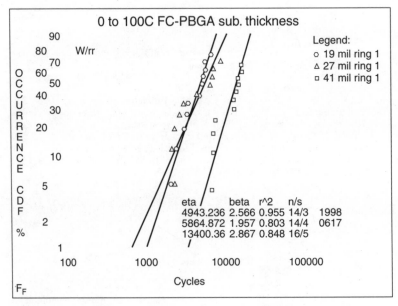

Figure 15.51 0 to 100°C reliability test data of FC-PBGA packages with varied substrate thickness.

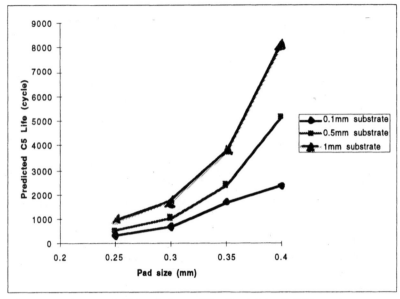

Figure 15.52   Impact of PCB pad size on C5 solder joint reliability.

16. Lau, J. H., and S. W. Lee, *Chip Scale Package, Design, Materials, Process, Reliability, and Applications*, McGraw-Hill, New York, 1999.
17. Lau, J. H., C. P. Wong, J. L. Prince, and W. Nakayama, *Electronic Packaging, Design, Materials, Process, and Reliability*, McGraw-Hill, New York, 1998.
18. Lau, J. H., *Flip Chip Technologies*, McGraw-Hill, New York, 1996.
19. Lau, J. H., *Ball Grid Array Technology*, McGraw-Hill, New York, 1995.
20. Lau, J. H., *Chip on Board Technologies for Multichip Module*, Van Nostrand Reinhold, New York, 1994.
21. May, A., J. H. Lau, R. Chen, L. Hu, and T. Chen, "Effect of Large Connector Insertion on the Solder Joint—Reliability of PBGA on PCB," *Proceedings of SMI Conference*, pp. 95–104, 1998.

Failure Analysis of Flip Chip on Low-Cost Substrates

16.1 Introduction

The past few years have witnessed explosive growth in the research and development efforts devoted to solder-bumped flip chip on low-cost organic substrates[1-28] as a direct result of higher requirements for package density and performance and of the cost advantages of organic over ceramic substrates. Just like many other new technologies, low-cost solder-bumped flip chip still faces certain critical issues. In the development of flip chip on board, it is reemphasized that the following (comparing with the conventional surface mount technology) must be noted and understood in order to obtain the full benefits of this technology:

- The infrastructure of flip chip is not well established.
- Flip chip expertise is not commonly available.
- Wafer bumping is still too costly.
- Bare die/wafer is not commonly available.
- Bare die/wafer handling is not easy.
- Pick and place is more difficult.
- Fluxing is more critical.
- Underfill slows down the SMT process.
- Rework is more difficult, if not impossible.
- Solder joint reliability is more critical.

- Inspection is more difficult.
- Flip chip assembly testability is not well established.
- Failure analysis is more difficult.
- Die shrinkage and expansion create system-level problems.
- There are known-good-die issues.
- Solder bump extrusion and chip cracking during solder reflow have been reported.

Failure analyses of 63wt%Sn-37wt%Pb solder-bumped flip chip on low-cost substrates with underfill encapsulant are presented in this chapter. Emphasis is placed on solder flow-out (also called solder bump extrusion), nonuniform underfill and voids, and delaminations. The x-ray, A-, B-, and C-mode scanning acoustic microscopy (A-SAM, B-SAM, and C-SAM, respectively), and tomographic acoustic microimaging (TAMI) techniques[16] are used to analyze the failure samples. Also, cross sections are examined for a better understanding of the failure mechanisms.

Mechanical shear tests are performed to characterize the shear strength at the underfill-solder mask interface and the underfill-passivation interface. The main objective of the present study is to achieve a better understanding of the mechanical behavior of FCOB assemblies with imperfect underfill encapsulants.

16.2   Failure Analysis of FCOB with Imperfect Underfills[27]

In this section, the design, materials, process, qualification tests, and failure analysis of solder-bumped flip chip on organic substrate with imperfect underfill encapsulants are presented. The assemblies are first subjected to the JEDEC level 1 and level 3 preconditions and then three repetitions of reflow. Then they are subjected to the temperature cycling test and a highly accelerated stress test. The failed assemblies are examined using the A-, B-, and C-SAM, TAMI, and x-ray techniques. Cross sections of the failure samples are also examined for a better understanding of the failure mechanisms.

16.2.1   Test chip

The dimensions of the test chip are $5 \times 4.2 \times 0.5$ mm. It has peripheral-arrayed pads (0.12×0.12 mm). The passivation opening is 0.1×0.1 mm. The 63wt%Sn-37wt%Pb solder bumps are deposited by the electroplating method. The average bump height is 94 µm (Fig. 16.1). The average shear strength is 63 g, as shown in Fig. 16.2. (The shear blade speed is 100 µm/s and the shear blade tip is 25 µm from the chip surface.) The failure location is in the solder, not at the terminal metals.

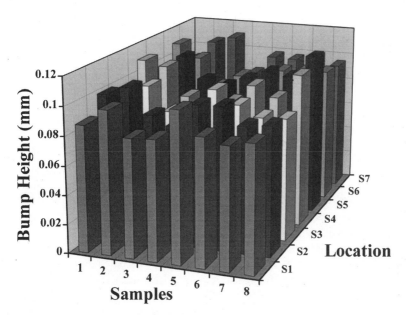

Figure 16.1   Solder bump height.

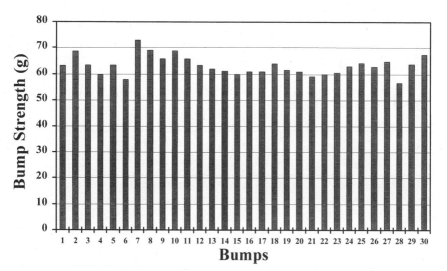

Figure 16.2   Solder bump shear strength.

16.2.2 Test board

The test board has 25 chip sites and is made of bismaleimide triazene (BT) material. Its thickness is 0.5 mm. The copper pad is 0.1 mm in diameter and the solder mask opening is 0.16 mm (i.e., non-solder mask defined copper pads). The finishing condition of the copper pads is Ni-Au.

16.2.3 Flip chip assembly

Siemens 80F[4] (1.6 mils at 4σ accuracy) with a spray fluxer (3 µl per dot) is used for this study. This spray fluxer can only dispense one dot of flux at a time on the copper pad of the test board instead of a peripheral-arrayed pattern. Heraeus Surf 26 no-clean flux is used for this study. The Fein Focus x-ray imaging system is used to check the alignment. The accuracy is more than 90 percent.

The Electrovert Omniflo 5 is used to reflow the solder bumps with a maximum temperature at 230°C. The x-ray apparatus is used again to verify the alignment and check for shorts. The alignment accuracy is 100 percent and there are no shorts. There are some voids in the solder joints; however, they are too small to become an issue. (Again, in this book, the bits of solder on the chip before it is joined to the substrate are called *solder bumps*. After the solder bumps have been reflowed on the substrate, they are called *solder joints*.)

The CAMALOT 1818 dispensing system is used in this study. The underfill material is HYSOL 4527. The BT substrate is placed on a hot plate set to 80°C. A 28-gauge nozzle is used to dispense the underfill in an L shape along two adjacent edges of the die. The underfill flow time on the hot plate is 60 s. The curing condition is 165°C for 35 min.

All samples are inspected by x-ray, A-, B-, and C-SAM, and TAMI techniques. At this stage, all samples are good.

16.2.4 Preconditions, reflows, and qualification tests

The JEDEC standard level 1 and level 3 preconditions are used in this study. First, all the samples are subjected to a temperature of 125°C for 24 h to bake out the moisture. Then, for level 1 precondition, the samples are subjected to 85°C/85 percent RH for 168 h, and for level 3 precondition the samples are subjected to 30°C/60 percent RH for 192 h. After moisture soaking (either level 1 or level 3), all the samples are subjected to three repetitions of reflow (after no less than 15 min and no more than 4 h of preconditioning).

After preconditioning and three repetitions of reflow, all the samples are inspected using x-ray, A-, B-, and C-SAM, and TAMI techniques. Most of the samples are good, and they are subjected to the qualification tests such as (1) HAST: 130°C, 85 percent RH, 96 h, and (2) temperature cycling: −65 to +150°C, 500 cycles. Thirty samples are used for both tests for level 1 and for level 3. All passed the tests. Cross sections of the tested samples are shown in Fig. 16.3. It can be seen that the solder joints are still in very good shape.

16.2.5   Failure modes and discussion

The focus of this study is on investigating the failure mechanisms of the few bad samples via A-, B-, and C-SAM, TAMI, and x-ray methods. Some important results are summarized as follows.

Solder flow-out. Figure 16.4 shows a series of pictures of a flip chip assembly that went through level 1 preconditioning and three repetitions of reflow. It can be seen from Fig. 16.4*a* (x-ray image) that some solder flowed out from the solder joints (solder bump extrusion). Figure

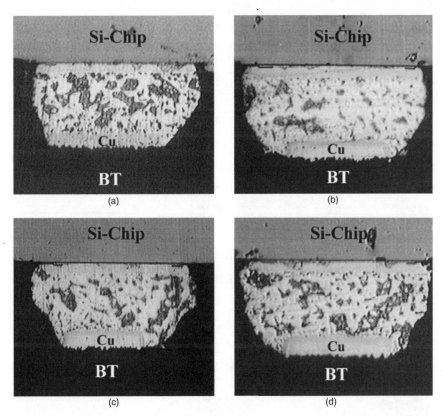

Figure 16.3 Cross sections of tested samples. (*a*) Level 1 + three reflows, then HAST. (*b*) Level 1 + three reflows, then TC. (*c*) Level 3 + three reflows, then HAST. (*d*) Level 3 + three reflows, then TC.

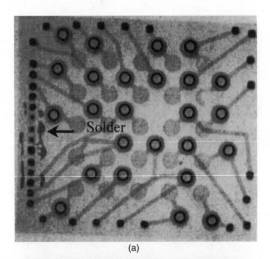

(a)

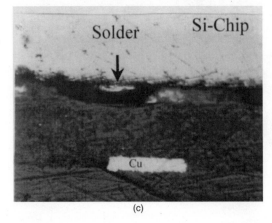

(c)

Figure 16.4 (*a*) X-ray image shows solder that has flowed out of the solder joint. (*b*) C-SAM image shows solder that has flowed out of the solder joint. (*c*) Cross-sectional view of the flowed-out solder.

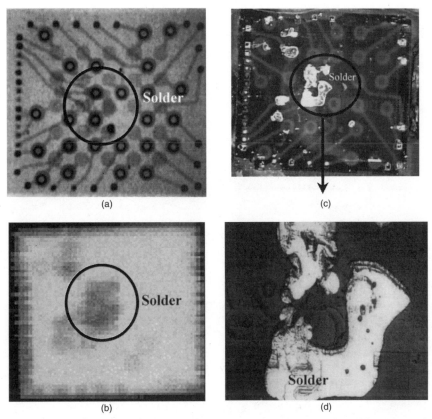

Figure 16.5 (*a*) X-ray image shows solder that has flowed out of the solder joint and moved to the center of the assembly. (*b*) C-SAM image shows a dark area of solder. (*c*) Solders lie on the solder mask of the BT substrate (with die removed). (*d*) Enlarged view of *c*.

16.4*b* is a C-SAM image that also shows that solder has flowed out from the solder joints. Figure 16.4*c* is a cross section of the flip chip assembly that shows that some solders have been pushed out of the solder joints and attached to the die surface. There is a gap between the die and the underfill.

Figure 16.5 shows a series of pictures of a flip chip assembly that went through level 3 preconditioning and three repetitions of reflow. Figure 16.5*a* is an x-ray photo showing a large dark area at the center. Figure 16.5*b* is a C-SAM image that also shows a large dark area at the center. Figure 16.5*c* shows a map of solder sitting on top of the solder mask of the BT substrate (the die has been removed). Figure 16.5*d* is an enlarged view of Fig. 16.5*c*. These pictures show that there are delaminations between the underfill and solder mask and that the solders flowed out from the solder joints to the center portion of the assembly.

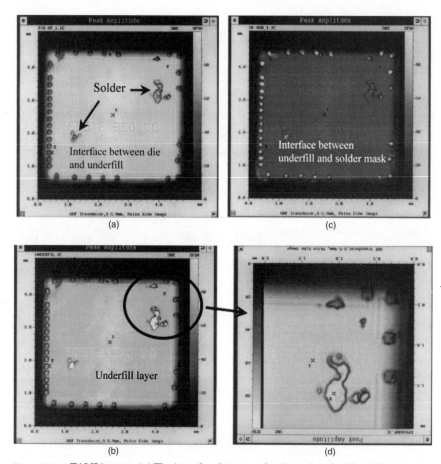

Figure 16.6   TAMI images. (*a*) The interface between the die and underfill. (*b*) The Underfill layer. (*c*) The interface between the underfill and solder mask. (*d*) Enlarged view of *b*.

The TAMI technique developed by Sonix is a very powerful tool used in performing failure analysis of solder-bumped flip chip assembly. This technique executes a number of TAMI scans at different focus positions and generates 30 TAMI images for each focal distance requested. These TAMI images can allow an observer to travel through the flip chip assembly layer by layer.

Figure 16.6 shows a series of TAMI images taken from samples that went through level 3 preconditioning and three repetitions of reflow. Figure 16.6*a* shows the interface between the die and underfill. There are some dark areas, which could be the flowed-out solder from the solder joints. This is evidenced by Fig. 16.6*b*, which shows the underfill layer. It can be seen that there are some white spots at that location, which means

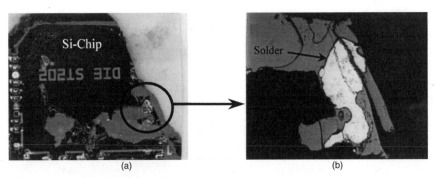

Figure 16.7   (a) Solders on die. (b) Enlarged view of a.

something (flowed-out solder) is blocking the transmission wave. Figure 16.6c shows the interface between the underfill and the solder mask. Figure 16.6d is an enlarged picture of Fig. 16.6b. From these TAMI scans it is possible to tell that the flowed-out solder is on the die surface and that there are delaminations between the underfill and passivation of the die. In order to verify this, the die is removed (Figs. 16.7a and b). It can be seen that the solder does attach to the die surface.

Based on the above, it can be concluded that solder flow-out is due to delaminations at the interface between the passivation of the die and the underfill, or at the interface between the underfill and the solder mask of the substrate, or both. In order to prevent solder flow-out, underfills with strong adhesion to the passivation and solder mask are needed. Also, controlling the moisture absorption and reflow temperature of the underfill FCOB assemblies would help. Finally, prebaking of the substrate at 125°C for 24 h is a solution to prevent solder bump extrusion.

Nonuniform underfill and voids.   It should be noted that not all the dark spots shown in the C-SAM or TAMI images are caused by solder flow-out. This should be verified by the x-ray images. For example, Fig. 16.8a shows a C-SAM image with many dark areas, while the x-ray image shows nothing. This means there is no metal in these areas and thus the dark spots are not solder. As a matter of fact, these areas are the fillers crowded together. This fact is shown on the BT substrate in Fig. 16.8b with the die removed. It can be seen that the dark areas in the C-SAM image are not caused by the solder flow-out but by a bulking up of filler (underfill nonuniformity). The white spots in the C-SAM images or TAMI are caused by voids in the underfill encapsulant. The void shown in Fig. 16.8b is visible as a white spot in Fig. 16.8a.

Delaminations.   Figure 16.9a shows a TAMI scan of the interface between the die and the underfill of a flip chip assembly that went through level 1

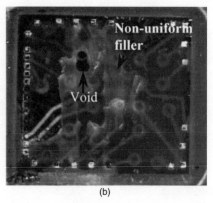

(a)   (b)

Figure 16.8   (a) C-SAM image shows nonuniform fillers. (b) Nonuniform fillers and voids on substrate (with die removed).

preconditioning and three repetitions of reflow. It can be seen that there are some dark areas, some of which are shown by x-ray imaging to be flowed-out solder. (The delamination between the die and the underfill also can be seen from Figs. 16.6 and 16.7.) Even without the x-ray image, the fact of solder flow-out is also evidenced in Fig. 16.9b for the underfill layer. This image shows some white spots in the upper right corner, meaning that something (solder) is blocking the transmission wave. Figure 16.9c shows the interface between the underfill and solder mask, and Fig. 16.9d shows the interface between the solder mask and the copper traces/pads. It can be seen that some of the copper traces have been lifted off. (The delamination between underfill and solder mask can also be seen in Fig. 16.5.)

Figure 16.10 shows the C-SAM and A-SAM images of a flip chip assembly that went through level 1 preconditioning and three repetitions of reflow. It can be seen from the C-SAM scan that there are two different types of solder joints, one darker and the other brighter. This means some solder joints are good (the darker ones labeled B) and some are not (the brighter ones labeled A). The brighter (bad) solder joints are shown in Figs. 16.11a through c. Figure 16.11a shows that there is a gap in the solder joint. Figure 16.11b shows a solder joint with a very tiny cross section. Figure 16.11c shows a solder joint with a narrow cross section.

16.2.6   Die cracking

For no-flow underfill with filmlike material,[6] considerable pressure is required in assembling the solder-bumped flip chip on board. Too low a pressure cannot make good solder joints. However, too high a pressure may crack the die. Figure 16.12 shows images of die crack obtained with C-SAM (on top of the chip) and B-SAM (from three sides of the chip). It can be seen that this crack did not originate from the die surface and is

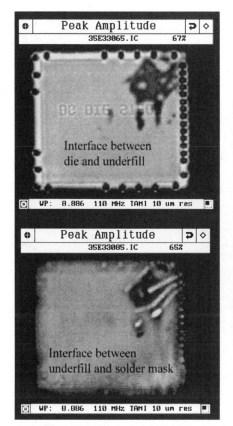 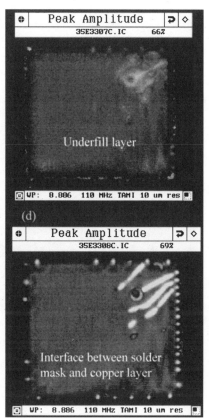

Figure 16.9 TAMI images. (*a*) The interface between the die and underfill. (*b*) The underfill layer. (*c*) The interface between the underfill and solder mask. (*d*) The interface between the solder mask and copper traces/pads.

propagated through the flip chip at 45°. The B-SAM image is below the C-SAM and is similar to the cross section from the side of the flip chip. On the left side of the B-SAM scan, two cracks are observed, while the right side of the B-SAM scan shows only one crack. This means that the two initial cracks are combined into one crack. It is noted that the underfill is capable of arresting the cracks.

16.2.7 Summary

The design, materials, and assembly of solder-bumped flip chip on organic substrate with imperfect underfills have been presented. These assemblies have been first subjected to the JEDEC level 1 and level 3 preconditioning and then three repetitions of reflow. Also, these assemblies have been subjected to the temperature cycling test and to a highly accelerated stress test. The assemblies that failed the precondi-

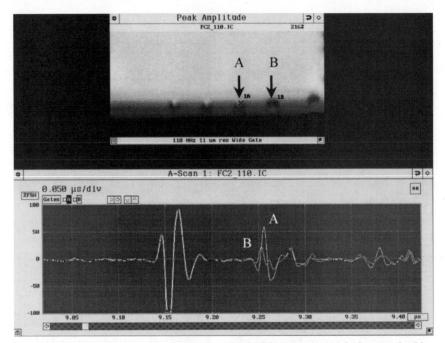

Figure 16.10   A-SAM image shows the lighter (bad solder joint A) and darker (good solder joint B) solder joints.

tioning and reflow tests have been examined by the A-, B-, and C-SAM, TAMI, and x-ray techniques. Cross sections of the samples that failed have been examined for a better understanding of the failure mechanisms. Some important results are summarized as follows.

- The present solder-bumped flip chip on organic substrate assemblies with the level 1 and level 3 preconditioning and three repetitions of reflow pass the HAST and temperature cycling tests.

- Solder flow-out from solder joints (solder bump extrusion) is due to delaminations at the interface between the passivation of the die and the underfill, or at the interface between the underfill and the solder mask of the organic substrate, or both.

- Solder flow-out from solder joints also could be due to underfill voids adjacent to the solder joints.

- Most of the underfills are not uniform. Local concentrations of filler may appear at some areas.

- Most of the underfill voids are randomly distributed.

- Solder joints with perfect underfill most likely will not fail under most operating conditions. Thus, underfill materials having strong adhesion with the passivation and the solder mask are one of the key

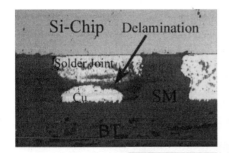
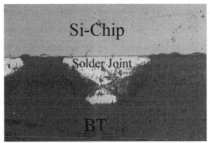
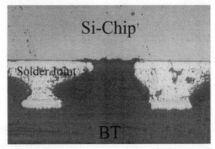

Figure 16.11   Cross sections of the bad solder joints shown in Fig. 16.10.

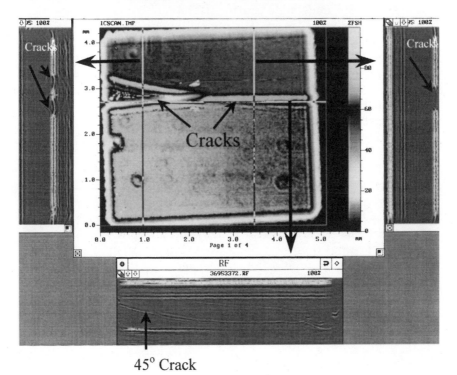

45° Crack

Figure 16.12   B-SAM and C-SAM images show the cracks on flip chip.

ingredients necessary for solder-bumped flip chip on low-cost substrate to become popular.

- X-ray, A-, B-, C-SAM, and TAMI methods are very useful and important for failure analysis of solder-bumped flip chip on low-cost substrate assemblies.

16.3 Interfacial Shear Strength[28]

In order to characterize the interfacial shear strengths between the underfill and the solder mask and between the underfill and the chip passivation, mechanical shear tests are performed. Cross sections of the failure samples are examined and discussed in this section.

16.3.1 Interfacial shear strength between solder mask and underfill

Figure 16.13 shows the test setup for determining the adhesion strength between the solder mask on the PCB and the underfill encapsulant. The specimen consists of a PCB, a flat, hard subject ($2 \times 1.5 \times 2.4$ mm), and an underfill encapsulant, and is subjected to a shear force (as shown in Fig. 16.13) in a Royce Instrument. The PCB is 0.5 mm thick and is made of bismaleimide triazine (BT) resin with a thin layer (less than 25.4 µm) of solder mask. The underfill material is made of epoxy resin with 60 percent silica filler content, and the filler size is less than 10 µm. The Young's modulus and the coefficient of thermal expansion (CTE) of the underfill material are 6 GPa and 30 ppm/°C, respectively.

The process of assembling the specimen is very simple. A drop of the underfill material is placed on top of the BT PCB, and then the hard subject is placed on top of the underfill. A lightweight metal is placed on top of the hard subject to ensure good contact and a standoff height of 50 µm. The whole configuration is put into a curing oven (150°C) for 30 min.

The test results are shown in Table 16.1, and the typical failure mechanism (top surface of the BT PCB) is shown in Fig. 16.14. It can be seen that the epoxy/glass of the BT PCB is obvious. That means the adhesion between the underfill and the solder mask is stronger than that between the solder mask and the epoxy/glass. In Table 16.1, the values in column 2 (Area) represent the areas where solder mask got peeled off. The values in column 4 (Shear Strength) represent the adhesion

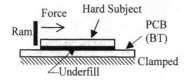

Figure 16.13 Setup of test for shear strength between solder mask and underfill encapsulant.

TABLE 16.1 Results of Shear Tests on the Interface
Between Underfill and Solder Mask

Number	Area (mm²)	Peak load (kg$_f$)	Shear strength (MPa)	Failure mode
1	9.76	28.8	28.9	Crack occurs
2	12.1	35.8	29	between the
3	11.2	36.6	32.1	solder mask
4	5.62	17.7	30.9	and the BT
5	8.19	23.8	28.5	substrate
6	7.02	15.5	21.6	
7	7.41	28.5	37.6	

strength between the solder mask and the epoxy/glass; the average value is about 30 MPa. Therefore, the shear strength between the underfill and solder mask should be much larger than 30 MPa.

16.3.2 Interfacial shear strength between passivation and underfill

A mechanical shear test, as shown in Fig. 16.15, is performed to characterize the interfacial shear strength between the passivation and the underfill materials. It is noted that the materials, assembly process, and dimensions of the specimen are exactly the same as those in the last case except that the PCB is replaced by a very large silicon chip. The silicon chip is 0.5 mm thick with a very thin layer (1 µm) of Si_3N_4 passivation. During the test, the peak loading force (P_u) at failure is recorded. After the test, an optical microscope is used to inspect the failed surfaces at both sides. Table 16.2 shows the test results and Fig. 16.16 shows the failure mechanism (top surface of the chip with passivation intact and some underfill encapsulant). In this case, the underfill encapsulant is broken into pieces. Some remain on the flat, hard subject and some remain on the passivation of the chip. Thus, the adhesion between the silicon and passivation is stronger than that between the passivation and the underfill encapsulant. The values in the column 2 (Area) represent the areas of underfill.

Epoxy-Glass Fiber

Figure 16.14 Fracture surface on BT after shear test.

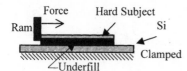

Figure 16.15 Setup of test for shear strength between passivation and underfill encapsulant.

For the calculation of interfacial shear strength, the failure surface is divided into three regions: the separation between the passivation and the underfill, the separation between the underfill and the upper substrate, and the voids. Obviously the voids cannot take any loading during the test. Therefore, the peak failure load is contributed by the shear strengths of the aforementioned two interfaces. It is extremely difficult to identify each interfacial strength with a precise close-form solution. However, an engineering estimation with a simplified model ($P_u = \tau_1 * A_1 + \tau_2 * A_2$) is feasible, where τ_1 and τ_2 are the shear strength of passivation-underfill interface and the underfill-top substrate interface, respectively. A_1 and A_2 are the areas corresponding to τ_1 and τ_2, respectively. P_u could be recorded during the test; A_1 and A_2 could be measured after the test. With two independent tests, τ_1 and τ_2 could be calculated ($\tau_1 \approx 52$ MPa). In fact, a third test is conducted to check the validity of this model. The results turn out to be consistent ($\tau_1 \approx 50$ MPa). Therefore, the shear strength between silicon and passivation should be larger than 50 MPa.

16.3.3   Load displacement response of a solder-bumped FCOB assembly

A flip chip assembly (chip size = 4.7 × 4.2 mm) including chip, underfill, and BT substrate is cross-sectioned to the center of a row of solder bumps. The cross section is finely polished for subsequent inspection. This procedure is necessary in order to clearly observe the crack path after the shear test. Figs. 16.17 and 16.18 illustrate the side view and the top view, respectively, of a cross-sectioned flip chip assembly. The Royce Instruments Model 550 is used to perform the mechanical shear tests. The shear wedge is placed very close to the substrate and against one edge of the solder-bumped flip chip with underfill on the BT substrate. The speed of the shear wedge is 0.001 in./s. Figure 16.19 shows

TABLE 16.2   Results of Shear Tests on the Interface Between Underfill and Passivation

Number	Area (mm²)	Peak load (kg$_f$)	Shear strength (MPa)	Failure mode
1	3.82	20.4	52.2	Crack occurs in the underfill
2	3.88	20.6	52.2	and between the underfill and
3	5.14	26.2	50.0	the chip passivation (Si$_3$N$_4$)

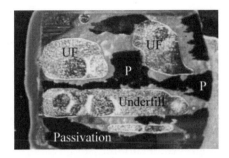

Figure 16.16 Fracture surface on Si after shear test.

the shear direction from the top view after cross-sectioning of the specimen. The experiment must be conducted with great care. Otherwise, the cross-sectioned sample will be shattered dramatically. A typical load displacement curve of solder-bumped flip chip on board with underfill is presented in Fig. 16.20. One minor and one major reduction in loading force are observed along the curve. The former should correspond to the crack initiation in the fillet of the underfill, the latter to the crack propagation causing the final failure of the whole assembly.

Figure 16.21a is a cross-sectional view of overall flip chip assembly after the shear test. The TCE of the underfill is 28.6 ppm/°C. To highlight the local failure mechanism, four zoomed pictures from the cross-section are made (see Figs. 21b through c). Figure 16.21b shows that the crack is initiated in the underfill outside the package and then propagated through the interface between the Si_3N_4 passivation and the underfill. Figure 16.21c reveals that the crack continues to extend through the interface between the Si_3N_4 passivation and the underfill, then passes

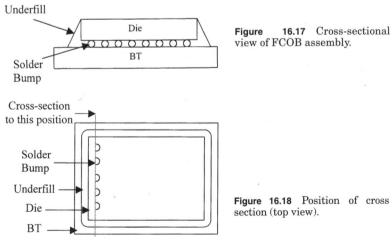

Figure 16.17 Cross-sectional view of FCOB assembly.

Figure 16.18 Position of cross section (top view).

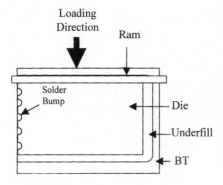

Figure 16.19 Shear loading direction of the specimen.

through a void in the underfill and causes the copper pad to be peeled off. Figure 16.21*d* indicates that the crack breaks the solder bumps. Figure 16.21*e* shows that the crack cuts through the underfill and goes all the way along the interface between the underfill and the solder mask.

From these observations of the shear test, failure may appear at (1) the interface between the Si_3N_4 passivation and the underfill, (2) the interface between the copper pad and the BT, (3) the solder bumps, or (4) the interface between the underfill and the solder mask.

16.3.4   Summary and recommendations

The general goal of this study is a better understanding of the mechanical behavior of FCOB assemblies with imperfect underfill encapsulants. An attempt has been made to characterize the shear strength at the underfill–solder mask interface and the underfill–chip passivation interface. A number of mechanical shear tests have been performed

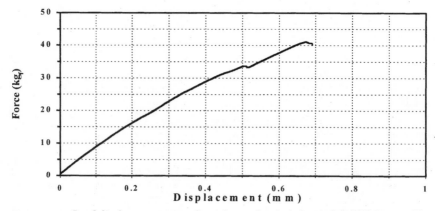

Figure 16.20   Load displacement curve from the mechanical shear of an FCOB assembly.

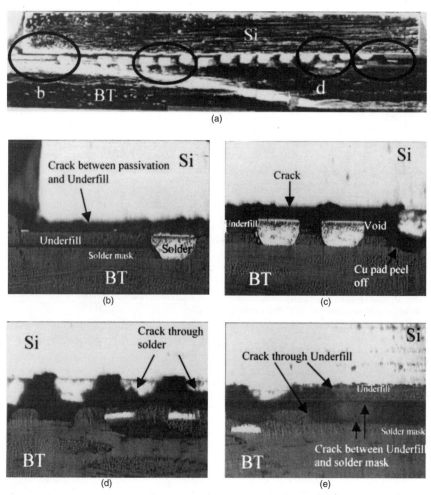

Figure 16.21 (*a*) Cross-sectional view of overall flip chip assembly after shear test. (*b–e*) Zoomed views of (*a*).

and some preliminary testing results have been obtained. It is found that the shear strength between the underfill and the solder mask should be much larger than 30 MPa and that the shear strength between the silicon and the passivation is no less than 50 MPa. This data should be valuable in future stress analysis on the FCOB assembly. Also, such efforts should be encouraged and will continue. After all, without interfacial strength and toughness, computational modeling can only serve the purpose of qualitative study. Therefore, in order to achieve predictive capability through numerical simulation, further research activities in material characterization are essential.

Acknowledgments

The author thanks Ricky Lee, Chris Chang, Arthur Chen, Jian Mire-madi, Joel Sigmund, Damon Rachell, and Jim William for their useful input and constructive contributions.

References

1. Wong, C. P., M. B. Vincent, and S. Shi, "Fast-Flow Underfill Encapsulant: Flow Rate and Coefficient of Thermal Expansion," *Proceedings of the ASME—Advances in Electronic Packaging,* vol. 19-1, pp. 301–306, 1997.
2. Nguyen, L., P. Fine, B. Cobb, Q. Tong, B. Ma, and A. Savoca, "Reworkable Flip Chip Underfill—Materials and Processes," *Proceedings of the International Symposium on Microelectronics,* pp. 707–713, San Diego, CA, November 1998.
3. Wang, L., and C. P. Wong, "Epoxy-Additive Interaction Studies of Thermally Reworkable Underfills for Flip-Chip Applications," *Proceedings of IEEE Electronic Components and Technology Conference,* pp. 34–42, San Diego, CA, June 1999.
4. Pascarella, N., and D. Baldwin, "Advanced Encapsulation Processing for Low Cost Electronics Assembly—A Cost Analysis," *The 3rd International Symposium and Exhibition on Advanced Packaging Materials, Processes, Properties, and Interfaces,* pp. 50–53, Braselton, GA, March 1997.
5. Lau, J. H., C. Chang, and R. Chen, "Effects of Underfill Encapsulant on the Mechanical and Electrical Performance of a Functional Flip Chip Device," *J. Electronics Manufacturing,* 7 (4): 269–277, December 1997.
6. Lau, J. H., C. Chang, and C. Ouyang, "SMT Compatible No-Flow Underfill for Solder Bumped Flip Chip on Low-Cost Substrates," *J. Electronics Manufacturing,* 8 (3,4): 151–164, September and December 1998.
7. Lau, J. H., and C. Chang, "Characterization of Underfill Materials for Functional Solder Bumped Flip Chips on Board Applications," *IEEE Trans. Components and Packaging Technology,* part A, 22 (1): 111–119, March 1999.
8. Lau, J. H., and C. Chang, "How to Select Underfill Materials for Solder Bumped Flip Chip on Low Cost Substrates?" *Int. J. Microelectronics and Electronic Packaging,* 22 (1): 20–28, First quarter 1999.
9. Qian, Z., M. Lu, W. Ren, and S. Liu, "Fatigue Life Prediction of Flip-Chip in Terms of Nonlinear Behaviors of Solder and Underfill," *Proceedings of IEEE Electronic Components and Technology Conference,* pp. 141–148, San Diego, CA, June 1999.
10. Wong, C. P., S. H. Shi, and G. Jefferson, "High Performance No Flow Underfills for Low-Cost Flip-Chip Applications," *Proceedings of IEEE Electronic Components and Technology Conference,* pp. 850–858, San Jose, CA, May 1997.
11. Wong, C. P., D. Baldwin, M. B. Vincent, B. Fennell, L. J. Wang, and S. H. Shi, "Characterization of a No-Flow Underfill Encapsulant During the Solder Reflow Process," *Proceedings of IEEE Electronic Components and Technology Conference,* pp. 1253–1259, Seattle, WA, May 1998.
12. Lau, J. H., *Flip Chip Technologies,* McGraw-Hill, New York, 1996.
13. Lau, J. H., and Y. Pao, *Solder Joint Reliability of BGA, CSP, Flip Chip, and Fine Pitch SMT Assemblies,* McGraw-Hill, New York, 1997.
14. Lau, J. H., C. P. Wong, J. Prince, and W. Nakayama, *Electronic Packaging: Design, Materials, Process, and Reliability,* McGraw-Hill, New York, 1998.
15. Lau, J. H., and S. W. R. Lee, *Chip Scale Package: Design, Materials, Process, Reliability, and Applications,* McGraw-Hill, New York, 1999.
16. Sigmund, J., and M. Kearney, "TAMI Analysis of Flip Chip Packages," *Adv. Packaging,* 50–54, July/August 1998.
17. Tsukada, Y., Y. Mashimoto, T. Nishio, and N. Mii, "Reliability and Stress Analysis of Encapsulated Flip Chip Joint on Epoxy Base Printed Circuit Board," *Proceedings of the 1st ASME/JSME Advances in Electronic Packaging Conference,* pp. 827–835, Milpitas, CA, April 1992.
18. Guo, Y., W. T. Chen, and K. C. Lim, "Experimental Determinations of Thermal Strains

in Semiconductor Packaging Using Moire Interferometry," *Proceedings of the 1st ASME/JSME Advances in Electronic Packaging Conference,* pp. 779–784, Milpitas, CA, April 1992.

19. Tsukada, Y., S. Tsuchida, and Y. Mashimoto, "Surface Laminar Circuit Packaging," *Proceedings of IEEE Electronic Components and Technology Conference,* pp. 22–27, San Diego, CA, May 1992.

20. Lau, J. H., "Thermal Fatigue Life Prediction of Encapsulated Flip Chip Solder Joints for Surface Laminar Circuit Packaging," ASME Paper No. 92W/EEP-34, ASME Winter Annual Meeting, Anaheim, CA, November 1992.

21. Lau, J. H., T. Krulevitch, W. Schar, M. Heydinger, S. Erasmus, and J. Gleason, "Experimental and Analytical Studies of Encapsulated Flip Chip Solder Bumps on Surface Laminar Circuit Boards," *Circuit World,* **19** (3): pp. 18–24, March 1993.

22. Tsukada, Y., "Solder Bumped Flip Chip Attach on SLC Board and Multichip Module," in *Chip On Board Technologies for Multichip Modules,* Lau, J. H., ed., Van Nostrand Reinhold, New York, pp. 410–443, 1994.

23. Wong, C. P., J. M. Segelken, and C. N. Robinson, "Chip on Board Encapsulation," in *Chip On Board Technologies for Multichip Modules,* Lau, J. H., ed., Van Nostrand Reinhold, New York, pp. 470–503, 1994.

24. Le Gall, C. A., J. Qu, and D. L. McDowell, "Delamination Cracking in Encapsulated Flip Chips," *Proceedings of IEEE Electronic Components and Technology Conference,* pp. 430–434, Orlando, FL, May 1996.

25. Lau, J. H., "Solder Joint Reliability of Flip Chip and Plastic Ball Grid Array Assemblies Under Thermal, Mechanical, and Vibration Conditions," *IEEE Transactions on Component, Packaging, and Manufacturing Technology,* part B, **19** (4): 728–735, November 1996.

26. Lau, J. H., E. Schneider, and T. Baker, "Shock and Vibration of Solder Bumped Flip Chip on Organic Coated Copper Boards," *ASME Trans., J. Electronic Packaging,* **118,** 101–104, June 1996.

27. Lau, J. H., C. Chang, and S. W. Lee, "Failure Analysis of Solder Bumped Flip Chip on Low-Cost Substrate," *IEEE Trans. Electronic Packaging Manufacturing,* January 2000.

28. Lau, J. H., S. W. Lee, C. Chang, and C. Ouyang, "Effects of Underfill Material Properties on the Reliability of Solder Bump Flip Chip on Board with Imperfect Underfill Encapsulants," *Proceedings of IEEE ECTC,* pp. 571–582, June 1999.

Index

ABOUT THE AUTHOR

John H. Lau is the president of Express Packaging Systems, Inc., in Palo Alto, California. His current interests cover a broad range of electronics packaging and manufacturing technology.

Prior to founding EPS in November 1995, John worked for Hewlett-Packard Company, Sandia National Laboratory, Bechtel Power Corporation, and Exxon Production and Research Company. With more than 29 years of R&D and manufacturing experience in the electronics, petroleum, nuclear, and defense industries, he has authored and coauthored over 150 peer-reviewed technical publications, and is the author and editor of 12 books: *Solder Joint Reliability; Handbook of Tape Automated Bonding; Thermal Stress and Strain in Microelectronics Packaging; The Mechanics of Solder Alloy Interconnects; Handbook of Fine Pitch Surface Mount Technology; Chip On Board Technologies for Multichip Modules; Ball Grid Array Technology; Flip Chip Technologies; Solder Joint Reliability of BGA, CSP, Flip Chip, and Fine Pitch SMT Assemblies; Electronics Packaging: Design, Materials, Process, and Reliability; Chip Scale Package: Design, Materials, Process, Reliability, and Applications; Low-Cost Flip Chip Technologies for DCA, WLCSP, and PBGA Assemblies;* and the forthcoming *Low-Cost and High-Density Interconnects.*

John served as one of the associate editors of the *IEEE Transactions on Components, Packaging, and Manufacturing Technology* and *ASME Transactions, Journal of Electronic Packaging.* He also served as general chairman, program chairman, and session chairman, and invited speaker of several IEEE, ASME, ASM, MRS, ISHM, SEMI, NEPCON, and SMI international conferences. He received a few awards from ASME and IEEE for best papers and technical achievements, and is an IEEE Fellow and ASME Fellow. He is listed in *American Men and Women of Science* and *Who's Who in America.*

John received his Ph.D. degree in theoretical and applied mechanics from the University of Illinois, an M.A.Sc. degree in structural engineering from the University of British Columbia, a second M.S. degree in engineering physics from the University of Wisconsin, and a third M.S. degree in management science from Fairleigh Dickinson University. He also has a B.E. degree in civil engineering from National Taiwan University.